The Parks Belong to the People

The Parks Belong

to the People

THE GEOGRAPHY OF THE NATIONAL PARK SYSTEM

JOE WEBER

SELIMA SULTANA

The University of Georgia Press
ATHENS

© 2024 by the University of Georgia Press
Athens, Georgia 30602
www.ugapress.org

Designed by Kaelin Chappell Broaddus
Set in 10.25/13.5 Bodoni Egyptian Pro Regular
by Kaelin Chappell Broaddus
Printed and bound by Versa Press
The paper in this book meets the guidelines for permanence
and durability of the Committee on Production Guidelines for
Book Longevity of the Council on Library Resources.

Most University of Georgia Press titles are
available from popular e-book vendors.

Printed in the United States of America
28 27 26 25 24 C 5 4 3 2 1

Library of Congress Cataloging-in-Publication Data

Names: Weber, Joe, 1970– author. | Sultana, Selima, 1965– author.
Title: The parks belong to the people : the geography of the National
    Park System / Joe Weber, Selima Sultana.
Other titles: Geography of the National Park System
Description: Athens : The University of Georgia Press, [2024] |
    Includes bibliographical references and index.
Identifiers: LCCN 2023024505 | ISBN 9780820365053 (hardback) |
    ISBN 9780820365725 (epub) | ISBN 9780820365718 (pdf)
Subjects: LCSH: National parks and reserves—United States—
    History. | United States. National Park Service—History. |
    United States—Historical geography. | National parks and
    reserves—United States—Maps. | National parks and reserves—
    United States—Management.
Classification: LCC SB482.A4 W43 2024 | DDC 363.6/80973—dc23/
    eng/20230522
LC record available at https://lccn.loc.gov/2023024505

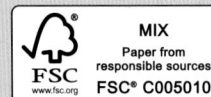

Published with the generous support of
Office of Research and Engagement at the
University of North Carolina at Greensboro

MIX
Paper from
responsible sources
FSC
www.fsc.org
FSC® C005010

Dedicated to my parents, who raised me in
Death Valley and gave me the opportunity to explore
and develop my lifelong interest in national parks.

—JOE WEBER

Dedicated to the memory of my beloved daughter,
Preeta Ananta Shaikh (October 11, 1991–October 17, 2020),
and to my son, Pantho Ananta Tsali, with whom I explored
and developed my love for America's national parks.

—SELIMA SULTANA

CONTENTS

FIGURES AND TABLES

Figures

Tables

ABBREVIATIONS

BLM   Bureau of Land Management

CCC   Civilian Conservation Corps

DST   Daylight Saving Time

FWS   Fish and Wildlife Service

IUCN   International Union for Conservation of Nature

NASA   National Aeronautics and Space Administration

NPS   National Park Service

TVA   Tennessee Valley Authority

UNESCO   United Nations Educational, Scientific and Cultural Organization

USFS   United States Forest Service

Selima Sultana

I grew up in Bangladesh, which is affectionately known as "the land of green" due to its endless cultivated rural vegetation, where anyone could run away from everyday crowded city life. At the same time, the wilderness was always a scary place because of the king cobra (*Ophiophagus hannah*), a ten-foot-long venomous snake whose bite can take your life within twenty minutes. There are eighteen national parks in the country, but they are small and look nothing like American parks. Most of them are quite recent; the country's protected area has quadrupled since I came to the United States in the early 1990s. Yellowstone in Wyoming was the first American national park I visited. I took a break from my graduate studies and went on a long summer road trip that also passed through Little Bighorn Battlefield National Monument in Montana and other units. One of my strongest memories of that first visit was how cold it was in the summer. After living in tropical-monsoon Bangladesh and then in humid-subtropical Georgia, I found it hard to imagine any place being so cold in July. It was also a strange feeling being in places where there were so few people and no one who really looked like me. This was not necessarily the case in Yellowstone, which was comfortably crowded, but it certainly was in the vast expanse of the northern plains and mountain West where towns and farms are so far apart.

After that first trip, I fell in love with America's national park system and started reading many books about it. The first book I read about national parks was *The National Parks: America's Best Idea*, published in 2009 by Day-

ton Duncan and Ken Burns, and then I watched the television documentary with my kids. That book was our inspiration when I took my children on the traditional American summer family vacation, showing them Yellowstone and the Grand Tetons, the Rocky Mountains, the Grand Canyon, the Great Smoky Mountains, the Blue Ridge Parkway, the memorials of the National Mall in Washington, D.C., North Carolina's Outer Banks, and other places. I've visited even more since my children went off to college, and I even became a novice hiker. Hiking on the Devil's Hall Trail and the McKittrick Canyon Trail in the Guadalupe Mountains National Park in Texas, I realized that people miss out on a lot of park experiences unless they hike through some of the most rigorous terrain. I would never have imagined what a beautiful treasure the Guadalupe Mountains are from only looking at the view from the visitor center. No wonder this park is so little known to Americans.

Acadia, Badlands, Big Bend, Carlsbad Caverns, Congaree, Death Valley, the Guadalupe Mountains, Shenandoah, Theodore Roosevelt, Yosemite, and Zion are all wonderful in their own unique way. The scenery in a national park is something that anyone can connect with, regardless of where they are from, but the landscapes in these parks could not be more different from what I grew up with. I prefer to visit the large national parks because of their vastness and serenity, but I also love historic sites and battlefields. Though I grew up with memories of wars different from the Civil War or the American Revolution, I can attest that someone who became a U.S. citizen as an adult might also be interested in learning the American history that most native-born Americans grew up with. I've only been able to visit a few dozen national parks, but I want to visit all of them.

On a family trip in the American West in 2006, my interest in national parks began to get more serious. I started paying more attention to the people who were there, as well as those who weren't there. Often, my family were the only nonwhite people present. Yet everyone I have ever met in whatever country has the same capacity to enjoy natural beauty. I started pondering: If the national parks were the best idea America ever had, why were there not more visitors like my family? As a geography professor and foreign-born American, I had to find the answers. After some research I quickly learned that there is relatively little information available about who goes to parks and why, yet there is no shortage of explanations. That's when I began working with my coauthor on national park topics. We both thought geography might be one of the reasons for the lack of visitors, yet geography had been overlooked in most studies. We began examining the role of where the parks are located compared to the U.S. population and found that geography does make a difference. Since then I've started collecting my own data: I've visited the Great Smoky Mountains with a research team to talk to visitors and staff, combed through the park's archives, and tried to better understand park visitors. I enjoyed talking to visitors and the hardworking NPS staff, who have one of the best jobs in the world. But I also came across old plans for segregated park facilities and heard from my own research team that they would never visit around Gatlinburg and Pigeon Forge, Tennessee, again, as "their people" weren't visible there, and they didn't feel comfortable. My work with my research team gave me many new perspectives on parks and opportunities to study materials that I hadn't considered in the past. I became a member of a Facebook group called "Hikers of Color," which has thousands of members. The future of the national park system is about more than the preservation of nature, wolves, and bears. All people should be able to enjoy and feel comfortable and safe in national parks and all visitors should see their history being preserved. Questions remain about what future changes can make all Americans want to visit and feel comfortable at the parks and enjoy America's best idea.

Joe Weber

My experience of the national park system began very early in life. My father worked for the National Park Service, and my first home was a house within Whiskeytown National Recreation Area in Northern California. (Sadly, in the summer of 2018 Whiskeytown was destroyed by the Carr Fire, which also burned down that house we once lived in.) After a short time there we moved to Death Valley (then a national monument), where I grew up. We lived in a standard 1960s ranch-style Mission 66 house right next to a stone CCC building from the 1930s. Among our neighbors were park rangers who had arrested Charles Manson a few years earlier. In 1983 I met Horace Albright, the second director of the National Park Service, who was in Death Valley for the park's fiftieth-anniversary celebration. Our family vacations were often to other parks, and I remember many camping trips to Lake Mead and Yosemite. On another vacation I first saw the national capital parks and was amazed to see that the buildings were real, not just pictures on TV. On regular trips to visit grandparents in Missouri we visited George Washington Carver National Monument several times, though my main memory is of the antique vending machine that sold Coke in glass bottles.

In addition to growing up with an interest in parks and a great appreciation for the hard work NPS employees do, I learned other lessons about parks along the way. I learned about the spatial organization in parks by observing the different areas that were open to tourists. I never worked for the NPS, but I did have summer jobs in Death Valley, and I got an inside view of much of what makes a park run: its water and sewer systems, trash collection, restroom cleaning, sign repair, employee housing maintenance, and road and trail repair. (My supervisor one year, Barbara Moritsch, wrote a book of her own, *The Soul of Yosemite: Finding, Defending, and Saving the Valley's Sacred Wild Nature* [2012].) I remember hearing talk about expanding Death Valley and making it a national park; this was an early lesson that parks could be changed.

After I left for college in Arizona I became a regular park visitor. Saguaro National Park outside Tucson became one of my favorites. I had time enough to hike almost every trail many times in all five Sonoran Desert seasons (winter, spring, dry summer, monsoon summer, and fall). Hiking the King Canyon Trail in the summer in the hours before sunset remains one of my favorite park experiences. I have fond memories of other southwestern park units as well. Chiricahua National Monument must surely be one of the most underappreciated park units in the country, but maybe it is best that way. I remember a visit in the summer of 1995 when, aside from a few park employees down at the visitor center, I seemed to have the monument all to myself. Zion and Arches National Parks have also become among my favorites, though finding solitude within them is much harder.

I also learned many lessons about parks. I heard about park units that had been eliminated from the system. When I heard that Death Valley had become a national park I learned how redesignating a monument to a park automatically elevates it above competing park units and draws visitors. I also learned about wilderness. In Death Valley wilderness seemed unnecessary; the mountains and many valleys were so remote that distance alone would protect them. My view changed when I encountered urban growth pressures on public lands in Arizona. I had taken it for granted that most parks were in the West because that was the best part of the country; only gradually did I realize the importance of public lands and timing to the distribution of parks. I've seen parks with overflowing trash cans, and I've visited parks where NPS employees are nowhere to be seen. I've heard NPS employees insulted for being lazy parasites rather than keepers of America's heritage.

There are many books written by current or former NPS employees, almost all of them on rangers, whose accounts predictably revolve around law enforcement and search and rescue. The best account of living and working in a park I have ever read is *Fire on the Rim: A Firefighter's Season at the Grand Canyon*, by Stephen Pyne (1995). He was a seasonal firefighter and not an NPS employee, but his account of what it's like to live and work in a park is very good. The nicknames for different places and groups of workers, the arrogance and uselessness of park rangers who, fortunately, rarely leave their air-conditioned offices and vehicles, and the clueless administrators who occasionally show up for a photo op but are otherwise never seen are well described. Other park books that I found particularly inspiring were John Ise's *Our National Park Policy: A Critical History* (1961) and Dwight Rettie's *Our National Park System: Caring for America's Greatest Natural and Historic Treasures* (1995), which is a behind-the-scenes look at how the park system functions.

Reading these books made me aware that there were opportunities to write about parks and even a need to do so that went beyond the usual histories or focus on photogenic wildlife. I became interested in studying parks because of the role geography plays in park visitation. That interest taught me to look at how individual parks and the system have changed over time, which is a topic that I have enjoyed digging into. Changing boundaries, internal development patterns, and especially places that used to be part of the park system but were removed fascinate me.

I always enjoy entering a park, seeing the entrance sign and a park ranger wearing the familiar uniform. It feels like coming home. There are many parks I want to see, some big, some small, and not always for the obvious reasons. I want to see Dinosaur National Monument in Utah, Klondike Gold Rush National Historical Park (Skagway Unit) in Alaska, the North Rim of the Grand Canyon in Arizona, and Uluru in Australia. Writing this book has helped me explore these parks vicariously.

The Parks Belong to the People

The National Parks and Geography

The U.S. national park system is a much-loved set of places that preserve scenic beauty and the nation's history (Fig. 1.1). Almost all of us have visited at least a few units, perhaps on dimly recalled family vacations, on an adventuresome trip, or while wandering through a city. You may have visited some parks that you didn't even know were parks. They are found throughout the country and attract hundreds of millions of visitors from all over the world. Parks are also often in the news due to congestion, fires, inadequate budgets, disappearing glaciers, border problems, rangers who have been shot, search and rescue missions, the creation of new units, and the lack of diversity among visitors. In 2016 there was news coverage of the centennial of the National Park Service (NPS).

There are many books on the national park system, but this one is different. It is not a book about bears or wolves or mountain scenery. Instead, we take a geographic perspective: Where are America's parks, and why are they in particular locations? What are they located near? Where do visitors come from? How does distance play a role in which parks are visited? How have park facilities developed to accommodate visitors? The national parks are geographic features: they have legally defined boundaries and are not located randomly or to serve the greatest number of people. They are affected by what is around them, as are shopping centers and neighborhoods. Their boundaries are even more gerrymandered than voting districts, and they are integrated into the nation's road network.

The study of geography can be differentiated into different fields, such as physical geog-

raphy (land, water, and climate) and human geography. Human geography naturally deals with people: our cities, economic practices, political systems, transportation, and cultures. We will focus on the human geography of parks, or parks as political and cultural features, sites for public facilities and economic activities. These park units were created at great cost to the federal government and with great benefit to the American people and environment. Therefore, they cannot be understood without considering whether all Americans have access and ensuring that the park system remains relevant in the future. Our perspective is quite different from others that focus on parks solely as nature preserves or wilderness.

This book is written principally from the spatial or locational tradition of geography, one of the four major philosophical tenets of geography introduced by geographer William D. Pattison in 1963. While all four traditions are interrelated or impossible to divorce from each other, the core concept of the locational tradition attempts to understand a topic in relation to location, distribution, and patterns over time utilizing the geographer's powerful tool of mapping (cartography). At the same time, this book also embraces other traditions of geography or concepts that emerged from criticism of spatial analysis. Given the eclectic and diverse nature of geography, our book discusses the varied geographic issues of national parks and is synchronized by a focus on geographic locations, distributions, and the processes of growth and development of park service units and park boundaries. This book also tries to explain the regularities that unite seemingly different parks, as well as the differences among them, such as

FIG. 1.1   Dawn at the Grand Canyon. For many people, this spectacular landscape represents the best of America's national park system. Photo by Murray Foubister, 2013. Wikipedia, Creative Commons Attribution–ShareAlike 2.0 Generic License.

the origins of visitors, visitation patterns in relationship to size and proximity to the parks, and how the formation of national parks, shaped by the sociopolitical structure of the locale, formed social, cultural, and economic relations in the surrounding communities. We have tried to answer these inherently geographic questions throughout the book while investigating different topics that relate to national parks.

We examine the national park system using several important geographic themes: concepts of place and location, spatial interaction, spatial organization, and the concept of cultural landscapes.

Geographic Concepts

Location may be absolute or relative; in the first case, it could refer to the latitude and longitude coordinates of a park feature, such as 44.460552, -110.82807, or to the street address of an urban unit, perhaps the one at 1600 Penn-sylvania Avenue NW in Washington, D.C. Several survey systems have been used to define park boundaries, and hikers rely on GPS coordinates to locate themselves. In the early years, locations were not always well known, but 150 years of mapping and surveying have usually eliminated these problems. Absolute location can change, as with the two-block move of the historic home that makes up Hamilton Grange National Memorial, an NPS site in St. Nicholas Park, Manhattan, New York City, or Death Valley's famous sliding rocks, known as "sailing stones," which move by themselves.

Relative location refers to a park's location compared to other features: What is the park near? What places does it lie between? The pioneer fort at Pipe Spring, Arizona, was midway between Zion and the North Rim of the Grand Canyon and became a national monument because that location made for a convenient stop for travelers between the larger parks (McKoy 2000). Relative location is also always changing: a park that was remote in 1900 may be within a big city today. Horseshoe Bend National Military Park in Alabama is located a few miles off Highway 280, once a busy highway for beach-bound northerners. The park catered to those travelers until the construction of faster roads drew them away from the park (Hebert and Braund 2016).

Distance and spatial interaction can be measured in many ways, including accessibility, which shows their relative location or distance to people. Distance is measured either in a straight line ("as the crow flies") or along a road, trail, railroad, or other network using miles or any other units. It can also be measured as the travel time between places (on foot, in a car, on a plane, or some combination). It could be measured as a cost, perhaps in dollars or in calories. Different measures will be appropriate depending on whether you are flying to a park or choosing which trail to take to your favorite peak. If you are using time as a measure of distance, it may be hard to predict and can change depending on the weather, the presence of a "bear jam" stopping traffic, what kind of a road you are on, or whether you are going uphill or downhill. No matter how distance is measured, it is related to the relative cost of movement, which has a profound impact on spatial interaction between people and parks. Some parks are easier for people to get to than others; they are a shorter drive or are located on faster roads or closer to an airport with more (or cheaper) flights. We can talk about the ease of reaching parks as their accessibility; Golden Gate National Recreation Area is in San Francisco, a city of over six million people that is far more accessible than Rainbow Bridge National Monument, which requires an all-day drive from the closest big city, followed by a three-hour boat ride across Lake Powell through the Utah desert. The varying accessibility of parks can be related to visitor numbers: more people tend to show up at those parks that are closer to more people and easier to get to. It is no wonder that Great Smoky Mountains National Park is the most visited national park, located as it is in the southeastern United States.

It has gotten easier to get to parks over time as travel has improved from horse-and-wagon days to cars; we call this space-time convergence or compression (Warf 2008). The South Rim of the Grand Canyon was rarely seen before 1882, when a railroad was built across northern Arizona (Rothman 1998). Then a rough three-day stagecoach ride from towns along the railroad allowed a few hardy travelers to see the canyon from the South Rim. Accommodations were limited in number and quality. In 1901 a railroad was built to within a few feet of the rim, and visitation exploded; the railroad made sure new hotels, restaurants, and other facilities were ready. The automobile and good roads created a new wave of visitors, with more and more facilities required for them. Mission 66 was an effort to keep ahead of this visitation growth.

Today few parks are as remote as the Grand Canyon was before 1882; residents of the Atlantic coast can reach even the most remote Alaskan park today in less time than they could have reached the Grand Canyon in 1881. These Alaskan parks have very low visitation and no development; neither outcome is inevitable and may be a temporary condition. If large numbers of people begin making their way to these parks, the demand for visitor facilities will grow. If new facilities are built, visitation will increase.

The opposite of space-time convergence is space-time divergence. It is less common, but examples can be found in the park system. The National Park of American Samoa is much harder to get to now than before the 1970s when airplanes crossing the Pacific would land at this island territory to refuel. Longer-range airplanes now fly between the United States and Australia without stopping.

In sum, the concept of distance and spatial interaction can explain how different places are connected and the movements of people, goods, or ideas between them. National park units are connected to the rest of the country and the world by well-developed transport networks, allowing the movement of people from all over the world to visit them. Visitors do not arrive from random locations but can be predicted by the concept of distance and spatial interaction.

Spatial organization describes how parks or larger areas are geographically organized or why facilities such as museums, campgrounds, and even roads are located where they are within parks. Geographers frequently refer to regions. These may be formal, as when boundary lines are drawn on a map to indicate a park or county or even natural features such as landforms. Conditions on one side of a boundary line may be quite different from those on the other, a situation with which any park visitor is familiar: once visitors pass the dramatic entrance sign, the road often changes, buildings may disappear, and visitors know they are in a special place. Parks are formal regions, but they are contained within larger formal regions such as counties, states, and congressional districts. Few parks will share the same combination of these three regions.

Regions may be functional when they indicate areas defined by their spatial interaction, such as a large metropolitan region to which workers commute and goods are shipped each day. Parks may be one element in a tourist-dominated regional economy, a small part of an annual animal migration cycle, or a link in a transport system moving big-city residents to and from their weekend recreation lands.

A *cultural landscape* is the combination of the natural world and human activities in an area. It could include settlement patterns, long-lived agricultural practices, architecture, and other human efforts that have created a built environment or shaped the natural one.

Large parks possess cultural landscapes in the form of carefully designed scenic roads and rustic park buildings, while in urban areas a cultural landscape might include architectural styles and even the details of a city's street pattern. Battlefield parks have become the most visible cultural landscapes in the park system, combining carefully manicured landscaping and countless memorials, markers, signs, and plaques. Even when there are no buildings or roads, a cultural landscape may exist, reflecting the meanings we give to a place, the names we bestow on natural features, and the ways people use the land. Cultural landscapes are always open to reinterpretation or disputes.

A related concept is the sense of place, shaped by people's interactions with a location, which vary with each person's experiences and sensitivities. Many park visitors have a strong attachment to the landscape of their favorite park unit. Our sense of place when we are in a park may be difficult to communicate to someone who has not experienced that location. For example, the description of national parks as embodying some of America's central values and experiences may not be understood by people from a different cultural background or with different experiences. A person's negative interactions with people in a national park might give them a feeling that they do not belong there. Places can be part of a social order, as in "knowing one's place."

Geographic Perspectives on the Parks

In addition to our geographic perspective, our book differs from other works on the park system because we examine the full range of the national park system, not just the large nature parks that often receive the greatest attention, such as Yellowstone. Most units in the national park system are not scenic mountains or canyons; instead, they are devoted to battlefields, historical events, or archaeological sites or people, and this focus is reflected in our approach. While each park is unique, all parks share the common imprint of NPS management. We discuss how these units differ from other places, such as national forests administered by the U.S. Forest Service, wildlife refuges overseen by the Fish and Wildlife Service, and wilderness areas found in all these areas. Toward the end of the book we also broaden the scope to examine protected places in other countries and Antarctica and even possibilities for parks on other worlds.

There is often a belief that what happens in Yellowstone affects all other units or sets a trend followed by others, sometimes known as the "crown jewels syndrome" (Benton 1998; Yochim 2009). This is not always the case: large wildlife and fires have been important to the park but are not relevant to many other units. While Yellowstone may be held up as the example of what a park should be, there are many ways in which it is the least representative unit in the park system, as we demonstrate throughout this book. There are many issues playing out in other kinds of units, and many recent trends are most apparent in newly created national monuments and historic sites (e.g., the historic sites that honor the civil rights movement). Our changing climate will have much larger effects on parks other than the early mountain parks.

Park units come in all shapes and sizes, and it is hard to define a typical park. Death Valley National Park is an example of a park with many physical and cultural resources. It preserves pristine desert scenery and climate, including the lowest elevations and hottest temperatures in the Western Hemisphere, the second greatest vertical relief of any valley in the United States, sand dunes, spectacular spring wildflower displays, mysterious moving rocks, and the endangered desert pupfish (Hunt 1975). The valley also contains a variety of historical resources, including early California mining history and the famous twenty-mule-team wagons hauling borax (Lingenfelter 1988). The labor of the CCC during the Great Depression is visible in many park buildings that are themselves now historical resources in their own right. Diverse personalities such as Death Valley Scotty, John Wayne, Charles Manson, and Ronald Reagan are associated with the park. In more recent years, the park has attracted *Star Wars* fans seeking locations where the movie was filmed and amateur astronomers enjoying some of the darkest night skies remaining in North America (Nordgren 2010). The park also offers a wide variety of motorized and nonmotorized recreational activities, and the Saline Valley Warm Springs might be the only clothing-optional place in the entire park system. The park also contains a variety of racial and ethnic themes. Several members of the first group of non–Native Americans to pass through the valley were African American slaves being taken to California in 1849. The park includes the site of an early Chinese settlement, was briefly home to Japanese American internees during World War II, is the first national park to contain an Indian reservation, and is one of the few parks to have had an African American superintendent (Burnham 2000; Schnayerson 2006). Classifying the park with a single theme or topic would be difficult, and most other park units are equally multifaceted.

Overview of the Book

Chapter 2 discusses what the national park system is, how many units there are, and what kinds of places are in it. Every park has a label, such as national park, monument, memorial, lakeshore, seashore, river, recreation area, preserve, reserve, parkway, historic site, historical park, national military park, battlefield park, battlefield site, and battlefield, and there are more than twenty additional site labels that do not fit into any of these categories, such as the White House. The National Park Service was created to oversee these units with an obligation both to protect the resources within them and to ensure that they remain available for public enjoyment. In addition to its parks, the agency has several important regional support centers and a headquarters a few blocks from the White House. The NPS has also become the overseer of the nation's history and is responsible for several programs designed to list historic sites both within and outside park units. The National Register of Historic Places lists over one million properties (National Park Service 2023b), while the national historic landmark designation has been applied to a select group of the country's most important sites (National Park Service 2023a).

Just as there are different kinds of park units, there are different ways of creating them. Most national park units exist because they were created by laws passed by Congress and signed by the president. This is a slow process that requires careful consideration of whether a place is of national significance and whether it is suitable and feasible as a park. There are many proposals for parks that never join the system. Park units may be added by presidential proclamation, by action of the secretary of the interior, or even as the result of agreements between the NPS and another federal agency.

Legally, these are all the same, though those based on presidential proclamation have often been controversial.

The 424 units of the national park system are only a small part of the protected places in the country. Other protected areas are operated by federal agencies such as the Bureau of Land Management, the Forest Service, or the Fish and Wildlife Service. These may have similar geographies and labels and may even look similar on the ground. Wilderness areas created by Congress provide a different but overlapping set of protected places that will be compared with the national park system. There are also national rivers and trails, which are sometimes part of the park system but usually are not.

The national park system did not emerge as a complete and fully planned system but grew out of the protection of Yosemite Valley in 1864. A few more parks were created in the vast western public domain, with their numbers growing at a faster rate in the twentieth century. In addition to mountain scenery, the park system began to include caves, geological oddities, and Native American ruins. In 1933 the system had one of its greatest expansions when the NPS took over battlefield parks from the army, many parklands in and around the District of Columbia, and national monuments operated by the Forest Service. Recreation parks centered on lakes, beaches, and even scenic roads were added. The parks also moved eastward in the 1930s in a move by the NPS to create a more national system with greater public and political support; in the 1970s similar goals caused the system to expand into large cities. The system took on a new mission of better representing the nation's diverse history, creating a growing series of units devoted to civil rights and culture. In 1978 the acreage of the national park system doubled thanks to the creation of many vast new units in Alaska, representing the greatest single expansion of the system. It has continued to expand since then, with little expectation that the country will soon run out of places worth protecting.

Chapter 4 provides an overview of the geography of the national park system in the early twenty-first century. This tour fittingly begins in Alaska, which contains 65 percent of the total area of the park system and nine of the ten largest parks. These vast parks, which contain glaciers, volcanoes, and high peaks, are almost a separate system, with few of the visitor amenities or crowds found in the forty-eight contiguous states. The Pacific Ocean is represented on maps of the park system with a few tiny insets of Hawaii and other island territories, but it is the largest region within the system. It is home to the southernmost, westernmost, and easternmost park units.

The western states are where the system began and where many famous parks are located. The Rocky Mountains and Sierra Nevada contain the towering mountain peaks that many people associate with the system, but the West embraces a wide range of environments, including deep canyons, deserts, archaeological sites, and fossil parks. Within the Four Corners states of Arizona, New Mexico, Utah, and Colorado lies the Colorado Plateau, the scenic heart of the park system, including the most spectacular canyons and Native American ruins.

Half of all park units are found in eastern states, though few are as large or scenic as those out west or in Alaska. Instead, the focus is often on recreation and history, whether that of the nation's presidents and leading figures of industry, art, or social movements or its battles and political struggles. Several cities have urban parks, and those of Washington, D.C., are the greatest single concentration of park units in the system.

Each park has a name that communicates its identity and purpose, though this is not always clear. Some names can be downright misleading, are easily confused, and have changed over time. They can have enormous political or even financial significance, which will likely produce more name changes in the future.

Chapter 5 examines the political geography of parks, especially their boundaries. There are several different ways they can be drawn, depending in part on location but also on the purpose of the unit. Natural features such as ridgelines and watersheds rarely make up park boundaries; instead, the squares and rectangles of the land survey system or urban street grids provide most boundaries. However defined, parks often contain private land that must be acquired, a long-running problem for the system. Many parks are in several sections, perhaps with two equal units or with many small outlying parcels of land. Park boundaries are frequently subject to change, whether to protect more features, eliminate some that are controversial, or fix problems with the original boundaries. Boundary changes may eliminate problems but introduce others in the form of more private land within the park.

Change in designation from one type of park unit to another (such as from a monument to a park) has been common and can have important consequences for a park's visibility and visitation. Designation changes are frequently pursued as a means of encouraging more visitors. Some park units have even been removed from the system.

National parks also overlie many other political boundaries, and a number are found along the nation's borders, which can make for management challenges. They fall within the districts of the 535 members of Congress, who differ considerably in their outlook on national park issues. Even the law enforcement power of park rangers varies among parks.

Chapter 6 looks within national parks, examining how they were planned and developed even before the National Park Service existed. From the earliest days, parks needed some form of transportation and accommodations for visitors, including roads, lodges, trails, campgrounds, restaurants, and museums or visitor centers. Early parks had facilities laid out in the horse-and-wagon era, with grand rustic lodges constructed to house eastern visitors on their multiday tours of the park. The coming of the automobile and growing numbers of visitors put increased pressure on the NPS to provide roads and inexpensive lodging for families. In the 1930s the CCC constructed roads, trails, campgrounds, housing, and other facilities in many parks, using a rustic architectural style. The layout of these facilities in these parks was guided by landscape architects working to showcase the parks' features.

In the 1950s many parks were deteriorating and badly in need of expanded facilities to meet the influx of visitors arriving by automobile. The Mission 66 program was launched in 1956 as a ten-year program to rebuild parks and accommodate crowds. New roads, trails, campgrounds, lodging, and the invention of the multipurpose visitor center were the result. As before, these facilities were located carefully to structure the visitor experience in the park and channel movement through the park.

Since those years there has often been an attempt at minimizing development within parks and moving visitor facilities away from scenic views, Native American ruins, and other attractions. This is most evident in the remote and undeveloped Alaskan parks but can be seen in many newer parks elsewhere. There has also been a partial return to the rustic style of the early years, though with a postmodern twist.

Chapter 6 also addresses the ways that different kinds of parks are developed. Large parks, monuments, and preserves are often divided into different visitor areas (or frontcountry) and backcountry areas for management (though those in Alaska are usually all backcountry). Some of the backcountry may be designated as wilderness, but this is surprisingly rare in the national park system; there is no wilderness in Yellowstone or the Grand Canyon. Recreation units are designed with many dispersed visitor facilities, often filling up much of the unit. Smaller units may be little more than a single building, requiring little spatial organization (aside from figuring out where visitors will park).

Visiting parks requires transportation to and within them. Chapter 7 focuses on the parallel development of parks and transport systems in the United States. In the early years, railroads were essential, not only bringing visitors but even developing tourist facilities within the parks and operating tours. Railroad companies built many enormous rustic lodges, some of which are still standing, and the visitor experience at Yellowstone and many other early parks is still based on the routes and lodging the railroads developed.

Americans gradually shifted from railroads to cars, and the NPS responded by building roads within parks to accommodate them. The NPS designed and built many scenic highways on the sides of mountains that remain popular today. It also built parkways, or parks that were little more than roads with carefully designed scenic views. Park roads are designed to follow the landscape with as few cuts or embankments as possible while still providing for safe movement of traffic.

Air travel is essential to reach remote destinations, especially in Alaska and the Pacific, while "flightseeing" has become a popular activity at parks such as the Grand Canyon. Boats are essential forms of travel to some units, perhaps most notably in the cruise ships that enter Glacier Bay every day. Buses are becoming more and more common as parks grapple with ever-increasing visitor numbers but a usually fixed mileage of roads and limited parking spaces, resulting in congestion. Buses are especially common in canyon parks, where linear routes can easily be operated and parking capacity is especially limited.

Visitors are essential to parks, and without them the park system would not exist. How many people visit, and how are they counted? Who are they, what do they do, and when do they come? The answers, discussed in chapter 8, can be quite varied, and it is remarkable how little information the NPS has about these issues.

Visitation to the entire system has almost always shown an upward trend, reflecting the growing popularity of parks, as well as the greater numbers and mobility of Americans. Since 1985 the pattern has been more varied, with some years and units seeing increases but not others. The most visited units are usually the closest to the most people, including urban recreation areas and the memorials of Washington, D.C. Parkways are also remarkably popular. Redesignating a national monument to a national park generally gives a modest rise in visitation, though this may only last a few years. Parks tend to be quite seasonal, with most parks having a summer peak, though some desert and Florida parks have a winter peak.

The NPS has relatively little data about who comes to parks, but it is evident that most visitors are local. Yet parks also attract visitors from great distances and from other countries; over one-third of foreign visitors to the United States visit a national park unit. These visits

are often to larger and well-known parks such as the Grand Canyon but also include some that are easy to get to. War in the Pacific National Historical Park on the island of Guam gets most of its visitors from Korea and Japan, which contain the closest big cities and have the most frequent airline service to the island.

We also look at what people do within parks and how these activities vary by park type and location. While enjoying the scenery is important, there are many possible activities, including hiking, cycling, camping, climbing, boating, learning about history, and even hunting in a few units. Not everyone is capable of hiking through rugged wilderness, and the NPS has been working to make parks more accessible to those with various mobility disabilities.

Park visitors are not a cross section of the American population; instead, they reflect particular age ranges, incomes, or racial and ethnic groups. Not the wealthiest, but older and higher income populations make up disproportionate shares of visitors (perhaps especially in Alaska), as they are more likely to have the time and money for long park vacations. Park visitors are also far whiter than the American population, leading to a growing movement to attract a more diverse crowd to the park system.

Parks need money to operate, and there are vast amounts to be made by those serving the millions of visitors who visit each month. Chapter 9 discusses park budgets, visitation, and maintenance backlogs of parks throughout the system, as well as the various sources of funds and what this money is spent on. Parks also contain many businesses or concessions that provide services to visitors, such as lodging, food, tours, river rafting, mule rides, and flightseeing, collectively earning more than $1 billion each year. NPS policies have long favored monopolies in parks, and several large corporations operate in the park system. Gateway towns outside a park serve visitors and may even eclipse the park itself. Parks have long been promoted for their economic benefits, and this remains a popular selling point for creating new parks.

Chapter 10 discusses threats to the park system. When the Park Service was created in 1916, it was charged with what has become known as the "dual mandate" to both preserve the parks and open them up to the public to the greatest extent possible. This mandate has produced an unending string of controversies over opening up parks to more visitors and activity versus keeping them as undeveloped as possible.

The parks have also been threatened by many outside forces. Remoteness was an early complaint about the parks that has largely receded. Resource extraction has been and remains a major threat to many parks, with many units containing important mineral or timber reserves whose value will not diminish in a world with growing appetites for products. Mining has been allowed in some parks, and the expense of cleaning up the messes left behind has been a headache for park managers. The country's growing population presents some of the largest threats to the integrity of the park system, including congestion; incompatible land uses crowding around park boundaries; air, noise, and light pollution; and long-term warming and drying of the climate. Another group of threats are those existing between the NPS and other government agencies, including recurring battles over whether the parks should be protected or used for national defense. More worrisome are changing cultural values that may endanger the visitor base parks depend on, as well as question the idea of parks altogether.

In chapter 11 we take a larger view of parks and other protected areas, comparing the U.S. system to those of other countries, some of which are quite different in function. Comparing American parks to others can reveal important truths about our system that are easily overlooked. Among these is the fact that American parks are fundamentally based on removing people from areas to be protected as a park. In many countries this would be unthinkable, and national parks in those countries may host substantial populations.

Just as chapter 4 included a tour of America's park system, in chapter 11 we take a quick tour around the world, looking at protected areas on each continent and in some countries. Many face threats far more serious than in the United States. We also look at protected areas in international waters and Antarctica. Finally, the chapter examines possibilities for protected areas on the moon and Mars, the final frontier for national parks.

Chapter 12 discusses the possible futures of the system. We discuss how the geographic forces at work may impact the future of the parks during the rest of the century.

A few final notes: throughout the book we generally refer to parks by their current name for convenience (Acadia rather than Sieur de Monts, Denali rather than Mount McKinley, Zion instead of Mukuntuweap, etc.). We also often omit the designation of a unit (such as national park or monument or historic site); these too have often changed, and these titles can become bothersome when repeated. Designations are also often abbreviated. And we frequently refer to the size of parks by their acreage. One acre contains 43,560 square feet, or nearly the size of a football field (minus the end zones), and there are 640 acres in a square mile.

TWO

America's National Park System

What is a national park? The Grand Canyon is one, a vast empty canyon with breathtaking views protected against development and serving as an inspiration to all who experience it, but there are many others that are vastly different. The Civil War battlefield of Shiloh, a crowded beach on the Gulf Coast, a museum devoted to a former president in downtown Cincinnati, a small battlefield nestled within suburban subdivisions in Greensboro, North Carolina, and the marble memorials in Washington, D.C., are all examples of units in the national park system.

This chapter explores exactly what the national park system and National Park Service (NPS) are, why they exist, and how parks are created. Many people might say they would have no problem spotting a national park when they see one, but the reality is that America's park system is made up of a variety of places, some of which look much like those of other federal systems. Those other agencies have employees who wear similar uniforms to a park ranger, and you may even visit a park without realizing which agency runs it. This chapter also examines state, local, and tribal parks, national forests, BLM recreation areas, wildlife refuges, and wilderness. It also compares their function and geography to those of national parks.

What Is the National Park System?

America's national park system comprises all the parks, monuments, recreation areas, historic sites, battlefields, parkways, lakeshores, seashores, and other units administered by the NPS, among them the Grand Canyon (Fig. 2.1).

It would be simpler just to refer to them all as parks, but there are more than twenty different designations in use for the national park system (Rettie 1995; Comay 2013) (Table 2.1). Eleven of these designations are found only in the Washington, D.C., area, where they make up part of the national capital parks (Comay 2013). Many of these labels are quite similar, such as national historic site and national historical park and the several kinds of battlefield parks. Further, the various designations often make little logical or functional sense. The Blue Ridge Parkway is a scenic road running through a linear park, while Shenandoah National Park is a linear park with a scenic road running through it. The visitor experience in the two park units is perhaps very similar.

There has been discussion of whether the number of designations should be reduced to better communicate the purpose of a park to the public. On the other hand, reducing the number could also limit the flexibility of the park system to encompass new types of places or themes. A simple classification is to divide the park system into three basic types of units: scenic, historic, and recreational; this approach will often be used in this book. The legislation that creates a park specifies its purpose, what activities are allowed or excluded, and how the park is to be operated; these are of greater importance than whatever label may be attached to the unit.

Another basic but difficult question about national parks is how many there are. As of December 2022 the official count is 424 units. Where does this number come from? What makes one park separate from others nearby? The Washington Monument, Lincoln, Jefferson, Martin Luther King, Jr., Vietnam, Korea, and

World War II National Memorials are all within sight of each other but are treated as separate units. They are all located on the National Mall, considered a separate park unit. Many parks are made up of multiple units (as many as thirty-eight) but are still counted as one park. Some parks are closed to the public, yet they are still listed as part of the official count.

Further, the official count of parks ignores the fact that some seemingly separate park units are jointly managed or operated as one unit. Sequoia and Kings Canyon National Parks in California's Sierra Nevada are adjacent to each other and operated as one park, with one budget and superintendent in charge. There are 364 distinct park units in the 2021 NPS budget (National Park Service 2020d), with small parks grouped together and operated as one. The Cape Hatteras Group includes Fort Raleigh, Wright Brothers, and Cape Hatteras, all located near each other in North Carolina's Outer Banks. The Flagstaff Area Parks is made up of three small monuments in Arizona. And most of the NPS sites in and near Washington, D.C., are in one of three groups (called the National Capital Parks East, the National Mall and

TABLE 2.1 National park unit designations, 2021

Designation	Abbreviation	Number
SCENIC PARKS		
National Park	NP	63
National Preserve	Npres	19
National Reserve	NR	2
RECREATIONAL PARKS		
National Lakeshore	NL	3
National Parkway	Nparkway	4
National Recreation Area	NRA	18
National River	Nriver	4
National Scenic Trail	NST	3
National Seashore	NS	10
National Wild and Scenic River	NWSriver	10
HISTORICAL UNITS		
International Historic Site	IHS	1
National Battlefield	NB	11
National Battlefield Park	NBP	4
National Battlefield Site	NBS	1
National Historical Park	NHP	61
National Historical Park and Preserve	NHPP	1
National Historic Site	NHS	73
National Memorial	Nmem	31
National Military Park	NMP	9
National Monument	NM	85
Other designation		11
TOTAL		**423**

Note: Some of these abbreviations are used by the NPS; others have been created by the authors.

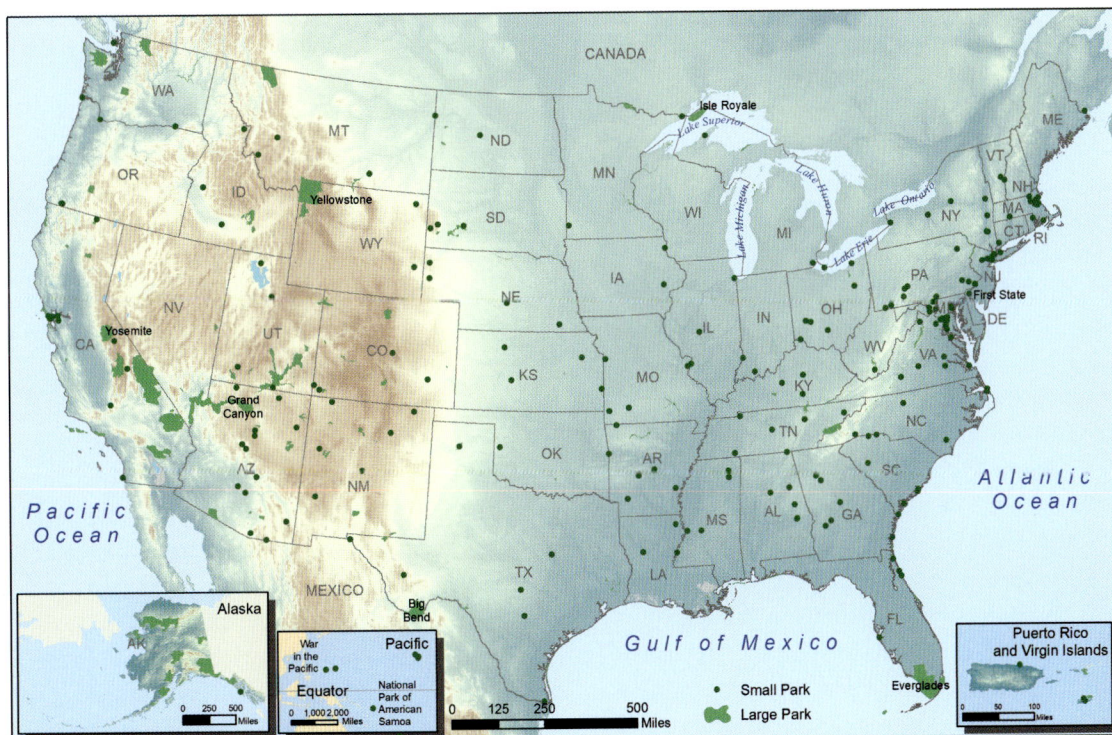

FIG. 2.1   Map of the national park system in 2021.

Memorial Parks, and the George Washington Parkway). The reverse is also possible. Klondike Gold Rush National Historical Park consists of two units over nine hundred miles apart in Seattle, Washington, and Skagway, Alaska. Each is operated independently and is officially a separate park. For convenience, this book uses the official NPS count of 424, but it should be remembered that this is a somewhat arbitrary number.

The National Park Service

For the first forty-four years of their existence, the national parks were overseen by the Department of the Interior. In 1916 President Wilson signed a law creating the National Park Service. The NPS was to manage the fourteen national parks, as well as twenty-one national monuments previously administered by the Department of the Interior (Mackintosh 1991). The law that created the National Park Service specified that "the service thus established shall promote and regulate the use of the Federal areas known as national parks, monuments, and reservations hereinafter specified by such means and measures as conform to the fundamental purpose of the said parks, monuments, and reservations, which purpose is to conserve the scenery and the natural and historic objects and the wild life therein and to provide for the enjoyment of the same in such manner and by such means as will leave them unimpaired for the enjoyment of future generations" (Dilsaver 1994a, 46). This statement has since become known as the "dual mandate" and in practice means that the NPS must preserve both nature and history while also opening parks up to tourist development, including the construction of roads, restaurants, trails, lodging, campgrounds, and cell phone towers (Foresta 1984; Sellars 1997; Runte 2010). The dual mandate has been the source of endless controversy.

Any new agency requires a director, and the first for the NPS was Stephen Mather. He was born into a wealthy family in San Francisco in 1867, became even wealthier in business, and by middle age was searching for new challenges (Shankland 1951). He loved camping in the mountains, and promoting America's na-

tional parks became his new cause. Upset at the poor condition of the parks he visited, he wrote the secretary of the interior and urged improvements. The secretary wrote him back and asked him to come and do the job himself (Mackintosh 1991). He did so as the first director of the National Park Service, remaining in that position until 1929. Mather was a tireless promoter of the parks and traveled over fifteen thousand miles in his first year of office, visiting all the parks and places proposed as parks and talking to politicians, business people, and locals to seek their support.

Mather suffered from health problems and had to step down in 1929; he died the following year. Fortunately, one of his first official actions in 1917 had been to recruit Horace Albright as his assistant. Albright was born in 1890 in Bishop, California, and was also a strong supporter of the growing system (Swain 1970). He took over as director of the NPS in 1929 and stayed in that position until August 10, 1933, a day that marked the first big expansion of the national park system. He died in 1987, having witnessed tremendous expansion in the system over his lifetime.

After Mather and Albright, the NPS has had seventeen directors up to 2021 (Table 2.2). Since 1964 a director's term in office has generally coincided with a presidential term, with each incoming administration appointing a new director. President Donald Trump's 2018 nominee was never confirmed by the Senate. "Chuck" Sams III was sworn in as the nineteenth director on December 16, 2021. The achievements and reputation of these directors among park employees vary; some worked their way up through the park service, others were political appointees with no understanding of the parks and how to manage them. Among them, Conrad Wirth was notable for directing the CCC program in the 1930s and initiating the Mission 66 program to rebuild parks in the 1950s. Several directors have written about their experiences while working in the NPS (Wirth 1980; Albright and Cahn 1985; Albright and Schenck 1999; Hartzog 1988; Ridenour 1994; Mengak 2012), which can provide an insider's view of the politics and pressures involved.

The park ranger is the iconic park employee, with his or her green uniform and flat-brimmed hat. These uniforms were derived from those of the U.S. Army cavalry, the first agency to patrol the national parks. The NPS arrowhead, adopted in its current form in 1951, is displayed on the uniform. The design includes elements that represent the major components of the park system: the tree and bison represent wildlife and vegetation, the mountain represents scenery and recreation, and the arrowhead represents history. The symbol has been trademarked to prevent its inappropriate commercial use.

In the early years, rangers took on all tasks related to protecting wildlife, fighting forest fires, and keeping tourists safe and informed. Today jobs have become specialized, and visitors encounter two kinds of rangers. Some work in visitor centers and give fireside talks or guided tours; these are interpreters, with a deep knowledge of the park and wildlife, archaeology, history, and other topics appropriate to each park. Others are federal law enforcement agents and carry guns; they undergo training at the Federal Law Enforcement Training Center in Glynco, Georgia, which also trains marshals, FBI agents, Drug Enforcement Agency personnel, members of the Bureau of Alcohol, Tobacco, and Firearms, and federal prison guards, among others.

Within a park, there are many other uniformed NPS employees. Many provide maintenance support for the park's roads, trails, buildings, water and sewer systems, and communications facilities. Others are scientists working to study and manage plants, wildlife, caves, and other geological resources; preserve the park's historic treasures; and carry out other tasks the park may require.

Despite its humble origins, the NPS has become a large bureaucracy based in Washington, D.C., with 12,363 employees in 2018. Its headquarters is within the Interior Building, at the northwest corner of C Street NW and 18th Street NW, a few blocks from the White House. While the White House is one of the 424 sites managed by the NPS, the Interior Building is not; it, like many others, is managed by the General Services Administration, a little-known fed-

TABLE 2.2   National Park Service directors

Name	Term of office
Stephen Mather	1917–29
Horace M. Albright	1929–33
Arno B. Cammerer	1933–40
Newton B. Drury	1940–51
Arthur E. Demaray	1951
Conrad L. Wirth	1951–64
George B. Hartzog Jr.	1964–72
Ronald H. Walker	1973–75
Gary Everhardt	1975–77
William J. Whalen III	1977–80
Russell E. Dickenson	1980–85
William Penn Mott Jr.	1985–89
James M. Ridenour	1989–93
Roger G. Kennedy	1993–97
Robert Stanton	1997–2001
Fran P. Mainella	2001–6
Mary A. Bomar	2006–9
Jonathan Jarvis	2009–17
"Chuck" Sams III	2021–

Note: There was no director between 2017 and late 2021, as President Trump's 2018 nominee was never confirmed.

eral agency that manages many federal buildings and vehicle fleets. In 2010 the building was named for Stewart Udall (1920–2010), a secretary of the interior who was an important friend of parks.

Each park has a superintendent in charge, with considerable freedom to carry out policies they find most appropriate. Like many federal agencies, the NPS has adopted a regional structure, with seven regions overseen by regional offices. Superintendents report to their regional office, which in turn reports to the headquarters in Washington, D.C. There are also several organizations within the agency that handle special needs. The Denver Service Center handles planning, design, and construction management. Some centers are within parks: the Horace Albright Training Center is at the South Rim of the Grand Canyon. The Harpers Ferry Service Center is in the West Virginia portion of Harpers Ferry National Historical Park and handles cartography and graphic design. The park maps and brochures familiar to every park visitor are designed here (National Park Service 2012). The Harpers Ferry Service Center also helps parks design museum exhibits. But the NPS has far more artifacts to display than visitors will see in those museums. The Western Archeological and Conservation Center preserves 14.7 million historic artifacts from dozens of western parks, ranging from prehistoric artifacts and early photographs to caretaker reports and even bullets from the Little Bighorn battlefield. Located in a large and secure nondescript building west of downtown Tucson, Arizona, the center houses the greatest museum collection you will never be allowed to see.

Where Do Parks Come From?

We tend to think of parks as naturally occurring and eternal, but they are government facilities just like a post office. They require staffing, a budget, and customers. Before looking at the development of the system we must look inside the often-messy process of creating and funding parks. There have been four ways a park unit can be created, though two have been rare.

Legislation

Most park units were created when Congress passed a bill, which was then signed into law by the president. This first happened on March 1, 1872, when President Grant signed a bill titled "An Act to Set Apart a Certain Tract of Land Lying near the Head-Waters of the Yellowstone River as a Public Park," known ever since as Yellowstone National Park (Dilsaver 1994a). Since that time hundreds of laws have been passed to create, eliminate, make changes to, regulate, and fund the national park system.

The process is more complicated today; since 1998 the NPS has been required to first investigate proposed park units to determine if they are worth adding to the system (Dilsaver 2008). The agency must evaluate whether the site meets three criteria: national significance, feasibility, and suitability. National significance has been a slippery concept. According to NPS guidelines, "A proposed unit will be considered nationally significant if it meets all four of the following standards: it is an outstanding example of a particular type of resource; it possesses exceptional value or quality in illustrating or interpreting the natural or cultural themes of our Nation's heritage; it offers superlative opportunities for recreation for public use and enjoyment, or for scientific study; it retains a high

degree of integrity as a true, accurate, and relatively unspoiled example of the resource" (National Park Service 2018a, 1). Suitability requires that a similar unit not already be included in the park system or not contained within the system of any similar land management agency such as the Forest Service. Feasibility is perhaps the most straightforward of the three, as the potential unit must be of sufficient size for resource protection and public use. A proposed park is not feasible if it would cost too much to purchase, would encounter tremendous opposition or is otherwise unlikely to be politically possible, or is in private ownership that is unlikely to part with it.

There are limits to what might be considered suitable. The guidelines specifically exempt "cemeteries, birthplaces, graves of historical figures, properties owned by religious institutions or used for religious purposes, structures that have been moved from their original locations, and reconstructed historic buildings and properties that have achieved significance within the past 50 years" (National Park Service 2018a, 1) from consideration, though exceptions may be made.

When a bill to create a park unit is being put together, Congress must decide what type of designation to use: Should it be a park, preserve, lakeshore, recreation area, or another type? There are no formal criteria for each of these, and Congress is free to create a new designation (Rettie 1995; Comay 2013). The purpose of the park should be identified in the bill, and restrictions on hunting, mining, or other activities must also be specified.

As does any other legislation, a park proposal starts with two bills, introduced in the House of Representatives and Senate. Each bill will be assigned to a committee for a hearing; important committees for national park issues in the House are the Committee on Natural Resources and the Subcommittee on Public Lands and Environmental Regulation and in the Senate the Committee on Energy and Natural Resources, the Subcommittee on National Parks, and the Committee on Environment and Public Works (National Park Service Office of Legislative and Congressional Affairs 2021). If the bills survive hearings in these committees, they will be voted upon by the full House and Senate. If the bills are successful in both houses, they will be reconciled and sent to the president for signing into law.

Very few bills pass all the way into law, and many parks were created only after several bills failed over several years or even decades. It took fourteen bills and sixteen years for Crater Lake National Park to be established despite a lack of any serious opposition in Congress (Hampton 1981). Many other parks were similar: Mesa Verde in southwestern Colorado was first proposed as a national park in 1891 but was not established until 1906, while Grand Canyon National Park was first proposed in the 1880s but not created until 1919. Big Thicket National Preserve in Texas may hold the record at twenty-eight bills submitted to create it before success (Cozine 2004).

Of the forty-one national parks created by 1980, only one-third passed Congress without serious controversy and delay. Yellowstone was a remarkable exception for its rapid passage through Congress; the bill was introduced on December 18, 1871; it passed in the Senate the next month and in the House in February. Among later parks, Wolf Trap may hold the record for swiftness: the bill was submitted in May 1966 and signed by President Johnson that October. The NPS has been regretting it ever since, as the northern Virginia park became mired in conflict with a private foundation and was in constant need of ever-larger amounts of money (Mackintosh 1983). Sometimes Congress creates a park against the wishes of the NPS; this happened when the NPS was opposed to making Pinnacles National Monument in California into a national park due to a lack of adequate scenery. Congress disagreed, and the monument became a park in 2013.

Because parks located in eastern states require that land be purchased before the park is created, these parks go through an additional step. A law was first passed authorizing the NPS to begin acquiring land, and another law was passed to establish the park once that goal was met. The Great Smoky Mountains went through these two steps in 1926 and 1934, Shenandoah in 1926 and 1936, Big Bend in 1936 and 1944, and Cape Cod in 1961 and 1966. This can create confusion over exactly when these parks came into being. There are several places that passed through the first step but not the second and so never became park units. Palm Canyon in California, Grandfather Mountain in North Carolina, Patrick Henry National Memorial in Virginia, and a Spanish-American War memorial park in Florida are among examples (Hogenauer 1991a). An ongoing case is the effort to establish the Ronald Reagan Boyhood Home National Historic Site in Dixon, Illinois. This was authorized in 2002 but has not yet been established, as the current owners have not agreed to sell the house to the NPS.

When a new park unit is created it must also be funded in a separate bill by the House and Senate Committees on Appropriations. It is entirely possible for a park to be created but not funded for several more years, as with Denali, which was created in 1917 but had no money until 1921 (Norris 2006).

In the early years, legislation usually involved one park at a time, but it has become common for parks to be created as part of a package within a large bill. The National Parks and Recreation Act of 1978 was one such bill. It marked a major expansion of the park system, though one that is still debated today. The law created or authorized nine new parks, modified the boundaries of twenty-six units, and renamed or redesignated four more. The new units were a mix of small historic sites and battlefields, as well as the Santa Monica Mountains National Recreation Area in Los Angeles. The act was seen at the time (and ever since) as an example of pork-barrel politics (or, in this case, park-barrel politics), with individual members of Congress working to get new units established in their districts, regardless of whether the NPS had any interest in these sites (Rettie 1995). This type of expansion into those areas where influential members of Congress hold sway is an unfortunate consequence of the American political system, as well as evidence of the growing politicization of the park system.

Parks may be created in bills that don't seem related to parks at all. The Defense Authorization Act of 2015 was an enormous $600 billion defense spending bill passed in December 2014. Hidden within it was legislation to

FIG. 2.2   Medgar and Myrlie Evers Home National Monument. This house in Jackson, Mississippi, is one of the newest park units. The Everses were civil rights activists, and Medgar was shot and killed in the carport. Photo by Jud McCraniem, 2018. Wikipedia, Creative Commons Attribution–ShareAlike 4.0 International License.

create eight national park units: Tule Springs National Monument was created north of Las Vegas; Valles Caldera appeared on the map in New Mexico; a World War I memorial was authorized for Washington, D.C.; and the new multistate Manhattan Project National Historic Site was created in Tennessee, New Mexico, and Washington. Several units were also expanded, including Oregon Caves. In 2019 White Sands National Monument was redesignated a national park by the National Defense Authorization Act for fiscal year 2020. The park was mentioned in section 28, part 51b of the 1,119-page law.

As of 2021 the most recent large park legislation also had a misleading title: the California Desert Protection and Recreation Act of 2019. It expanded Death Valley, Joshua Tree, and Mojave National Preserve in the California desert but also dealt with a variety of public lands in other states. It authorized the new Medgar and Myrlie Evers Home National Monument in Mississippi (Fig. 2.2) and Mill Springs Battlefield National Monument in Kentucky (both later established in 2020). It eliminated the World War II Valor in the Pacific National Monument and replaced it with three separate monuments:

Pearl Harbor Memorial and Tule Lake along with a third, non-NPS site in Alaska. It also redesignated several units, adjusted the boundaries of others, and authorized special resource studies of several potential new park units.

Proclamation

Another pathway to a national park unit exists due to the Antiquities Act of 1906, a law that allows the president to create (or proclaim) a national monument to protect a resource, with southwestern archaeological sites the original motivation (Rothman 1994a, 1994b; Harmon, McManamon, and Pitcaithley 2006). Using this method, a short statement is written describing the feature to be protected and its boundaries; it is then signed by the president. A national monument now exists, though a separate bill must still be passed by Congress to provide funds for the new monument.

President Theodore Roosevelt signed the Antiquities Act into law in 1906 and then used it to create Devils Tower National Monument in Wyoming, along with seventeen more monuments during his time in the White House. Presidents who followed him also made frequent use of the act: Presidents Taft, Wilson, Harding, Coolidge, and Hoover had created fifty-four more national monuments by 1933. These presidents are not well known today, but their use of the Antiquities Act was vital to the growth of the park system. Hoover's use of it was particularly noteworthy for expanding the system into the southwestern deserts with the addition of Death Valley and White Sands. Hoover was also responsible for the dam that created the first large national recreation area, Lake Mead, a shimmering blue lake in this desert region (Dodd 2007).

President Franklin Roosevelt was an enormous supporter of the national park system and proclaimed eleven more monuments during his thirteen years in office. But during his extended stay in the White House, he also signed legislation allowing for new kinds of park units to be created, such as recreation areas, parkways, memorials, battlefields, and historic sites. This allowed an expansion of the park system but reduced the importance of monuments; they were no longer a miscellaneous category for any site not worthy of being a park. The Antiquities Act was used only once during World War II and only sparingly afterward; the next four presidents used it only six times, and Presidents Nixon and Ford did not use it at all. This was a time of tremendous growth in the park system, but it was accomplished through legislation, reflecting a consensus among Americans and their elected representatives of the value and importance of the park system.

From President Carter's tenure onward, this consensus has diminished. In 1978 Carter created fifteen monuments in Alaska, the single biggest expansion in the history of the national park system. Since then, proclamations to create new national monuments have been issued almost exclusively by other Democratic presidents: Clinton proclaimed twenty-one (though not all were under the administration of the NPS) and Obama twenty-eight, the most of any president. In contrast, President George W. Bush used the act twice, President Trump used the act to create a new monument once, while Presidents Reagan and George Bush did not use the act at all.

Because they require only a presidential signature, monuments can be created quickly and without submitting a park proposal to debate in Congress. The Antiquities Act sidesteps the questions of natural significance, feasibility, and suitability that must be addressed by any park unit created by Congress. While park guidelines rule out cemeteries as a possibility for park legislation, proclamations are not so limited. African Burial Ground National Monument, an enormous cemetery beneath New York City, was created this way in 2006.

But there are disadvantages. The Antiquities Act has been controversial due to the size of many monuments, the vast range of features to be protected, and the absence of any immediate threat that would warrant creating a national monument to protect a place, along with opposition based on the potential negative effects on land use (Vincent and Baldwin 2004). As members of Congress are unlikely to vote for a park that their constituency and campaign donors strongly disapprove of, a successful park bill is one that has gained widespread approval and support. The use of the Antiquities Act may ignore the wishes of Congress and local citizens, who may resist or undermine the new units in a variety of ways. The Antiquities Act is also not subject to review under the National Environmental Policy Act. Establishing a new national monument can be accomplished much easier than building a road to the monument, which would require such a review.

The first strong opposition to monuments occurred in 1943. After a small Grand Teton National Park was created in Wyoming in 1929 the NPS began a secret campaign to expand the park eastward into the Jackson Hole valley (Righter 1982). The NPS recruited John D. Rockefeller Jr., the richest man in America, to work behind the scenes to buy up much of the valley's land, which he then donated to the NPS. President Roosevelt created Jackson Hole National Monument in 1943 out of this land, enraging many remaining landowners around the valley, who worked to ensure that no funding would go to the new monument. After a few years, tempers cooled enough for Congress to join the monument with Grand Teton National Park to create a larger Grand Teton National Park in 1950, but Congress's support for this action came with the price of banning the Antiquities Act from further use in Wyoming without congressional approval.

President Carter's proclamations of several large monuments in Alaska provoked more outrage over the act. In 1980 many of these monuments were made into national parks, but again, congressional approval came with the cost of banning the Antiquities Act from Alaska. In 1996 President Clinton proclaimed the first of a series of monuments, including the controversial Grand Staircase–Escalante in Utah (not under the administration of the NPS). This and the later Bears Ears National Monument (also not within the park system) were reduced in size in 2017 due to strong opposition by some people living near them, though Grand Staircase–Escalante was restored to its original size in 2021. Bills continue to be introduced to further reduce the freedom of the president to issue proclamations, and this path to parkhood may therefore be further diminished in the future.

The status of monuments is further complicated by the fact that Congress can also pass legislation authorizing a national monument and has done so several times since 1929, beginning with Badlands in South Dakota. Monuments proclaimed by the president can also later be made into national parks by Congress. Many large national parks went through this process (e.g., Zion, Grand Canyon, Death Valley, Joshua Tree, Bryce Canyon, and most large parks in Alaska). For these parks, monument status was a stepping-stone to full parkhood. This has most recently occurred with Camp Nelson National Monument in Kentucky, which was created by proclamation in October 2018 and again by legislation in March 2019.

Other Methods for Creating Parks

National park units have been created in other ways that do not require either the president or Congress. The Historic Sites Act of 1935 gave the secretary of the interior the authority to designate national historic sites, and several secretaries have used this power to create thirteen of these sites. Most of these have since been redesignated into other kinds of units by Congress, but two (Vanderbilt Mansion in New York and Pennsylvania Avenue in the District of Columbia) remain in the system because of their original interior secretary proclamations.

Several reservoir-based national recreation areas were created through interagency agreements between the NPS and either the Bureau of Reclamation, the Army Corps of Engineers, or the International Boundary and Water Commission (a joint U.S.-Mexico agency that manages the Rio Grande). The NPS was essentially a subcontractor handling the recreational management of a reservoir created by these agencies. The first of these was Lake Mead in 1936. The Bureau of Reclamation had built Hoover Dam on the Colorado River (Fig. 2.3) and brought in the NPS to develop beaches, campgrounds, and marinas. This unit was expanded to include Lake Mohave in 1947, again through an interagency agreement. The unit is the sixth most visited in the entire system but came about not through the actions of Congress or the president but by bureaucrats working out details of construction contracts. Not until 1964 was a law passed by Congress regarding this unit. Eleven other reservoir recreation areas were created this way in the 1950s and 1960s, though all but two (Washington's Lake Roosevelt and Curecanti in Colorado) have been either formalized by legislation or removed from the system.

It can be difficult to figure out exactly how a park came to be created and the steps it went through. Records of successful legislation on the national park system and proclamations have been collected by the National Park Service Office of Legislative and Congressional Affairs (2021) and displayed on its website. The documents don't always make for easy reading, but they can be fascinating glimpses into the development of park units.

Table 2.3 shows the original legal basis for the creation of the 423 park units that existed in 2021, as well as the 29 delisted units. Legislation accounts for almost three-quarters of all parks created, with presidential proclamation accounting for most of the rest. Aside from large parks such as Yellowstone, Yosemite, Sequoia, Kings Canyon, and Grand Tetons, most western parks were created through presidential proclamation. In contrast, most eastern units started with the "traditional" route via

FIG. 2.3   Hoover Dam on the Colorado River, 1941. This dam resulted in the creation of Lake Mead, the country's first national recreation area. Photo by Ansel Adams. National Archives and Record Administration, NAID 519837.

TABLE 2.3   The legal basis for creating park units, 2020

Legal basis for creation	Number	Type	Percent	Status of basis
Legislation	317	All	74.07	Valid
Presidential proclamation (Antiquities Act of 1906)	105	NM	24.53	No longer valid in Wyoming and Alaska if more than five thousand acres
Interior secretary proclamation (Historic Sites Act of 1935)	13	NHS	3.04	Valid, last used in 1987
Interagency memorandum of agreement	12	NRA	2.80	Valid, last used in 1965

congressional legislation. A small number of units along the East Coast were proclaimed by the president; most of these are coastal forts. Other small urban units were proclaimed by the secretary of the interior, and several western recreation areas were created by agreements between the NPS and other agencies.

Other Responsibilities of the National Park Service

The mission of the NPS has been expanded far beyond managing the nation's scenic and historical treasures; the NPS has also become the nation's historic preservation office as well.

The Historic Sites Act of 1935 was designed to add historic sites to the national park system (Mackintosh 1985). The act required a survey to identify historic sites for inclusion in the national park system. This survey would systematically examine locations associated with U.S. history with the goal of including representative sites from each era or major historical event. In 1936 a list of potential historical park units was created, but few of them became parks. One of the sites identified was the Blair House, then a private residence in Washington, D.C. The owner did not want to sell the house but worked with the NPS to have a plaque put on the house describing its historical status, the beginning of a tradition of marking historical properties (Mackintosh 1985).

After World War II the inventory process started again, still with the goal of adding to the park system. The issue of privately owned properties being considered nationally significant but not being added to the NPS led to the idea of designating these as national historic landmarks in 1960. A national historic landmark would remain in private ownership and receive no federal funds, but the owner would be allowed to purchase a plaque and certificate. National historic landmarks were therefore sites of national historical significance; the structure is of suitable quality for inclusion in the park system but for one reason or another is not part of the park system. Designating a property as a national historic landmark proved to be a useful means of heading off political pressure to add a location to the national park system. The program was also expanded to allow national historic landmark designation of buildings or sites already in the park system. National historic landmark status was removed if a property was destroyed or significantly altered. As of 2021 there are 2,596 national historic landmarks, predominantly in large cities.

A similar program to identify the nation's most significant natural features was launched in 1962 but has never had the visibility or support of historical landmarks. There have been 599 places designated as national natural landmarks, including Arizona's Meteor Crater, California's Mount Shasta volcano, and even a freeway cut through a mountain in Birmingham, Alabama.

The National Register of Historic Places was begun in 1966 as a list of all historical properties in the country (Mackintosh 1985). These include homes, churches, synagogues, mosques, bridges, hotels, theaters, mills, factories, schools, hospitals, sculptures, battlefields, ships, shipwrecks, archaeological sites, and even entire neighborhoods. There are over ninety-five thousand properties currently listed on the national register distributed throughout the fifty states, every territory, and even four foreign countries. Only 44 of America's 3,243 counties or their equivalent are without a national register site.

The NPS does not own these places or have anything to do with their care or management; almost all are privately owned. The owners of these buildings or sites may get tax benefits for preserving them, but there are otherwise no protections for these places or restrictions on what owners can do with them. State historic preservation offices must begin the process of documenting the history of a place and nominate it to the national register.

Other Federal Land Agencies

The NPS is only one of several federal agencies that manage and protect the nation's public lands, meaning those owned by the federal government (R. Wilson 2016). These agencies have their own identities and purposes but are easily confused with one another.

The Bureau of Land Management

The Bureau of Land Management (BLM) is one of the least known public land agencies, yet it administers more land than any other and is today the closest competitor to the NPS. It was created in 1946 from a merger of the General Land Office and the Grazing Service (Skillen 2009) and oversees mining, ranching, and energy development in western states. The land the BLM manages is left over from the settlement of the frontier; this land was not homesteaded and did not contain resources warranting a national park unit, forest, or wildlife refuge. The agency has almost twelve thousand employees and earns $13 billion a year in revenues (against a $1.1 billion budget), making it one of the very few public agencies that turn a profit.

Although environmentalists frequently deride the BLM as the "Bureau of Livestock and Mining," it manages an enormous collection of spectacular public lands for recreation, scenery, and environmental protection (Wilkinson 1992). This management began in 1967 outside Las Vegas when the BLM decided to develop Red Rock Canyon for its scenery and recreation (C. Williams et al. 2015). This area had been inspected by the NPS but turned down as a potential national park unit; the BLM has built a visitor center, roads, hiking trails, and picnic ar-

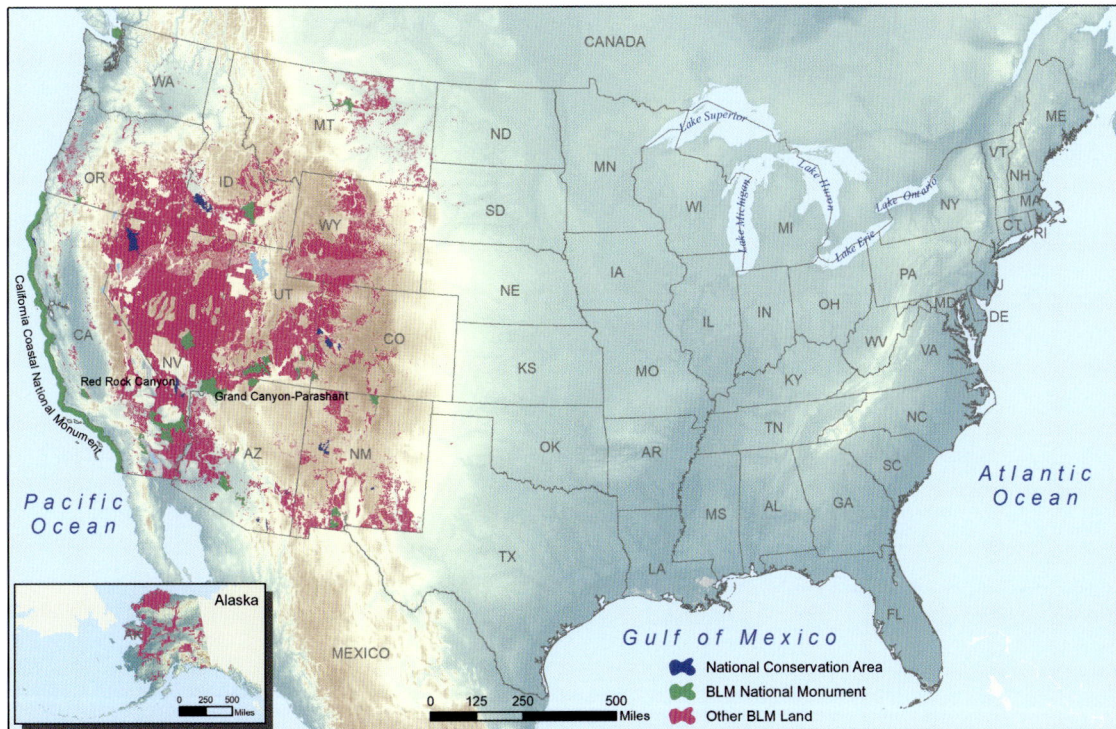

FIG. 2.4   BLM national conservation lands, 2021. These lands are all found in the West, where they contain a wide variety of landscapes and uses.

eas worthy of any national park unit. The Las Vegans who fill its parking lots and trails every weekend care little who manages it; they know a spectacular place when they see it. Other national conservation areas followed in the western states.

In the 1990s the BLM received control of several new national monuments, a break with tradition, because since 1933 all had been administered by the NPS. The first of these was Grand Staircase–Escalante in southern Utah (Vincent and Baldwin 2004; Squillace 2006). This monument was 1.8 million acres of spectacular canyons and cliffs, bigger than all but a few national park units in the forty-eight contiguous states. It filled in much of the land between Bryce Canyon, Capitol Reef National Park, and Glen Canyon National Recreation Area.

This spectacular monument was followed by many others, including Ironwood Forest and Agua Fria in Arizona, Mojave Trails in California, Basin and Range in Nevada, and Grand Canyon–Parashant in Arizona, all containing vast undeveloped desert valleys and mountains. In 2000 the BLM organized its national conservation areas, national monuments, and other

lands into the national conservation lands (Fig. 2.4). These now comprise 25 national monuments, 221 wilderness areas, national historic and scenic trails, national wild and scenic rivers, and 16 national conservation areas in western states. The BLM asserts that the national conservation lands are not a separate park system but a set of areas managed to preserve environmental conditions and landscapes, but in many ways, these areas function as parks. It will be interesting to see how this system is developed in the future. One certainty is that it will not move east as the NPS did, because the BLM has no land in the eastern United States and is unlikely to ever have any.

The United States Forest Service

The first national forest reserves were created in 1891 (Steen 2004). The number of these reserves grew rapidly in the following years and came to be known as national forests. The United States Forest Service (USFS) was created in the Department of Agriculture to manage these forests in 1905. Until 1910 these forests were directed by Gifford Pinchot, who provided strong leadership for the new agency

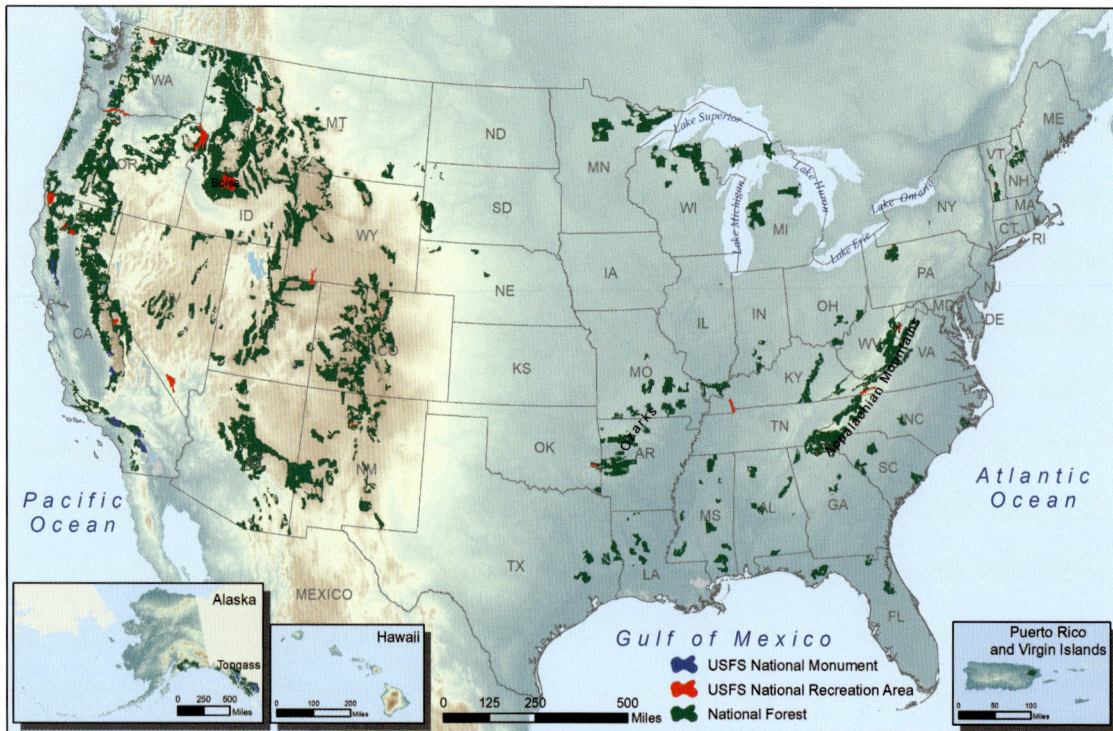

FIG. 2.5   National forests in 2021. Like the national park system, these national forests moved eastward and preserve a range of landscapes.

but often clashed with the Department of the Interior over control of national parks. The USFS managed forests for multiple uses, allowing logging, cattle grazing, mining, and, in later years, recreation.

At the present time, there are 154 national forests, along with 20 national grasslands (Fig. 2.5). Forests are found at higher elevations in western mountain ranges and the Appalachian and Ozark Mountains (as were early national parks), along with the northern Midwest, scattered locations in the South, the Black Hills, and the Tongass forest in the verdant Alaska Panhandle. Grasslands are found in the western Great Plains. Although there it has fewer units, the USFS is a substantially larger organization than the NPS, with over twenty-eight thousand permanent employees.

Over the years, recreation has become an important function for the agency, and many forests (or substantial portions of them) are now managed exclusively for recreation with campgrounds, picnic areas, trails, and ranger stations. These forests are functionally very similar to national park units. The similarities between them have become further blurred, as

the USFS has been given management of seven national monuments and several national recreation areas in recent years. Washington's Mount Saint Helens was made a national monument and developed for tourism after its 1980 eruption. It is quite possible for a visitor to spend a day at the park without realizing it is not part of the national park system (although a movement is under way to bring it into the national park system). The Giant Sequoia National Monument in California is adjacent to Sequoia National Park and easily confused with it; however, it offers few facilities to casual visitors.

Unfortunately, the USFS is undergoing another transition as forest fires become more common and destructive. Over half of the USFS budget is now spent on firefighting, and this amount is expected to grow in coming decades. What will happen to recreation or other uses remains to be seen.

A few words about Smokey Bear. The beloved bear warning against forest fires was part of a USFS campaign that began in 1940; in 1950 an actual bear was rescued from a fire in New Mexico's Lincoln National Forest. The bear was first named Hotfoot Teddy, but the name was

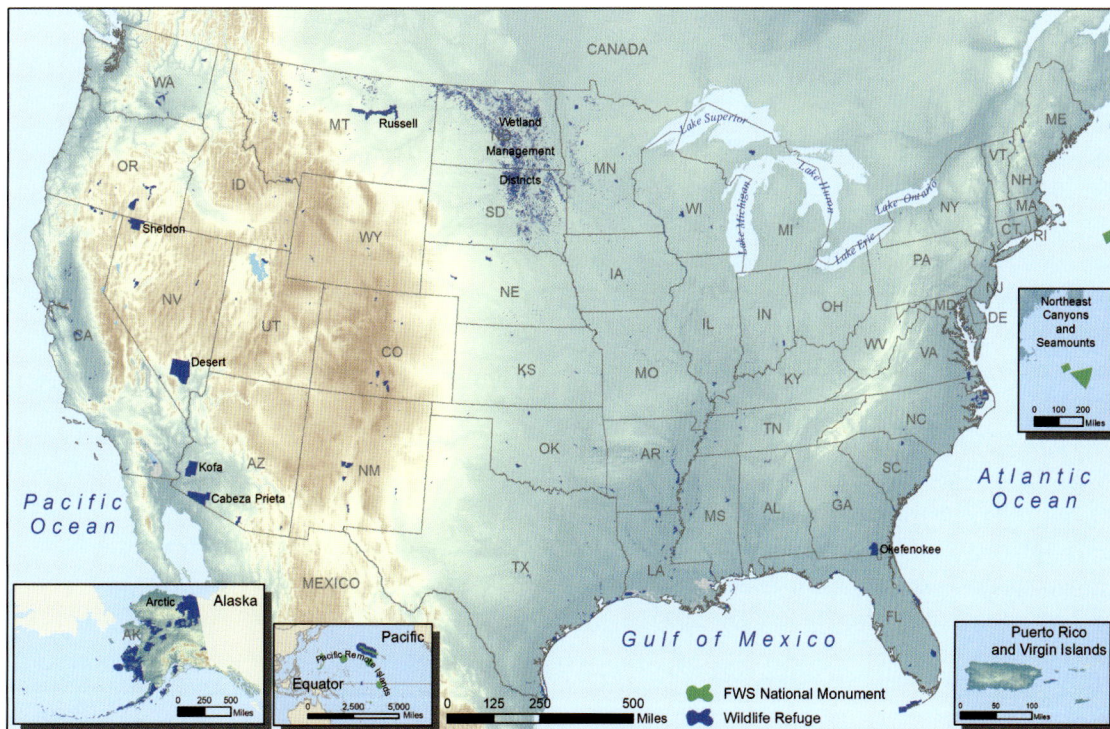

FIG. 2.6   Wildlife refuges in 2021. While most refuges are tiny, there are several notable ones in western states and especially in Alaska.

later changed to Smokey, and the bear became a mascot for the Forest Service's fire prevention campaign. He remains well known today but often to the chagrin of NPS employees, who must explain to tourists that he had nothing to do with the park system but is instead a mascot for the forest system. And park rangers may also correct your spelling: there are lots of bears at Great Smoky Mountains National Park, perhaps the best place to see them outside Alaska, but here Smoky isn't spelled with an *e* (and the bears don't wear pants).

Fish and Wildlife Refuges

Concern for preserving the nation's wildlife became a political force in the late nineteenth century after several species were carelessly exterminated and more were threatened. One solution was to create national wildlife refuges to allow plants and animals enough habitat to survive. The first was created in 1903, and there are now 568 across the country (Fig. 2.6). These refuges and monuments are managed by the Fish and Wildlife Service (FWS), which was created in 1940 and now has about nine thousand employees. The agency is contending

with much the same sort of pressure the NPS is struggling with, including climate change, urban sprawl, and invasive species.

Wildlife refuges encompass a wide range of ecosystems and environments throughout the country. In addition to wildlife and plant protection, refuges are managed for recreation, including hiking, birdwatching, and even hunting and fishing. Like that of the national park system, most refuge acreage is in Alaska; the Arctic National Wildlife Refuge is the biggest (and it is larger than any national park unit or forest). Within the forty-eight contiguous states, the Desert National Wildlife Refuge in Nevada is the largest, and it is larger than all but two national park units outside Alaska. The Kofa and Cabeza Prieta in Arizona, Sheldon in northern Nevada, Russell in Montana, and scattered wetland management districts in the Dakotas are also evident on Figure 2.6.

A substantial portion (37 percent) of the nation's refuges are found scattered across the Pacific. These take the form of seven vast marine national monuments, including Papahānaumokuākea, which encompasses 140,000 square miles of ocean and small islands northwest of Niʻihau and Kauaʻi. The Mariana Trench Marine National Monument was established in January 2009 and includes almost ninety-seven thousand square miles of ocean, including a strip almost eleven hundred miles long that contains the Mariana Trench, the deepest spot in the world. Pacific Remote Islands Marine National Monument includes tiny Howland Island, briefly famous in 1937 when Amelia Earhart disappeared while trying to reach it after a twenty-five-hundred-mile flight from New Guinea. No trace of Earhart, her copilot, Fred Noonan, or their aircraft has ever been found. Today the lonely island is visited only once every two years by an FWS ship.

Added together, the various Pacific Ocean national monuments total over 330,000 square miles, or about half the size of Alaska. The enormous size of these marine national monuments stems from their usually filling the entire exclusive economic zone surrounding each island; according to international law, the zone extends two hundred miles from shore. The actual territorial boundary of the United States extends to twelve miles from shore; most of the area of these marine national monuments is therefore outside the United States and in international waters.

The FWS, along with the National Marine Fisheries Service, administers the Endangered Species Act. This law, passed in 1973, was inspired by the extinction or near extinction of several species in North America. While hunting was a problem in the nineteenth century, in the twentieth the culprit was often the destruction of the habitat a species might need for nesting, feeding, or hiding from predators. Suburbanization, agriculture, reservoir construction, and recreation development have caused the destruction of habitats. The goal of the act is to protect species facing extinction due to habitat loss by protecting the remaining habitat. Once a species is determined to be facing extinction, the FWS is required to identify the critical habitat needed to support the species and develop a recovery plan for it. Bald eagles, grizzly bears, and wolves are among the biggest success stories of the act.

Endangered species might be within a national wildlife refuge but very often are not;

some are within national park units. The Devils Hole pupfish in Death Valley is one example; they live in only one cave and may be the world's rarest fish. Government actions on federally owned land may need to be changed to prevent further harm to an endangered species; in the case of the Devils Hole pupfish, the government has regulated groundwater pumping over a large area to prevent water levels falling below a critical elevation. The act allows the creation of an overlay zone where special regulations apply on federal, state, or private land separate from others discussed in this chapter.

The Smithsonian Institution

Another federal government agency that has substantial overlap with the NPS is the Smithsonian Institution, although it is not a land management agency. The Smithsonian was created in 1846 and became a national history and science museum, as well as carrying out scientific research. Like the NPS, the Smithsonian has expanded far beyond its original location and has many affiliated sites throughout the country, as well as in Panama, but the Smithsonian is best known today for its museums in Washington, D.C. The National Air and Space Museum, National Museum of American History, National Museum of the American Indian, and National Portrait Gallery, among many others, hold many historical treasures and are visited by many of the same tourists who visit the Washington Monument, the Lincoln Memorial, and the Vietnam Veterans Memorial a short walk down the National Mall. Most visitors to the mall are likely not even aware of which facilities are part of the NPS and which belong to the Smithsonian.

The two organizations have substantial overlap in their coverage of historical events: the National Park Service has a site for the Wright brothers' first airplane flight, while the Smithsonian has the airplane. The NPS preserves the Gettysburg battlefield and the site of Lincoln's immortal address but none of the five known copies of it (the Smithsonian has one). But neither agency has the Declaration of Independence, the Constitution, or the Emancipation Proclamation; these treasures are under the control of the National Archives and Records

Administration and on display elsewhere on the National Mall.

Other Agencies

Several other federal agencies manage lands that may be confused with the national park system. In the eastern part of the country, the Tennessee Valley Authority (TVA) and the Army Corps of Engineers manage many large reservoirs as recreation areas, with campgrounds, picnic areas, beaches, and boat ramps. These are quite similar to western national recreation areas developed by the Bureau of Reclamation and the NPS. The National Oceanic and Atmospheric Administration includes the National Weather Service and is not a land management agency; it does, however, manage (jointly, with the FWS) five marine national monuments in the Pacific Ocean. Finally, President Lincoln and Soldiers' Home National Monument in Washington, D.C., is under the jurisdiction of the Armed Forces Retirement Home and the National Trust for Historic Preservation, a nonprofit organization. The monument was created in 2000 and includes a house used as summer quarters by several presidents, including Abraham Lincoln.

Wilderness Areas

"Wilderness" is a common term for describing the natural landscape within many parks, but it also refers to a type of land use created by Congress (Vale 2005). The Wilderness Act of 1964 defines wilderness:

> A wilderness, in contrast with those areas where man and his own works dominate the landscape, is hereby recognized as an area where the earth and its community of life are untrammeled by man, where man himself is a visitor who does not remain. An area of wilderness is further defined to mean in this Act an area of undeveloped Federal land retaining its primeval character and influence, without permanent improvements or human habitation, which is protected and managed so as to preserve its natural conditions and which (1) generally appears to have been affected primarily by the forces of nature, with the imprint of man's work substantially unnoticeable; (2)

has outstanding opportunities for solitude or a primitive and unconfined type of recreation; (3) has at least five thousand acres of land or is of sufficient size as to make practicable its preservation and use in an unimpaired condition; and (4) may also contain ecological, geological, or other features of scientific, educational, scenic, or historical value. (United States Congress 1964)

Wilderness areas differ from most other federal protected lands in that they have no specific agency to manage them; they are administered by the agency that controls the land within them. So while the country's 803 wildernesses are grouped within the National Wilderness Preservation System, there is no unified management. The NPS has 40 percent of the acreage of wilderness areas, the USFS has 33 percent, the FWS has 19 percent, and the BLM has the remaining 8 percent. The majority of designated wilderness areas are in the West, where public lands predominate (Fig. 2.7). Wilderness can be found throughout the Rocky Mountains, the Cascade Range in the Pacific Northwest, and the Sierra Nevada in California. But the heart of the wilderness system is in the southwestern desert ranges and basins. Much of the Mojave and Sonoran Deserts in California and Arizona is included within wilderness lands belonging to the NPS, the FWS, and the BLM.

Wilderness has also spread eastward: the five-thousand-acre requirement noted in the act was subsequently relaxed in 1975 to allow wilderness areas in the eastern United States. The Sipsey Wilderness in Alabama was the first of these. There are now more than 109 million acres of wilderness in the country, making up about 4.5 percent of total land area.

Although the NPS has the largest share, the amount of wilderness in the national park system is surprisingly low. Only about half of national parklands are within wilderness. Park managers often feel that the way parks are operated already provides them with a level of protection that is as high as or even higher than that given to wilderness areas. For that reason, no designated wilderness exists in Yellowstone, Glacier, the Grand Canyon, and many other parks. However, most of Death Valley National

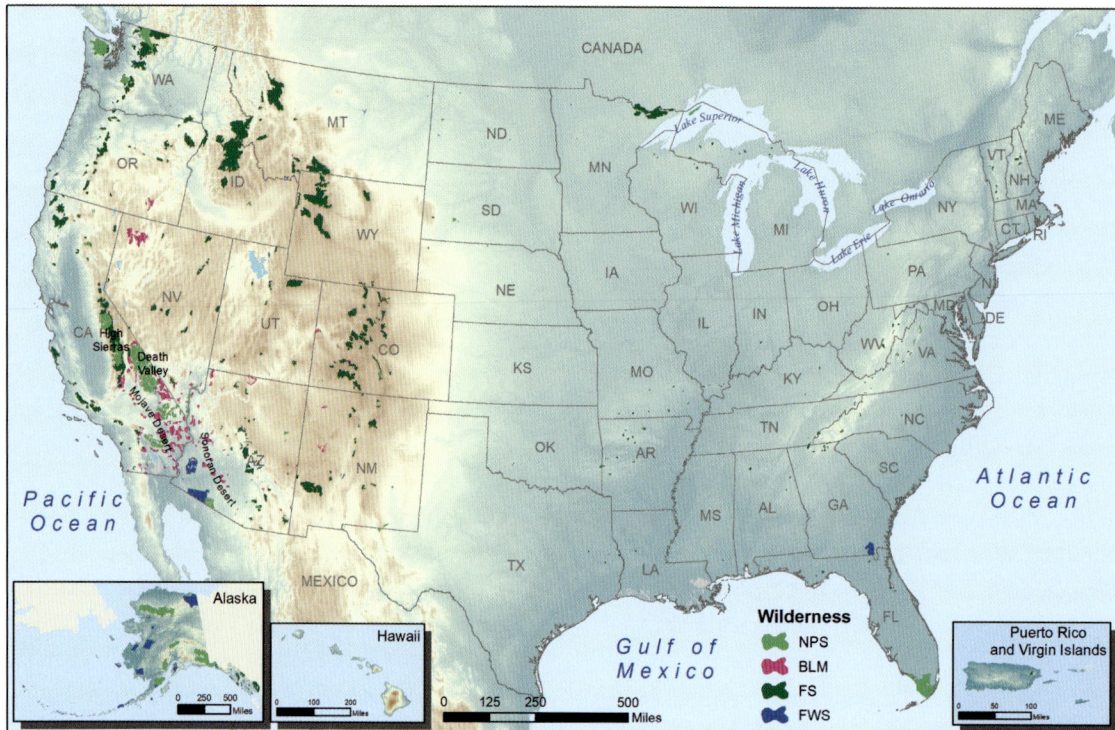

FIG. 2.7 Wilderness areas in 2021. These are concentrated in the western states and especially in the southwestern deserts.

Park is wilderness, the largest such land in the park system.

Rivers and Trails

The nation's rivers are also shared by different agencies. The Army Corps of Engineers manages twelve thousand miles of navigable rivers as parts of the nation's transportation system, including most of the Mississippi River and its tributaries, which must be dredged and dammed to allow safe barge travel. The Bureau of Reclamation and the TVA have built thousands of dams on other rivers for electricity generation and flood control. These developments spurred Congress to pass the Wild and Scenic Rivers Act of 1968 to protect many of the remaining rivers from development. The act currently includes 12,754 miles of waterways on 209 rivers. Most of these are smaller rivers not well known outside their areas; the Chattooga River, which runs through Georgia, North Carolina, and South Carolina, may be the most famous due to the movie *Deliverance* (1972). Rivers located in deserts may be quite small but are still important to their ecosystems. California's Amargosa Wild and Scenic River can be

stepped across but is the only reliably flowing water for many miles.

These rivers are administered by various federal agencies, with the Forest Service having the majority. Seven of those under NPS control are considered units of the park system, while the BLM and Fish and Wildlife Service have others. Some are even under the supervision of a state or local agency.

The National Trails System Act of 1968 created another shared resource: long-distance trails. These are administered by the NPS, the Forest Service, or the BLM. As of 2021 there are eleven national scenic trails and nineteen national historic trails. Of these, the Appalachian National Scenic Trail is by far the most famous, a twenty-two-hundred-mile trail from Georgia to Maine. The Appalachian, Natchez Trace, and Potomac Heritage National Scenic Trails are units of the national park system.

Other Park Systems

There are many other park systems across the country. Several Indian reservations have established tribal parks. The best known is surely Monument Valley (Tsé Bii' Ndzisgaii), on the

Navajo Reservation on the Arizona-Utah border, seen in countless movies. The nearby Ute Mountain Tribal Park in Colorado is another notable example.

All states have some sort of state park system, though these vary tremendously in size and function. The first was at Niagara Falls in New York in 1885, followed later that year by one in the Adirondack Mountains (McClelland 1998). The Adirondack Park is the largest state park in the nation, although 52 percent of the land within it is still in private ownership, and it has a resident population of several hundred thousand people.

Other states followed this lead, and by the 1920s most states were interested in creating parks. The NPS helped create many of these systems (Foresta 1984) and provided advice on park design and operation. Many state parks began as temporary national park system units. These were recreation demonstration areas, developed under NPS supervision in the 1930s. Several of these units later joined the national park system, but most became state parks. Alabama's Oak Mountain State Park, the first and largest in the state, was one of these units.

Today there are over 10,200 state parks. California has the most, with 280, including Anza-Borrego Desert State Park, one of the largest in the country. Big Bone Lick is one of forty-six state parks in Kentucky, Hard Labor Creek one of sixty-three in Georgia, Hocking Hills one of eighty-three in Ohio, Tubac Presidio one of thirty-one in Arizona, Icelandic one of thirteen in North Dakota, and Route 66 one of eighty in Missouri. Many parks are recreational, while others are scenic or historic; the exact mix of these types varies among states.

Counties and cities also have parks, some of them substantial in size. New York's Central Park is perhaps the most famous, though at 843 acres it is only the fifth largest in the city. Golden Gate Park in San Francisco and Griffith Park in Los Angeles are other well-known examples. Forest Park in St. Louis is not much known outside that city anymore but was the site of the 1904 summer Olympics. The biggest city park of them all is South Mountain Park in Phoenix; at 16,283 acres, it is bigger than many national park units. Nearby McDowell Mountains Regional Park is the largest county park in the country. These are rugged Sonoran Desert landscapes offering picnic areas and hiking trails very much like those of nearby national park units. California's Vazquez Rocks Natural Area Park is a much smaller county park but is one of the most recognizable park landscapes, familiar to movie and television fans, especially those of *Star Trek*.

Conclusions

America's national park system consists of many spectacular and inspirational places that are much better known than the government agency created to oversee them. The NPS carries out a wide range of activities within these parks, including wildlife caretaking, law enforcement, highway construction and maintenance, and many other tasks. The NPS is also the nation's history and preservation department; there are many sites near you on the national register.

The creation of many park units is often tied up with individuals who worked tirelessly to see a landscape they loved protected. Arizona's Saguaro National Park had two influential friends. The first was Frank Hitchcock, a former postmaster general who was instrumental in getting Saguaro National Monument created in 1933. He lived nearby and was convinced of the value of protecting the cactus forest, and he had connections within the administration of President Hoover that made the idea possible (Burtner 2011). In 1961 Saguaro was expanded with the addition of a west unit; in this case, Secretary of the Interior Stewart Udall, who had lived nearby, was instrumental in convincing his boss, President Kennedy, to sign the proclamation (Einberger 2018).

Not every potential park has friends like those, and for every park that is created there are many that weren't. An exact count of failed park proposals does not exist, but there are many glimpses into how difficult this process is. There were 103 national monument nominees in 1933, of which only 9 became monuments (Schneider-Hector 1993). One count found over two thousand places that had been unsuccessfully proposed (Dilsaver 2008). California includes 167 places that have been proposed for addition to the park system, but only 24 have made it. Of the 143 rejected areas, almost half were natural areas, about 40 percent were historical and archaeological sites, and the remainder were recreation oriented. These include Lake Tahoe, Ancient Bristlecone Pine Forest, Mono Lake, Mount Shasta, La Brea Tar Pits, Santa Catalina Island, the ghost towns of Bodie and Calico, Spanish missions, and Sutter's Fort. For those proposals for which information was available, a lack of national signifi-cance accounts for 62 percent of the rejections, while 25 percent were rejected for lack of feasibility and 11 percent for lack of suitability. Most rejected sites are protected: forty-eight are managed by other federal agencies, twenty-six are state parks, fourteen are local parks, and many of the remainder are protected by private organizations. Active park proposals often run for years or even decades before either culminating in a park unit or being abandoned by proponents (Hampton 1981).

THREE

The Geographic Evolution of the National Park System

This chapter discusses the geographic and thematic evolution of America's national park system, beginning with the protection of Yosemite Valley in 1864 and followed soon after by Yellowstone. Since then many more locations have been protected, and a truly national park system was created with 424 places across the United States. The system grew from its origins in the western mountains eastward, to lower elevations and deserts and into cities. The actions of President Carter in 1978 doubled the size of the national park system. The scope of the system has also grown as new types of units have been added, including battlefields, historic houses, memorials to presidents, beaches, lakes, scenic roads, and other places unimaginable in 1864.

The geographic and thematic growth of the system was not thought out in advance or even carefully planned; it was haphazard and often in response to specific threats or opportunities unforeseeable to planners. A 1933 expansion was the result of the transfer of land from the military and the Forest Service, along with many parks in the District of Columbia, to the National Park Service. A huge dam being built in Nevada led to the NPS getting involved in recreation parks oriented around reservoirs and later beaches. The discovery of oil in the new state of Alaska led to the greatest expansion of all in 1980 when government agencies scrambled to grab as much of the state as possible. On many occasions, the system was declared complete only to be expanded or even doubled in size; there is no reason to believe it will ever be complete.

There are many books on the history and growth of the park system. Among the most useful is one written by the NPS: *The National Parks: Shaping the System* (Mackintosh, McDonnell, and Sprinkle 2018). It has gone through four editions, with the most recent published as a special issue of the *George Wright Forum*. The 2005 edition is available online from the NPS (Mackintosh 2005). This chapter is heavily indebted to this book.

The First Parks

Many have discussed the origins of the national park idea and the revolutionary concept of setting aside land not for any productive purpose but merely for scenery or inspiration (e.g., Ise 1961; Mackintosh 1991; Runte 2010). An important component of this new idea was geography, as America's national park system was a part of the settlement of the vast western frontier. When the Revolution ended in 1783 the new U.S. government looked westward. The Northwest Ordinance of 1787 provided a plan for how the western lands would be settled. The Far West became the nation's public domain, under the control of the federal government. New states added to the Union gave up all unclaimed land to the federal government; that land was then added to the public domain. At this time, much of the nation's wealth and labor was tied up in land, and the public domain was a key resource for the nation's development. Homestead laws gave individuals the opportunity to claim land for agriculture; mining laws gave prospectors the ability to search out mineral resources and profit from their development; vast land grants were given to railroads to help them finance new transportation corridors, opening the frontier and binding the coun-

FIG. 3.1   *Looking Down Yosemite Valley*, by Albert Bierstadt, 1864. This view is looking down the canyon to the west, with El Capitan on the right. This is one of several similar paintings of Yosemite by Bierstadt. In all of them he tried to depict the valley as an uninhabited paradise. Birmingham Museum of Art.

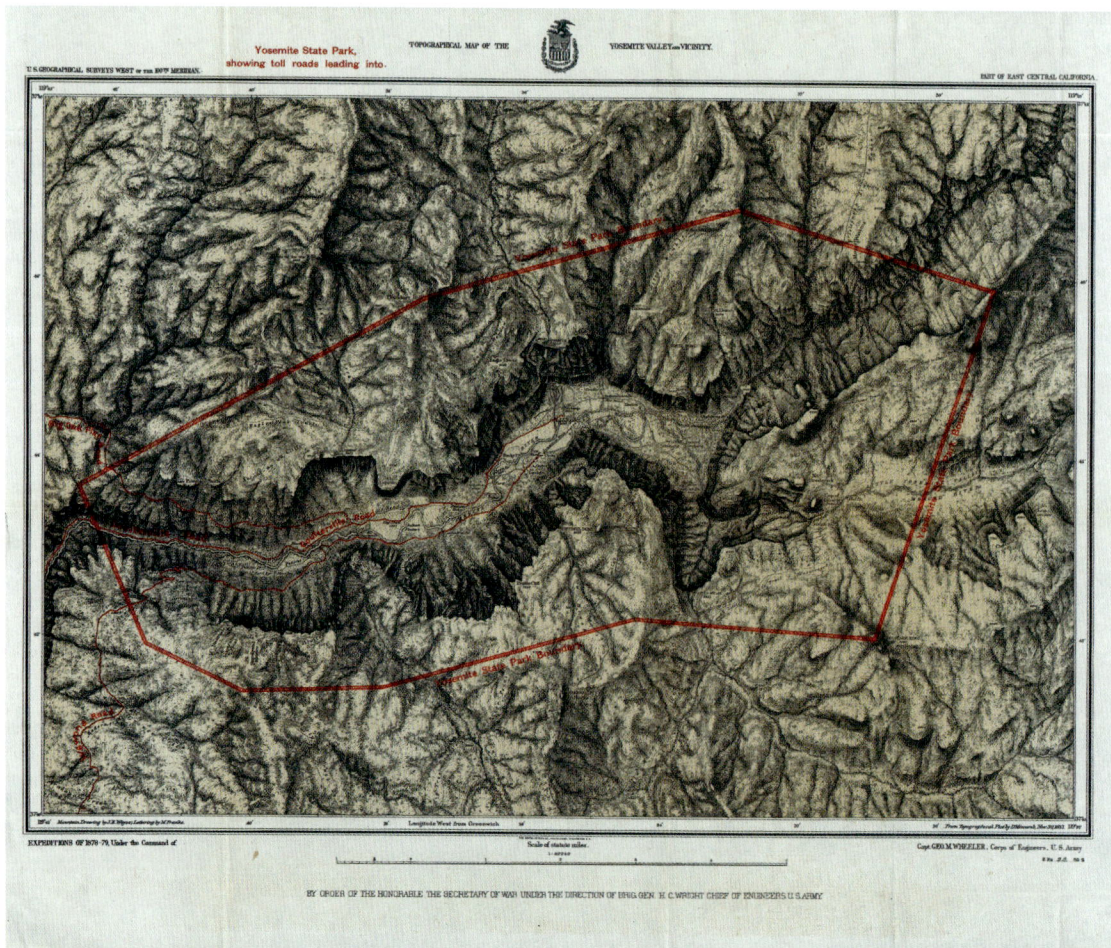

FIG. 3.2   Map of Yosemite Valley, by George Wheeler, 1893. The thick red lines show the boundary of the 1864 Yosemite state park, much smaller than today's national park. David Rumsey Map Collection, https://www.davidrumsey.com/.

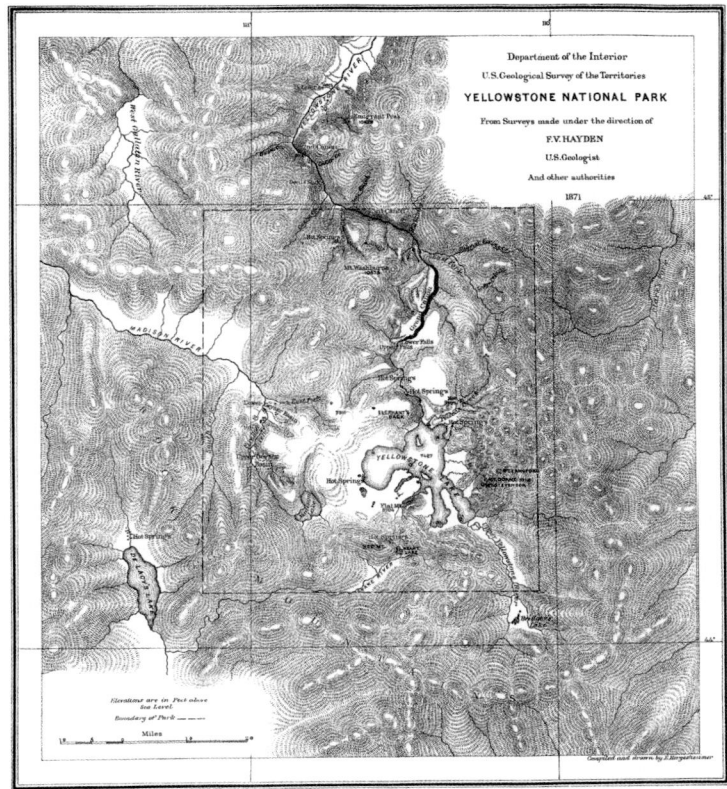

FIG. 3.3 Map of Yellowstone, by the Hayden Geological Survey, 1871. Little was known of this mysterious region at the time, but this map captures many landmarks still familiar to tourists today. United States Geological Survey.

try together. Many of these grants or claims were speculative in nature, as much of the frontier had not been mapped or explored, and the values of new lands were not fully known.

It was soon apparent that the western frontier contained not just valuable lands but also scenic wonders. Perhaps nowhere was this truer than in California, which became a state in 1850. The state's mineral wealth attracted the earliest attention, but the vast agricultural and timber resources of the state quickly became important. This led to the discovery by white explorers in 1851 of one of the most spectacular sights in America, Yosemite Valley (Fig. 3.1). This canyon, along with nearby giant sequoia trees, quickly became a tourist attraction. The laws of the land allowed individuals to claim the canyon's lands, trees, water resources, and mineral wealth for their own use and enrichment, which quickly destroyed much of the canyon's beauty. This process had begun by the 1860s when it was interrupted by an unprecedented event. Members of Congress, now down to the representatives of twenty-five states due to the treasonous secession of the remainder, voted to preserve Yosemite Valley. This was the first effort by the U.S. government to preserve a natural area for its scenic characteristics. This legislation was signed into law by President Lincoln on June 30, 1864, when Yosemite Valley was granted to the state of California as a public trust (Runte 2010). The grant included only the valley, a much smaller area than would later be identified with Yosemite (Fig. 3.2). While this situation would today be interpreted as the creation of a state park, it represents the origins of the national park system.

Eight years later another wonder was set aside for its scenery when the Yellowstone plateau was preserved (Fig. 3.3). But this time there was no state to grant the land to; Wyoming, Montana, and Idaho were all territories. The federal government instead retained control over it (Runte 2010). The law that created Yellowstone did not use the term "national park" anywhere in the text. Yellowstone is referred to only as a public park or pleasuring ground, but it was the first such park owned by the nation.

This Yellowstone park was much larger than

FIG. 3.4 *The Grand Canyon of the Yellowstone*, by Thomas Moran, 1872. Perhaps the most iconic work of art involving the national parks, this painting was seen by many easterners and helped promote the nation's western wonders. Smithsonian American Art Museum.

Yosemite Valley, but this was because the Yellowstone region remained little explored and mapped at the time (Fig. 3.4). The boundaries had to be made large and simple because there was little idea exactly what was there and where it was located and whether there were more features to be found (Ise 1961; Mackintosh 1991; Runte 2010; Sheail 2010). If the area had been better mapped, it is quite likely that it would have been much smaller and perhaps in several small sections around specific geothermal features or canyons (Sheail 2010). While we may think of Yellowstone as the prototype of what a park is or should be, it is a major exception to almost all later park units in its vast size and simple boundaries.

An enduring story about the creation of Yellowstone National Park is the "campfire myth" (Schullery and Whittlesey 2003). In 1870 the Washburn Expedition traveled through the Yellowstone region. The group camped in a meadow near the confluence of the Firehole and Madison Rivers on September 19, 1870. By this time the group had seen many of the features of Yellowstone and was quite impressed. The conversation around the campfire turned eventually to the possibility of preserving the area as a park for the nation. Upon returning home, the members began to lobby for the

park, with Yellowstone National Park the eventual result. On closer inspection the story becomes questionable at best. The account of the campfire discussion was first published in 1905 by one of the participants and is riddled with inconsistencies and unsubstantiated claims. The diaries of others who were present do not mention any such discussion. By the 1960s the campfire story had been largely discredited, except by the NPS, which had found the story useful from the beginning. The lovely meadow remains an inspiring place, regardless of what happened there in 1870.

With these two places protected by the federal government, it was perhaps inevitable that more would follow. Sequoia and General Grant National Parks were established in the California Sierra Nevada in 1890 to protect giant sequoia trees, along with the large Yosemite National Park surrounding the state grant (returned to federal authority in 1905) (Fig. 3.5). The next two parks were Cascade Range volcanoes: Mount Rainier in Washington was created in 1899 and has the distinction of being the first national park to be called a national park in the legislation that created it. The first park of the twentieth century, Crater Lake in Oregon, was designated in 1902. In 1903 Wind Cave National Park in the Black Hills of South Dakota became

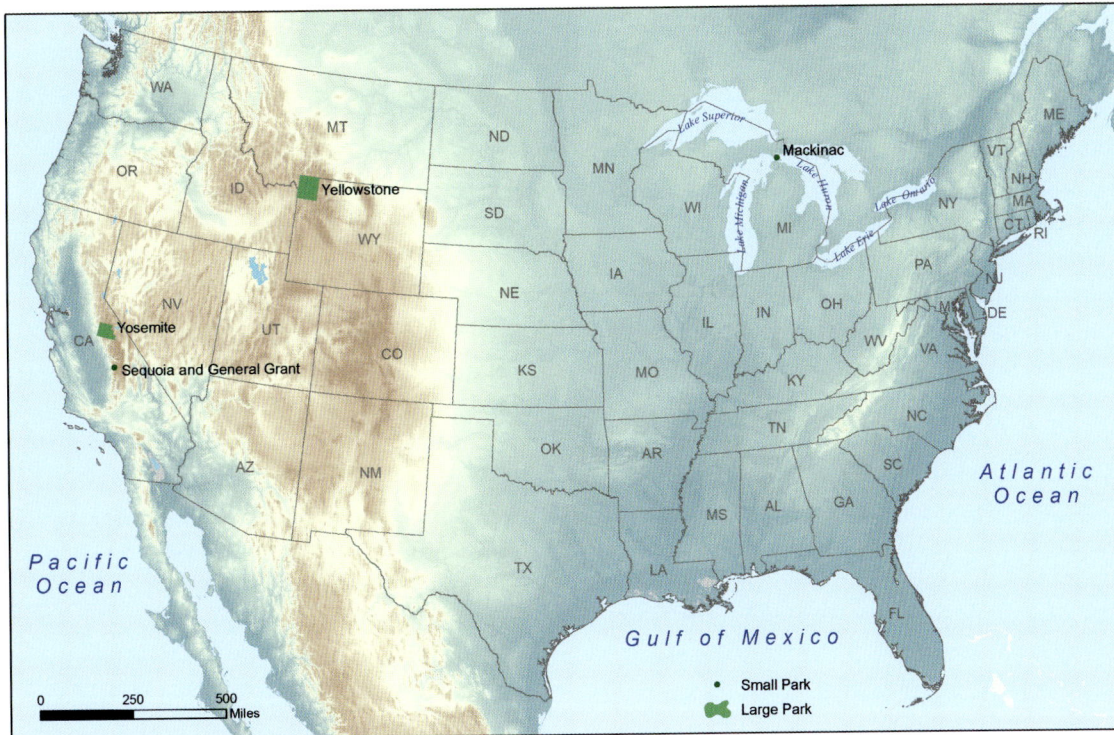

FIG. 3.5 Map of the national park system in 1890.

the easternmost park, as well as marking the extension of national parks underground. Glacier National Park in Montana (1910), Rocky Mountain National Park in Colorado (1915), and Lassen Volcanic National Park in California (1916) continued the focus on high mountain scenery.

The Antiquities Act of 1906 and National Monuments

In the late nineteenth century, Americans had become increasingly aware not just of the spectacular scenery to be found in the varied West but also of the natural resources and spectacular prehistoric ruins found there. Just as national parks were created as an attempt to prevent the loss of scenery in areas that were newly explored, other attempts were made to protect natural and prehistoric resources (Sellars 2007). The General Land Office, which managed the public domain, could withdraw an area from mining, logging, or homesteading, though this provided no protection against trespassing, vandalism, or hunting (Rothman 1994a, 1994b). After 1895 the president could proclaim a forest reserve to protect timber resources in the West; these reserves were carved out of

the public domain and managed by the General Land Office until a new federal agency, the Forest Service, was created to manage them.

Another set of resources increasingly seen as needing protection were spectacular cliff cities and other prehistoric dwellings in the Southwest (Rothman 1994a, 1994b). By the late nineteenth century, an "archaeological frontier" had reached the Southwest, causing scholars in eastern universities and organizations to become aware of the spectacular ruins, while these same sites also became attractive to plunderers (D. Smith 2002; Sellars 2007). The Four Corners area of southwestern Colorado, southern Utah, northeastern Arizona, and northwestern New Mexico included the richest collection of ruins in North America. Among the first of these were the ruins on the Mesa Verde plateau, which had been abandoned for over six hundred years when they were rediscovered in 1888. This collection of ruins was made a national park in 1906, the first park to be created for reasons other than scenery.

But in central Arizona, another spectacular prehistoric ruin offered a different approach. Casa Grande, the remains of a large three-story adobe building, had first been seen and named by Europeans in 1694 (Clemensen 1992). It was the largest example of structures created by the Hohokam culture about seven hundred years ago, along with large canal networks to divert water from nearby rivers to support irrigated agriculture in the desert (Doolittle 2000). In the late 1870s American farmers began moving into the area, drawn by the same fertile soils and water that supported the Hohokam. There was some interest in preserving the ruins, which were much too large to plow under. Easterners also took an interest in the ruins. Archaeologists from Massachusetts investigated the ruins in 1879, which led to a call by influential Bostonians to preserve the site. In 1892 the Casa Grande Reservation was set aside under the authority of the General Land Office. The reservation meant little in practical terms, just that the land was not available for homesteading and farming. No facilities existed, no fences kept wandering cows out, and a single caretaker was appointed to check on the site.

Ultimately, a new law was considered necessary, with penalties for vandalizing the protected site. The Antiquities Act, signed into law in 1906, allowed the president to set aside land to protect archaeological sites or other features of scientific interest to be called national monuments (Rothman 1994a, 1994b; Harmon, McManamon, and Pitcaithley 2006). However, from the very beginning, the designation was used for features other than those relating to antiquities; the act essentially provided a second and separate means of adding virtually any feature or place to the park system. The first national monument was proclaimed in Wyoming to protect a towering rock formation known to white settlers as Devils Tower; this had been proposed as a national park in 1892 but not considered worthy (J. Rogers 2007). Many more geological wonders, such as Devils Postpile, Pinnacles, Natural Bridges, Rainbow Bridge, Arches, and Cedar Breaks, were preserved as national monuments in coming decades. Antiquities, the original purpose of the act, were also protected. These were mainly Native American ruins in the Southwest (Arizona and New Mexico) such as Wupatki, Walnut Canyon, Navajo, Tonto, Gila Cliff Dwellings, Montezuma Castle, Tuzigoot, Pecos, Hovenweep, Aztec Ruins, Yucca House,

and Bandelier. Some sites preserved both, such as Canyon de Chelly, which features a spectacular sandstone canyon and cliff ruins (Fig. 3.6). Casa Grande eventually became a national monument itself in 1918. Many historic sites were also protected this way, including Spanish missions (Tumacácori), forts (among them Marion, Matanzas, Stanwix, Laramie, and Pulaski), and a scattering of other places.

National monuments were created to protect a wide range of scientific wonders and curiosities. Among these were locations where fossil remains of extinct plants and animals were being unearthed. The late nineteenth and early twentieth centuries represented a golden age of fossil hunting, and it should be no surprise that many of these locations later ended up as national monuments. Agate, Fossil Cycad, Fossil Butte, John Day, Hagerman, Florissant, Petrified Forest, and Dinosaur National Monuments all were proclaimed to protect important fossil resources. Though excavations at most of these sites had ended before they were proclaimed as monuments, fossil discoveries have continued in the park system.

Caves were another fascination of the time and were a new type of feature to be protected. Although Wind Cave was a national park, it was the Antiquities Act that allowed the park system to greatly expand underground. Jewel Cave, Carlsbad Caverns, Lehman Caves, Timpanogos Cave, and several others were soon added. These caves were protected before their full extent became known; Mammoth, Jewel, and Wind Caves turned out to be among the world's longest. Some are still being explored.

Other resources protected by the Antiquities Act were desert plants at Saguaro, Organ Pipe Cactus National Monument, and Joshua Tree National Park. Glacier Bay National Monument was added in 1925 to preserve an outstanding example of ecological succession following the retreat of glaciers (Catton 1995). Mount Olympus National Monument (later Olympic National Park) and Mount McKinley (later Denali) were created to preserve habitat for elk herds, though this purpose was forgotten in favor of scenery and environmental preservation. A 1952 proclamation that protected the endangered Devils Hole pupfish, which live in a

FIG. 3.6   *Canyon de Chelly*, by Ansel Adams, 1942. This canyon complex is both a scenic wonder and home to ancient Ancestral Pueblo ruins and recent Navajo history. National Archives and Records Administration, NAID 519852.

Nevada desert oasis, was perhaps the first that was specifically and entirely directed toward wildlife, though as an addition to Death Valley National Monument rather than as a separate monument.

Several early national parks were volcanoes, and many more were protected as national monuments. Cinder Cone National Monument and Lassen preserve Lassen Volcano in California's Sierra Nevada; Capulin Volcano in New Mexico and Sunset Crater in Arizona are recent and striking cinder cones; and Craters of the Moon in Idaho and Lava Beds in California preserve recent volcanic lava flows. Katmai and Aniakchak are (and remain) little-known active volcanoes in southwestern Alaska.

Monuments were initially seen as small reservations, but that changed in 1908 when the 808,120-acre Grand Canyon National Monument was proclaimed (Sellars 2007). Many other large scenery-based monuments were established over the years, among them Canyon de Chelly, Colorado, Zion, Death Valley, and Bryce Canyon, though many of these were later promoted to national parks. The Antiquities Act was later used to protect even larger areas, including Katmai in Alaska, which in 1931 became the largest unit in the national park system.

The Creation of the National Park Service

The growing collection of national parks and monuments was haphazardly managed by the government. The Department of the Interior received control of many, but those monuments

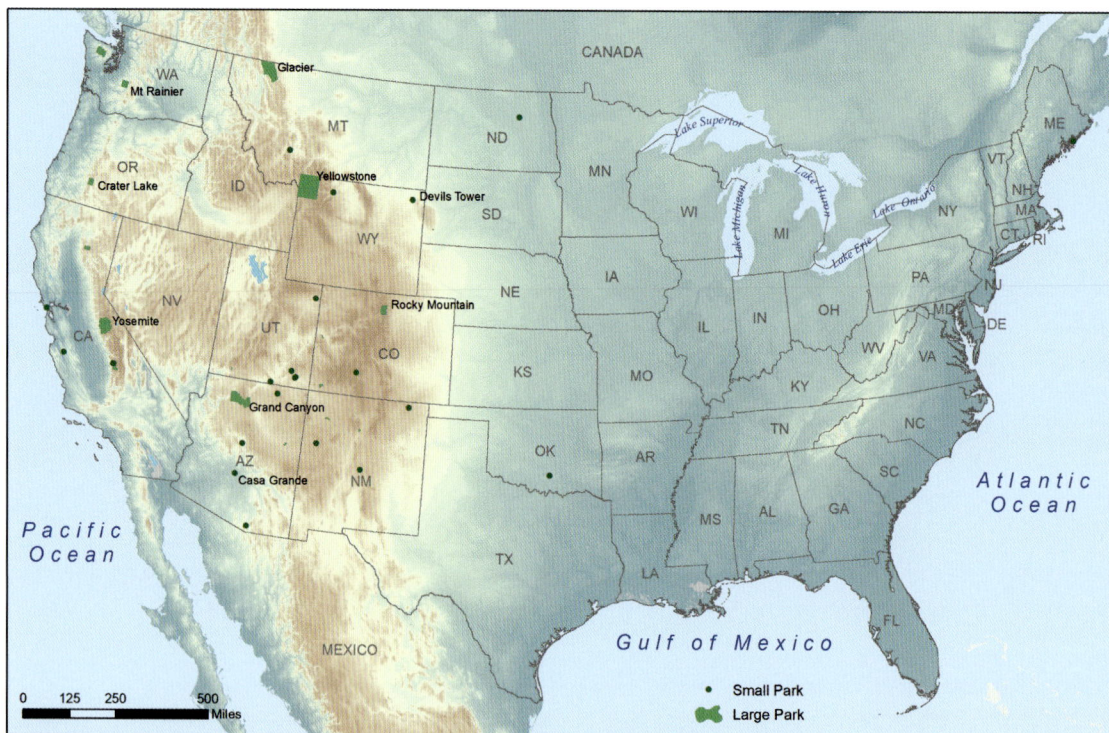

FIG. 3.7   Map of the national park system in 1916. This depicts the system at the time the National Park Service was created to manage the parks.

created in national forests were administered by the Forest Service within the Department of Agriculture. This agency was formed in 1905 to manage forest preserves. Many monuments were declared on military reservations and were assigned to the War Department (later called the Defense Department).

In 1916 the National Park Service was created within the Interior Department to manage the national parks, as well as those national monuments previously within the Department of the Interior. The NPS was made up of fourteen parks, twenty-one monuments, and two reservations (Mackintosh 1991) (Fig. 3.7). It was around this time that the word "system" was first applied to the collection of parks; the first use of that word may have been by Horace Albright in 1918 (Rettie 1995).

The NPS sought more parks; Mount McKinley (today's Denali) was the first new one after the agency came into being. Others were the Grand Canyon (1919), Bryce (1924), and Grand Teton (1929). Zion became a national park in 1919, but it had already been in the system since

1909, when it was Mukuntuweap National Monument. The agency liked it (but not the name) and was able to persuade Congress to pass a law enlarging and renaming the monument, as well as making it a national park.

As the NPS moved toward a policy of developing national parks for tourism, the parks became the focus of the agency's limited funds and personnel. National monuments were considered second-class sites, and little money was spent on them (Rothman 1994a, 1994b). But the NPS also pursued a policy of having promising areas proclaimed as monuments and then working to have them redesignated as national parks when the opportunity and funding were available. Many new large scenic monuments, later to become parks, were added this way. Among them were Katmai, Glacier Bay, Carlsbad Caverns, Badlands, Arches, White Sands, Great Sand Dunes, Death Valley, and Black Canyon of the Gunnison. Many small historical or archaeological monuments were added early on as well: Scotts Bluff, Yucca House, Aztec Ruin, Hovenweep, and Pipe Spring, followed years later by

Bandelier. These remain national monuments today.

Parks in the Territories

The first unit created outside the contiguous states was Sitka National Monument in the Alaska Panhandle. The town of Sitka had been founded in 1799 as the capital of Russian America but had lost its status as the capital of the territory. The national monument there preserved Russian history and Native Alaskan culture, the first park unit created to protect a living culture. Mount McKinley National Park was created in 1917 to protect the big game in what was known as "America's Serengeti" (Norris 1996b). The park would later become more famous for its high peak, though the wildlife is still an attraction. The next year Katmai National Monument was proclaimed to protect Mount Katmai, a volcano that had erupted in 1912 and produced a tremendous ash cloud that affected temperatures throughout the Northern Hemisphere (Norris 1996a). After the eruption scientists found what they named the Valley of Ten Thousand Smokes, a vast landscape of steam vents. In 1931 the monument was expanded to 2,697,590 acres, becoming the largest park unit in the nation. Glacier Bay was created in 1923 to preserve an outstanding example of ecological succession where rapidly retreating tidewater glaciers were being replaced by thick forest.

Hawaii was annexed by the United States in 1898 and quickly became famous for its many natural wonders, with active volcanoes foremost among them. Hawai'i Volcanoes National Park was established in 1916 to preserve the active volcanoes of Kīlauea and Mauna Loa on the big island. A separate unit included the Haleakalā volcano on the island of Maui, split off to become a separate park in 1961.

The National Park System Moves East

That the first and still today the majority of large national park units were in the western United States is often explained by the spectacular scenery found in this part of the country and the absence of similar landscapes in the East. But another explanation is also important:

land in the settled eastern portion of the country was in private hands, while that in the West was more commonly part of the public domain (land owned by the federal government). All the early park units were created out of the public domain; since the federal government already owned and controlled the land, it was a simple manner to change its status. Creating a national park in the eastern United States was a different story. The NPS had no mandate or funds to buy land if a suitable feature could be found. A new model for geographic expansion was developed in the East. Citizens could purchase land and donate it to the federal government for use as a park. This was used in 1916 when the first national park unit east of the Mississippi River was created: Sieur de Monts National Monument, to be renamed Lafayette National Park in 1919 and eventually known as Acadia. Wealthy landowners concerned about threats to an area they held dear sought a way to protect the land from development; a national park unit proved to be the means to do so.

Stephen Mather and Horace Albright, overseeing the new National Park Service, were also motivated to acquire land in the East for other reasons. Aside from Acadia, all the parklands were in the West and often in areas that even today are remote and far from population centers, giving them a poor relative location. Congressmen representing eastern states questioned why public money should be spent where so few lived. The NPS was aware of this and sought to expand eastward in part to increase its base of support among the public.

In the 1920s NPS staff went looking for a location for an eastern park. After visiting several areas, park staff settled on the Great Smoky Mountains, straddling the Tennessee–North Carolina boundary. This mountain range was, however, occupied by six hundred people farming in mountain valleys, while most of the range had been clear-cut by logging companies. A wealthy donor provided half the money needed to purchase this land, and ordinary citizens provided the rest. Many families did not go willingly but were forced out by eminent domain. The Great Smoky Mountains National Park was finally created in 1934. The NPS chose well, as the park has become one of the most popular in the system.

A similar story occurred in Virginia when the state government purchased the Blue Ridge Mountains from private landowners and donated it to the NPS to create Shenandoah in 1935. Five hundred people left their land, some against their will. Mammoth Cave in Kentucky was established in 1941 after the land was purchased by donated money or taken by eminent domain. Over one thousand people are said to have been displaced, and, as they had at other eastern parks, resentments lingered.

Due to the circumstances in which Texas was admitted to the United States, it has no federal public domain (Hubbard 2009). For that reason, creating parks within it followed the same model as in eastern states: the state of Texas first acquired land from private owners and then transferred it to the NPS. Big Bend went through a lengthy process of land acquisition before it was officially created in 1944.

New Units and Themes in the 1930s

The 1930s was one of the most eventful decades in the geography of the national parks. It saw a tremendous increase in the size and scope of the park system, as well as massive construction of visitor facilities and conservation projects by the Civilian Conservation Corps (CCC).

The Reorganization of 1933

When the NPS was created in 1916 it received control of all national monuments that had been under the jurisdiction of the Department of the Interior. Some monuments proclaimed before 1916 had been assigned to the Forest Service or War Department and remained under their control after the National Park Service was created. This created an awkward situation for managing these units, especially from the point of view of the NPS. This situation was resolved on August 10, 1933, when President Roosevelt issued an executive order transferring all national monuments under the control of the Departments of Agriculture and War to the NPS.

From the Agriculture Department, the park agency received fifteen monuments, all in western states and mountainous settings. Most were small geological oddities or archaeological sites, but they included Mount Olympus (later Olympic National Park), Lehman Caves (later Great Basin National Park), Saguaro (later a national park), and Chiricahua. The army contributed ten monuments, all but two in eastern states. These included several Gulf and Atlantic coastal forts, the Statue of Liberty, and the Little Bighorn battlefield in Montana. The army also contributed two national parks: Abraham Lincoln's Birthplace and Fort McHenry (later redesignated into a national historical park and a national monument, respectively).

Roosevelt's order also gave the NPS a system of battlefield parks created by the War Department. These had been established from the 1890s onward as efforts to commemorate the Civil War. Chickamauga and Chattanooga National Military Park and Antietam National Battlefield were the first in 1890, followed by Shiloh in 1894, Gettysburg in 1895, and Vicksburg in 1899 (Sellars 2005, 2007). These battlefields contained large cemeteries along with collections of monuments erected by veterans' organizations. Besides their commemorative purpose, they provided educational opportunities for officers to learn military tactics. Later battlefields were added to the system, including sites representing the War of 1812 (beginning with the Chalmette Monument in New Orleans in 1907) and the American Revolution (beginning with Guilford Courthouse National Military Park in 1917).

The army's battlefield park system stretched from Pennsylvania through an arc to Vicksburg and New Orleans. Adding it to the national park system brought the parks a huge boost in visibility; Shenandoah or Mammoth Cave might be known to a few, but all Americans knew of Gettysburg. These battlefield parks also introduced new designations to the park system, such as national military park and national battlefield; the NPS did not change these labels to match its other parks. The NPS was also content with these battlefields, and relatively few were added after 1933, and these have been usually quite small.

Roosevelt's executive order also gave the NPS responsibility for many parks, memorials,

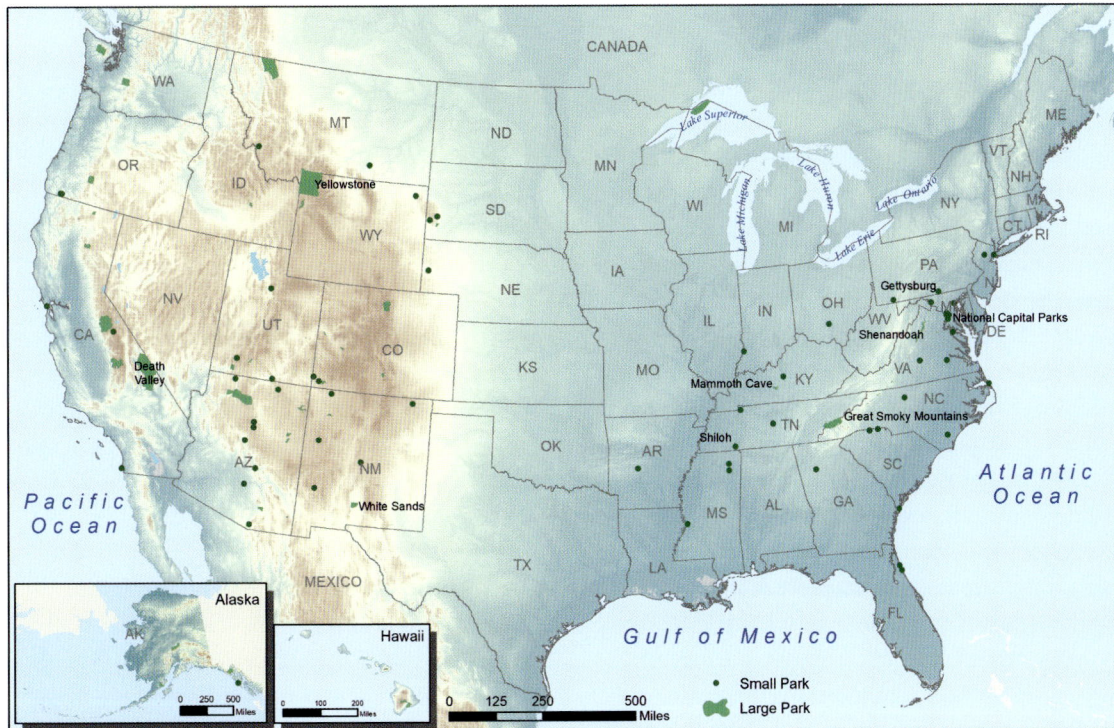

FIG. 3.8   Map of the national park system in 1933. This depicts the system after it was dramatically expanded by the reorganization of 1933.

and roads in the Washington, D.C., area (Mackintosh 1991). The District of Columbia was created in 1790 as a federal territory to house the capital, and its unique political geography meant that it contained buildings and parks that did not match municipal or county parks elsewhere. Among them were the Washington Monument and recently completed Lincoln Memorial.

This event greatly enlarged the size of the national park system, as well as greatly expanding the scope of the system (Fig. 3.8). At this point, the national parks became a truly national system of America's most spectacular and historic sites (Foresta 1984). More changes came in the 1930s when new types of parks emphasizing history and recreation were created.

National Memorials

Mount Rushmore is surely the most famous place in South Dakota. It began as a local project to promote tourism in the state, but as costs grew, the never-finished memorial was transferred to the federal government (R. Smith 1985). This was not the last time the NPS would end up with a park this way. Though it is a fa-

mous and popular example of the national park system, it has no direct connection to any of the four presidents, none of whom ever visited the mountain. It is a national memorial, which commemorates a person, event, or idea. This type is unique among those in the park system in that there need not be any geographic connection between the person or event being commemorated and the chosen location.

The first memorials were those for Washington and Lincoln on the National Mall. Washington's column was completed in 1885 and Lincoln's temple in 1922. Both were transferred to the NPS during the 1933 reorganization. Mount Rushmore was added to the system in 1939, followed by a Washington, D.C., memorial for Thomas Jefferson in 1943. Many more followed later.

Historic Sites

The preservation of historic sites has long been the initiative of individuals and private organizations in the United States. The preservation of George Washington's home on Mount Vernon by the Mount Vernon Ladies Association after 1860 and the reconstruction of Colonial

Williamsburg in the 1920s by John Rockefeller Jr. were well-known models for how historic preservation could be carried out (Mackintosh 1985). The 1906 Antiquities Act came into being to preserve Native American ruins and then was applied to other features. In 1930 it was used for historic preservation when Colonial Williamsburg and George Washington's birthplace became national monuments. The federal government had gotten into the historic preservation field.

In 1933 the transfer of national monuments and battlefields to the NPS greatly increased the number of historic sites the agency supervised. Even more historic sites began to enter the system following the Historic Sites Act of 1935, which created the categories of national historic site and national historical park. Many of these new units were created in the eastern states and became an important component of the park system's eastern growth.

Recreation Areas

In the 1930s the NPS became involved with parks built around reservoirs and seashores, beginning a push into recreation for the park system (Gonzales 2017). The construction of Hoover Dam on the Colorado River near the small town of Las Vegas, Nevada, in the 1930s resulted in a new type of park unit (Dodd 2007). The dam created Lake Mead, the world's largest artificial lake. Although the dam was being built by the Bureau of Reclamation for agricultural water storage and power generation, it was clear that there would be significant recreational opportunities. But the Bureau of Reclamation was unwilling and unprepared to manage these opportunities, so it turned to the NPS. Park Service staff had previously battled to keep reservoirs out of national park units but now found themselves in possession of a new unit consisting of an artificial reservoir. This presented many in the NPS with a serious dilemma: how to accommodate reservoirs without opening the park system to more. A compromise of sorts was reached when the stretch of river between Grand Canyon National Park and the new lake was proclaimed as Grand Canyon National Monument in 1932 and managed as

part of Grand Canyon National Park (the monument was eventually merged into the park in 1975). The lake itself was to be managed as a new category of park unit (Dodd 2007) and in 1936 was established as Boulder Dam Recreation Area, later Lake Mead National Recreation Area.

The NPS next applied the idea of a recreation park to the northern end of North Carolina's Outer Banks, preserving them as Cape Hatteras National Seashore in 1937 (Gonzales 2017). The idea of parks centered on beaches would become a popular one after the park system resumed growth following World War II.

Parkways

The Blue Ridge (1933) and Natchez Trace (1934) Parkways were a new kind of national park unit; rather than drive to an attraction, the road and driving experience themselves would be attractions. Not all within the NPS were pleased to have them in the system, but they quickly became among the most popular with the public. The Blue Ridge Parkway is 469 miles long, running from Shenandoah National Park to the Great Smoky Mountains. It was not complete until 1987. The Natchez Trace Parkway is of similar length, 444 miles, but much different in character. Instead of connecting national parks, it connects cities: Nashville to Natchez, Mississippi. The primary emphasis is history rather than scenery. Like the Blue Ridge Parkway, it was completed only recently, in 2005.

Continued Growth

Aside from the War Department's Abraham Lincoln National Park in Kentucky (now Abraham Lincoln Birthplace National Historic Site), the first presidential site outside Washington, D.C., was Andrew Johnson National Monument, created in 1935 in eastern Tennessee. It set a precedent of creating sites for presidents. Theodore Roosevelt was particularly popular, receiving four units. Adams National Historical Park (1946) preserves the home of father and son presidents John Adams and John Quincy Adams. Abraham Lincoln has been another park favorite, with two units complementing his Kentucky birthplace park and Washington, D.C., memorial. Presidents Van Buren, Hoover, Taft, and Garfield are not among the most famous presidents, but all received sites (which are themselves little known).

Franklin Roosevelt's home became a national historic site while he was in office, but it has since been customary to create a site for a president after he leaves office. A unit commemorating President Kennedy was added just four years after his death, while Eisenhower received one two years before he died in 1969. Since then it has become common for each president to receive a park unit: Johnson (1969), Truman (1983), Carter (1987), and Clinton (2010). At the present time, the NPS is authorized to create a Ronald Reagan boyhood home unit, but this will depend on the NPS's ability to purchase the home (there is currently a dispute with the owner over an appropriate compensation). Interest in a site for George W. Bush is growing, and this may be the next presidential site.

Expansion into New Environments

At the beginning of 1933 most large parks were still mountainous and forested. This began to change when large parks were created in places that were quite different from Yellowstone, Acadia, and Crater Lake. In the months before the 1933 reorganization, the first large desert parks were created: Death Valley in California and Nevada and White Sands in New Mexico. Saguaro joined with the reorganization, and in 1936 Joshua Tree was added in California, followed by Organ Pipe Cactus in Arizona. Texas's Big Bend was authorized in 1935 and established in 1944. Vast desert basins, sand dunes, dry lake beds, cacti, and other desert plants were probably not most Americans' ideas of spectacular scenery at the time, but they reflected a growing interest in the country's deserts.

The desert parks at least had mountains, the vast wetlands of southern Florida had none. Yet the Everglades began to be seen not just as the home of mosquitos and snakes to be drained for productive purposes but as an area worthy of preservation. When Everglades was made a national park in 1934 it was like nothing else in the park system and is still one of a kind.

Not all sites are accessible by road. The Channel Islands, an archipelago off the coast of California, have largely escaped development and are sometimes referred to as "America's Galápagos" due to their abundant marine life and relative isolation from the mainland. In 1938 two of the islands in the group were made into a national monument. This was expanded to include three other islands in 1980 when it became a national park.

Expansion in the Postwar Era

World War II brought a temporary end to the growth of the 1930s, as well as the steadily increasing visitation. Two-thirds of park employees went into the military (T. Davis 2016), as did the majority of the CCC workers who had been building facilities within many parks. The parks went through a dormant period, but the needs of the war put new burdens on the system; Fort Pulaski was taken over by the army. Agency staff resisted opening Olympic to logging but was not successful in preventing it from being carried out in Glacier Bay (McDonnell 2015). The NPS was more amenable to the military using parks as recreation areas for soldiers or sailors, even allowing the military to take over Yosemite's Ahwahnee Hotel. At some parks, soldiers made up a substantial portion of visitors during the war years. Parks were also useful for training, with winter warfare training carried out at Mount Rainier and Mount McKinley.

After the coming of peace in 1945, the visitors returned in ever greater numbers, and the park system began a period of tremendous and continuing expansion in the geography and type of parks (Fig. 3.9). New types of physical features, such as coral reefs, have been added to the park system. New themes have been added, new types of units have been created, and the park system has geographically expanded in several areas. By the early twenty-first century, the national park system has evolved into a large and tremendously varied set of places.

The Growth of Recreation Units

The number of these units exploded after peace returned in 1945. The demand for outdoor rec-

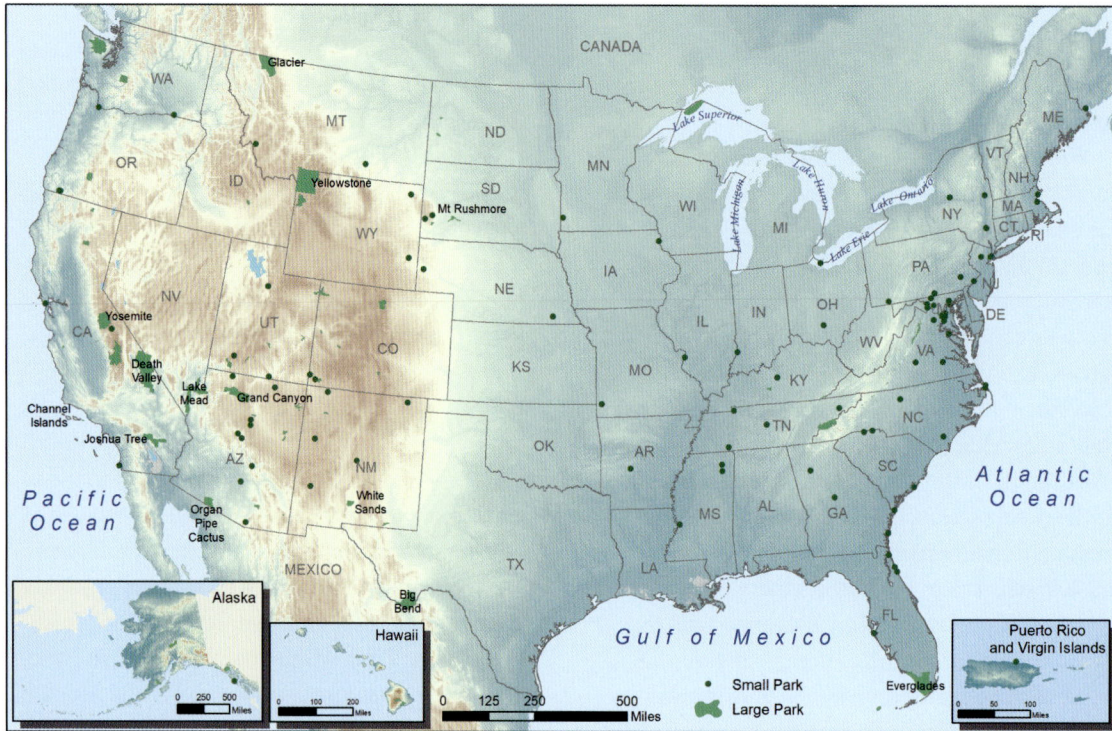

reation increased steadily after the war, and the NPS undertook surveys to find new outdoor recreation resources (National Park Service 1964). Many new seashores were added: Cape Lookout (1966), also on the Outer Banks, Cape Cod (1961), Fire Island (1964), Assateague Island (1965), Cumberland Island (1972), and Canaveral (1975) elsewhere on the Atlantic coast; Padre Island (1962) and Gulf Islands (1971) on the Gulf of Mexico; and Point Reyes (1962) on the Pacific. Pictured Rocks (1966), Indiana Dunes (1966), Apostle Island (1970), and Sleeping Bear Dunes (1970) National Lakeshores ring Lakes Michigan and Superior.

Reservoir-based recreation areas were established in western states, among them Lake Roosevelt (1946), Glen Canyon (1958), Whiskeytown (1962), Bighorn Canyon (1964), Lake Meredith (1965), Amistad (1965), and, the very last to be added, Ross Lake and Chelan Lake (1968). One reservoir unit that didn't happen was at Delaware Water Gap National Recreation Area. This site was to have been the first eastern reservoir unit, planned as a heavily developed recreation area between New York

City and Philadelphia. Park planners envisioned ten million visitors a year at a time when the most popular units only received eight million annually (Albert 1987). It was predicted that the unit would be the "Central Park of Megalopolis." The reservoir was eventually canceled, and the national recreation area was instead developed around a free-flowing river, a new concept in the park system.

Other river-based units were soon added as part of the National Wild and Scenic Rivers System, though few of them are managed by the NPS. These include national rivers, national scenic rivers, and national wild and scenic rivers, all intended to protect free-flowing rivers throughout the country. The Buffalo River in Arkansas (1972), Saint Croix National Scenic Riverway in Minnesota and Wisconsin (1968), Niobrara National Scenic River in Nebraska (1991), Upper Delaware Scenic and Recreational River in New York and Pennsylvania (1978), Obed Wild and Scenic River in Tennessee (1976), and New River Gorge National River (1978, now a national park and preserve) and Bluestone National Scenic River (1988) in

West Virginia are examples of units under NPS control. These units reflect growing environmental awareness, as well as growing interest in whitewater rafting.

Preserves are like parks, with a focus on outdoor scenery, but they allow economic activities such as mining, farming, and hunting, normally banned in parks. Big Thicket and Big Cypress were the first preserves, in 1974; they had originally been planned as national parks, but a broadening of allowed activities was required to create the necessary political compromise for their creation (Cozine 2004). Since 1980 most preserves have been in Alaska; Bering Land Bridge, Noatak, and Yukon–Charley Rivers are among these. While preserves allow new kinds of activities in parks, they are the only kind of NPS area where the agency acknowledges earlier settlement and activities.

There are also many affiliated areas, those with some NPS involvement but not counted as part of the system (Rettie 1995). National scenic trails and national historic trails are the outcomes of the National Trails System Act of 1968. Eleven are national scenic trails, though only three (Appalachian, Natchez Trace, and Potomac Heritage) are counted as NPS units. The Appalachian is the most famous, a 2,200-mile trail from Georgia to Maine added in 1968. Natchez Trace (1983) is within the Natchez Trace Parkway and consists only of several unconnected trails. Potomac Heritage (1983) is a set of trails between the Chesapeake Bay in Virginia through Washington, D.C., up the Potomac River that will eventually connect to Pittsburgh.

Nineteen of these trails are national historic trails; none of them are counted as NPS units, but the service assists with planning and promotion and has official websites for many. Most of these trails will be driven rather than walked. The Selma to Montgomery National Historic Trail was established in 1996 and commemorates the 1965 civil rights march along U.S. Highway 80, though most visitors will drive rather than walk the route. The NPS has built two visitor centers and has placed several signs along the road. The Juan Bautista de Anza National Historic Trail (1990) commemorates a Spanish expedition between Nogales, Arizona,

and San Francisco in 1775 and 1776. The expedition pioneered an overland route between Mexico and California.

National heritage areas are not national park units but are instead historic areas or districts that have received congressional designation. The NPS provides some assistance to these areas and has carried out numerous studies but does not own or manage any. Over four dozen have been created since 1984. The Gullah Geechee Cultural Heritage Corridor is the best known of these. Created in 2006, it runs along the Atlantic coast from northwestern Florida to North Carolina to help preserve a distinct cultural region of the country.

Parkways

The Blue Ridge and Natchez Trace Parkways were joined postwar by several others. The George Washington Memorial Parkway, Rock Creek Parkway, and Suitland Parkway opened in 1944, and the Baltimore–Washington Parkway was a freeway built between 1947 and 1954. (The remaining parkway, named for John D. Rockefeller Jr., is simply a parcel of land in Wyoming connecting Yellowstone and Grand Teton.) Many park units have spectacular and well-known roads, but these are part of a much larger park and not the focus of the unit, as they are in parkways.

Given the importance of the automobile in American life, it is striking that no additional parkways were created by the NPS. Other parkways along the lines of the Blue Ridge and Natchez Trace were considered but never built. As late as 1966 a study was conducted to identify possible new parkways and even a national scenic highway network (United States Department of Commerce 1966). The Natchez Trace Parkway shows that the parkway concept can be extended to areas lacking in scenic splendor, and there is no inherent reason why parkways could not have been created across the country. Nonetheless, this remains a road not taken by the Park Service.

Expanding into History and Culture

Building a park in a new part of the country is one way to expand the system geographically;

another is to create units representing the varied social geography of the country, especially its struggles over equality. The first park unit directly centered on an African American was George Washington Carver National Monument, created at his Missouri birthplace in 1943 (Toogood 1973; Krahe and Catton 2014). Other units followed, including Booker T. Washington National Monument (1956), Tuskegee Institute National Historic Site (1974), Carter G. Woodson Home National Historic Site (2006), Tuskegee Airmen National Historic Site (1998), Harriet Tubman Underground Railroad National Monument (2014), and Freedom Riders National Monument (2018). This remains an active and growing area of the geography of the park system, though one so far found mainly in the southeastern United States. It will likely spread across the country.

Pacific Islanders were represented in the national park system beginning in 1955, when the City of Refuge National Historical Park was created in Hawaii (renamed Puʻuhonua o Hōnaunau National Historical Park in 1978) to preserve aspects of Native Hawaiian culture. This was followed by Puʻukoholā Heiau National Historic Site in 1972 and Kaloko-Honokōhau National Historic Site in 1978. The National Park of American Samoa was created in 1988 and preserves Polynesian culture, as well as forests and coral reefs. The visibility of Asians in the national park system is dominated by several World War II prison camps created to house Japanese American civilians. Manzanar National Historic Site was created in 1992, Minidoka Internment Camp was created in 2001, and additional sites are under study for possible park units. No other Asian populations are represented by park units.

Although the remains of Native American and Native Alaskan ancestral cultures are preserved in many sites, there are few units directly preserving or commemorating living Native American and Native Alaskan cultures. Hubbell Trading Post National Historic Site (1965) was established as an active trading business within the Navajo Nation. Canyon de Chelly is owned by the Navajo Nation, Death Valley National Park includes the Timbisha reservation within

it, and Pipestone National Monument (1937) remains an active quarry for Native Americans. Many park units also contain locations considered sacred to Native Americans, raising a number of issues yet to be resolved.

Another group increasingly visible in the national park system is women. Several national park sites devoted to women and women's rights have been created as well, among them Clara Barton National Historic Site (1974), Maggie Walker National Historic Site (1978), Women's Rights National Historical Park (1980), and First Ladies National Historic Site (2000). Most recently, the first park unit commemorating Hispanics was established in 2012 when César E. Chávez National Monument was created in California to commemorate the labor leader and civil rights activist. The gay rights movement was included with New York City's Stonewall National Monument in 2016.

Expanding the Memorial Landscape

National parks are among the country's best-known and most important public spaces, and they are used to commemorate events important to the country. The USS *Arizona*, destroyed in the December 7, 1941, attack on Pearl Harbor, remains at its mooring spot as a national memorial. After the war's end, twenty-six more memorials were added to the system, commemorating presidents, leaders, and disasters but mainly wars. The completion of the Vietnam Veterans Memorial in 1982 set off a flurry of new memorials on the National Mall, including ones for World War II, the Korean War, and eventually World War I.

Urban Parks

The roots of the national park idea lie in the western wilderness, and to many, this remains the ideal. But it is an increasingly inaccurate depiction of the geography of the system. Many units of the system, including many of the largest and most visited, are urban. There are several geographic dimensions of urban parks: a collection of memorials, monuments, small parks, and other odds and ends in the Washington, D.C., area that are known collectively as the national capital parks; small historic sites added to the system since the 1930s that often lie in urban areas; and large recreation areas added in the 1970s during a time when an urban recreation crisis was considered acute.

The National Capital Area

Because of the unique political geography of the District of Columbia and historical happenstance, the NPS has also been assigned duties here that elsewhere fall to local government. Since 1933 the NPS has had the responsibilities for many local parks in the city, including Anacostia and Rock Creek Parks, which have ball fields and golf courses, as well as countless small parks and squares throughout the city. The NPS is also responsible for maintaining parts of the Washington, D.C., highway and freeway systems. The George Washington Memorial Parkway, Rock Creek Parkway, and Suitland Parkway opened in 1944, and the Baltimore–Washington Parkway is a freeway built between 1947 and 1954. The Wolf Trap National Park for the Performing Arts was created in 1966 as a venue for musical acts and theater in northern Virginia.

Historic Sites and Battlefields

Jefferson National Expansion Memorial (1935) in St. Louis and Salem Maritime National Historic Site (1938) near Boston were early examples of new kinds of parks authorized by the 1935 Historic Sites Act. Many small historic sites have been added to cities since the 1930s, and many large cities in the United States now have one or occasionally more small historic sites commemorating an individual or event. These are typically quite small and have limited facilities, often with a narrow appeal. These sites may be famous, such as Independence Hall (1939) in Philadelphia, which houses the Liberty Bell, or they might be considered as minor attractions in a city, such as William Howard Taft National Historic Site (1973) in Cincinnati.

In other cases, cities came to parks. Several battlefields, such as Kennesaw Mountain near Atlanta and Guilford Courthouse in Greensboro, North Carolina, were established in rural areas but have since become surrounded by urban neighborhoods. They are heavily utilized by residents and serve as city parks in addition to their national park status. This has created conflicts between use and preservation. Battlefield parks are managed according to goals different from those of other park units; picnicking, sunbathing, or similar innocuous activities are not permitted in battlefields.

The Urban Recreation Movement

In the 1960s concern about the lack of outdoor recreation opportunities became important, in part reflecting a growing interest in urban and minority interests in the federal government (Foresta 1984). Much of this interest was directed at state and local governments, but the NPS became involved with the idea of national recreation areas within large urban areas. The country was becoming increasingly metropolitan in the postwar era, and much of the park system was considered to be poorly located relative to this population. As it had with the earlier eastward expansion of the park system, the NPS also sought to broaden its base of support by moving parks closer to the population centers.

The first national recreation areas to be created were Gateway National Recreation Area in New York City and Golden Gate National Recreation Area in San Francisco, both in 1972. These sites are large and scattered, and they contain a vast range of facilities for outdoor recreation, scenery, performing arts, and history. The creation process for Golden Gate began in the 1960s when the military began to discard lands around the Bay Area and the prison on Alcatraz Island was shut down (Rothman 2004). Various proposals for these sites were put forth, including both parks and new commercial development. The park proposals carried the day, and in 1972 Golden Gate was created. Many new management problems arose as the park was developed for which the NPS was not prepared. Problems arising from dogs and mountain bikes were particularly difficult; existing national park regulations banned them, but they are very popular activities in urban parks.

These two sites were seen as a pair, one for each coast, but after they were created, pressure emerged for more sites (Foresta 1984). Several more were soon created: Cuyahoga Valley National Recreation Area in 1974 near Cleveland, Ohio (later to become a national park), Chattahoochee National Recreation Area in 1978 near Atlanta, and Santa Monica Mountains that same year in the Los Angeles area. In addition to these five, many other sites were established near large metropolitan areas, making the distinction between urban and traditional recreation areas one of degree. Cape Cod, Fire Island, and Point Reyes National Seashore, Indiana Dunes National Park, and Delaware Water Gap National Recreation Area are all near major metropolitan areas and justified in part by the need to serve urban populations (Foresta 1984). While these sites are near cities, they do not contain the collection of resource types and locations that made Gateway and Golden Gate so different from existing sites.

Geographic Expansions

Several big expansions took place after World War II. A new and larger Grand Teton National Park was created in 1950, and other large parks appeared, including Canyonlands (1964), Guadalupe Mountains (1966), Biscayne and the North Cascades National Park group (1968), Great Basin (1986), El Malpais (1987), and Valles Caldera (2014).

But the postwar era has seen the greatest growth in the number of park units and size in outlying territories, as well as greatly extending geographic coverage. The territorial expansion of the United States did not end with Alaska and Hawaii: Guam, Puerto Rico, and American Samoa were added in 1898, the U.S. Virgin Islands were purchased in 1916, and several scattered groups of Pacific Islands were added after World War II. National parks were created in all these areas during the postwar expansion: Puerto Rico (San Juan National Historic Site in 1949), the U.S. Virgin Islands (Virgin Islands National Park in 1952, followed by others), Guam (War in the Pacific National Historical Park in 1978), the Northern Mariana Islands (American Memorial Park in 1978), and American Samoa (National Park of American Samoa

in 1988). But the greatest expansion of all was far to the north, in Alaska.

The Alaska National Interest Lands Conservation Act of 1980

When Alaska became a state in 1958 a complicated set of events was initiated (Andrus and Freemuth 2006). The federal government was willing to give over substantial portions of the public domain to the state and Native Alaskan groups. This process, which was different from the government's process in other states, did not move quickly, and it was not until oil was discovered in the North Slope Borough of the state that settling the land claims became urgent.

The Alaska Native Claims Settlement Act of 1971 called for the government to identify lands for parks and wildlife refuges within two years and then gave Congress five years to enact legislation to designate these areas, setting off what has been termed the "scramble for Alaska." Lands were identified, but congressional politics froze the legislation. With the five-year limit days from running out, President Carter proclaimed ten new national monuments in Alaska in December 1978. This action was intended to be temporary to give Congress time to complete the task of designating new parks. The move was denounced by Alaska's congressional delegation, but many of these monuments were converted to national parks by the 1980 Alaska National Interest Lands Conservation Act, which led to the near doubling of national park acreage and the creation of what some have termed a second park system.

Mount McKinley National Park was greatly enlarged and renamed Denali National Park and Preserve, though the mountain officially remained Mount McKinley. The new Wrangell–St. Elias National Park and Preserve, the largest single unit in the park system, was created farther east around North America's second tallest peak and one of its largest icefields. Glacier Bay National Monument was expanded into a park and the new Kenai Fjords National Park was created to preserve tidewater glaciers on the Kenai Peninsula. The volcanic lands of Katmai National Monument became a national park, and the nearby Lake Clark National Park and Preserve was created to preserve other active volcanoes, as well as rain forest and tundra. Aniakchak National Monument and Preserve contains another volcano and is remarkable for being the least visited national park unit. Farther north, in the Brooks Range, Gates of the Arctic, Kobuk Valley, Noatak, and Alagnak Wild River were created. Bering Land Bridge, Cape Krusenstern, and Yukon–Charley Rivers were in the vast interior of Alaska between these two mountain ranges.

The California Desert Protection Act of 1994

The United States contains four deserts: the Mojave in Southern California, the Great Basin in Nevada and Utah, the Sonoran in southern Arizona, and the Chihuahuan in West Texas. Several large national monuments were proclaimed protecting the landscapes and vegetation of these deserts in the 1930s: Death Valley (1933) and Joshua Tree (1936) in the Mojave, Saguaro (1933) and Organ Pipe Cactus (1937) in the Sonoran, and White Sands (1933) and Big Bend (1944) in the Chihuahuan Desert. After World War II visitors to the desert parks increased tremendously; what was once remote became easily accessible. Areas that were untouched and protected by isolation came under the increasing influence of urban visitors with four-wheel-drive vehicles. At the same time, the desert received far more respect from American culture. Writers such as Joseph Wood Krutch (1952, 1955, 1958, 1961) and Edward Abbey (1968) wrote passionately about the desert and the beauty to be found there. After several decades of struggle, the California Desert Protection Act was passed in 1994. It expanded both Death Valley and Joshua Tree National Monuments and redesignated them as national parks (Wheat 1999). After this expansion Death Valley became the largest national park in the contiguous forty-eight states. The act also created the new Mojave National Preserve, located between the two parks.

Conclusions

On March 25, 2013, First State National Monument was established in Delaware. The small unit, later made into a national historical park, preserves the early history of the state at several locations. Delaware was the last state or territory of the United States to have an NPS unit, and the monument was created largely for that purpose. The national park system now extends throughout all fifty states and territories. There is probably a park unit much closer to you than you realize.

Since 1872 the possibilities for expanding the park system have been thought to be exhausted on numerous occasions, but new places have continually been added. Many reasons are evident for this. New environments, such as deserts and swamps, have been found beautiful and worthy of protection by new generations; the range of possible topics has been continually widened, and new features worthy of protection have been found. The NPS also used geography as a strategy to build support for the new park system by moving the system eastward and later into big cities.

FOUR

The Geography of the National Park System

In 1980 Alan Hogenauer, a university professor, became the first known person to visit every national park unit (Kuznia 2012). At that time, the system only numbered 348 units, but he continued to visit new units when they were added and had reached all 401 that existed when he died in 2013. Few of us will have the opportunity to follow in his footsteps, but we can take a quick tour of the 423 units of the system that existed in 2021.

Where are America's national parks and other units of the park system? There are many ways to approach the geography of the park system, and here we will do so by looking at different regions, among them Alaska, the Pacific, and the Southeast, as well as several subdivisions of the West based on physiographic or landform categories. Geographer Nevin Fenneman (1865–1945) pioneered this approach by creating the first national maps of landform categories (Fenneman 1916). His goal was to use terrain to divide the country into homogeneous subdivisions or formal regions (Hunt 1967). The Great Plains and the Rocky Mountains are among these, but many will not be so familiar. These basic landform categories remain in use today to describe the United States (United States Geological Survey 2003) and have even been used by the National Park Service to examine whether any potential parks have been missed.

Two NPS books provide some of the best guides to the park system: the fourth edition of *The National Parks: Shaping the System* (Mackintosh, McDonnell, and Sprinkle 2018) and *The National Parks: Index 2012–2016* (National Park Service 2016b). This chapter also draws heavily on the wonderful maps and information contained in the websites of individual parks.

A Tour of the Parks

Alaska Parks

It is appropriate to begin the tour in Alaska, which is home to over half of the park system by acreage and has the biggest parks, tallest mountains, biggest canyons, deepest depths, biggest animals, and most bears. It is the heart of America's park system, the land of fire and ice, with the most active volcanoes and the largest glaciers. Nowhere else in America comes close. Unfortunately, most Americans never come close either, and its parks remain little-known treasures, with some of the lowest visitation of any park unit (Fig. 4.1).

Most Alaskan parks are part of the state's two mountain ranges (Hunt 1967). The Alaska Range extends the length of the Aleutian Islands, across southern Alaska, and down the panhandle. The Aleutian Islands are home to hundreds of volcanoes, many of them active. Alaskan volcanoes and those in the contiguous forty-eight states are far more dangerous than the more famous Hawaiian versions. These volcanoes create enormous explosions and ash clouds that can disrupt life for hundreds of miles downwind.

Aniakchak National Monument and Preserve contains an active volcano with a six-mile-wide caldera formed during an eruption thirty-five hundred years ago. It was little known until a minor 1931 eruption drew attention to the area (Norris 1996a). Aniakchak is also noteworthy for being quite possibly the least visited unit in the entire park system; only

FIG. 4.1   Map of Alaska national park units in 2021. Over half of the land area of the system is in Alaska and is located in two belts along the Alaska and Brooks Ranges.

about one hundred people a year make it here. According to the park's website, more people successfully climb Mount Everest each year than visit Aniakchak. Much of this is due to its remoteness and bad weather; visitors arrive by chartered floatplanes that land on a lake within the caldera, but the nearly constant clouds, fog, rain, and wind often make this impossible. The monument may well have the most reliably bad weather of any park unit. There are no facilities within the monument, no communication with the outside world, and many grizzly bears. It is not for the average park visitor.

To the east, Katmai National Park and Preserve is also volcanic and includes a volcano that had a massive eruption in 1912, considered to have been the biggest eruption of the twentieth century. Scientists who arrived after the eruption found what they named the Valley of Ten Thousand Smokes, where steam wafted out of countless fumaroles throughout the valley. It became a national monument in 1918, and it was expanded to include bear habitat in 1931, giving it the distinction of being the nation's largest national park unit until 1978.

Lake Clark National Park and Preserve is just north of Katmai and has two active volcanoes, one of which, the 10,197-foot-tall Mount Redoubt, last erupted in 2009. The park also in-

cludes its namesake, Lake Clark, which is one of the largest lakes in the state. It is also noteworthy for its range of environments, including rain forests, alpine tundra, and rivers filled with salmon, and especially for its large bear population. Fishing and bear watching are two of the most popular activities, with visitors arriving by boat or floatplane; the park may be the very best place to watch bears in the entire park system.

The Alaskan range has a long arc through southern Alaska; here ice takes over from fire as the main force on the land. The southern summit of Denali is the highest point in the park system (and North America) at 20,320 feet above sea level. On a spectacularly clear day, it may even be seen from Anchorage, 130 miles away. Unfortunately, these high peaks also create many clouds, and only about one-third of visitors will ever glimpse it from any distance. Several lower peaks stand around Denali, including 17,400-foot-tall Mount Foraker. Although it is the third tallest mountain in the United States, it gets little attention next to the great one.

The park also contains hundreds of glaciers covering one-sixth of the park's area (Sherwonit 2013; Snyder 2013). The Kahiltna Glacier is forty-four miles long, and several others run over thirty miles. The most spectacular is Ruth Glacier, which runs through the Great Gorge, a mile-wide canyon on the south side of Denali. The canyon's cliffs tower five thousand feet over the Ruth Glacier, which is several thousand feet deep. If the glacier were to disappear, the resulting canyon would be deeper than the Grand Canyon, with granite cliffs twice as tall as those of Yosemite.

One unusual characteristic of Denali is that 36 percent of visitors arrive by train, and another 26 percent arrive on organized bus tours (Norris 2006). The park is one of several with a train station, but railroads are nowhere this important for any other park. The Alaska Railroad brings many visitors from either Anchorage or Fairbanks or directly from cruise ships docked at the port of Seward. Another characteristic is that a majority of visitors to this park are over the age of fifty; Alaska attracts many retirees with time and money to visit this vast and very expensive state.

Most visitors see the park from its one road, which runs eighty-nine miles into the park (Sherwonit 2013), one of the longest distances one can drive into a national park unit outside the southeastern parkways. Others see it from the air, while a hardy few experience it on foot. Around a thousand people a year spend several weeks attempting to climb the peak, with slightly more than half succeeding. It is not an easy climb, and each year brings several search-and-rescue missions and more additions to the one hundred people who have so far died on the mountain.

Denali also holds the record for having the state's highest national park visitor center, at 3,733 feet; most Alaskan park visitor facilities are at or near sea level. The reason is simply that at higher altitudes the climate becomes more extreme; an elevation of twenty-five hundred feet in Alaska is like one of twelve thousand feet in Arizona (Hunt 1967), far higher than any visitor center or town in that state.

Perhaps the most amazing fact about Denali is that it was originally created not to preserve the mountain but as a game preserve amid what was known as "America's Serengeti." Caribou, moose, Dall sheep, wolves, and bears can be seen there and still hunted today within the national preserve section of the park. Only when the park was expanded in 1980 was the entire mountain protected within its boundaries.

Hundreds of miles to the east is another towering range and an even vaster park. Wrangell–St. Elias is the largest single park unit in the country and is home to the second tallest peak in the nation (Mount Saint Elias, at 18,008 feet) (Bleakley 2002). It is bigger than Switzerland, has mountains thousands of feet taller, and has far more glaciers. It is tempting to refer to Wrangell–St. Elias as America's Switzerland, but this does the Alaska park a disservice; it would be more appropriate to refer to Switzerland as a lesser and far more crowded version of Wrangell–St. Elias. It is also perhaps the only park unit in the state that most visitors can drive into. The McCarthy Road leads to the well-preserved Kennecott ghost town, the most spectacular such attraction in the park system.

Kenai Fjords, Glacier Bay, and Wrangell–St. Elias are famous for their massive glaciers (Molnia 2008; Snyder 2013). From Glacier Bay to Cordova is the glaciered coast, where glaciers run into the ocean and icebergs can be seen calving off into the sea. The Malaspina Glacier in the southeastern corner of Wrangell–St. Elias is the world's largest tidewater glacier, bigger than Rhode Island. Most visitors see these glaciers from the decks of cruise ships, especially in the aptly named Glacier Bay. Such a view would not have been possible two hundred years ago, as Glacier Bay was entirely filled with ice. The bay exists because its glaciers have shrunk by sixty-five miles since 1800 and continue to retreat rapidly. The shrinking ice produced the deepest water in the entire national park system. The fourteen-hundred-foot-deep waters are also home to the biggest animals found anywhere in the park system: migrating humpback whales.

To the southeast the Alaska Panhandle is the wettest and warmest part of the state and covered in dense forests. Two small sites are nestled within the towering mountains and valleys of the Alaska Panhandle: Klondike Gold Rush National Historical Park and Sitka National Historical Park. Although located far away from most of the state's population today, they were once the most accessible and populous places in the state (Borneman 2003). Sitka commemorates early contacts between Russian and American cultures with Native Alaskans at what used to be the territorial capital. Klondike Gold Rush was named after the nineteenth-century gold rush to Dawson City, Canada, which brought both nations' Arctic territories to prominence. But except for these small parks, the densely forested, cloud-shrouded islands of the panhandle are largely the domain of the Forest Service. This may be the one area where the NPS has clearly lost the most spectacular places to a sister agency.

Far to the north, there are still more parks. The northern Brooks Range includes Bering Land Bridge, Cape Krusenstern, Noatak, Kobuk Valley, and Gates of the Arctic (Hunt 1967). All but Bering Land Bridge are north of the Arctic Circle, with continual daylight in summer months. Gates of the Arctic National Park and Preserve holds the distinction of being the northernmost park in the country, extending out onto the state's vast North Slope. The Continental Divide runs through several of these and forms the northern boundary of Noatak, the only place in the entire system where the divide makes up a park boundary. Despite their high latitude, these mountain parks possess only a few scattered glaciers (Molnia 2008). This is due not to temperature but to the dryness of the Arctic; Alaska has wider variations in precipitation than any other state, from the wet forests of the panhandle to the dry North Slope with less than five inches of rain per year, half that of Phoenix, Arizona. There is no snow out of which glaciers can form here. Instead, Kobuk Valley National Park contains sand dunes, along with vast caribou herds.

Cape Krusenstern preserves a large section of coastline, with many barrier islands. It is America's northernmost beach park, though swimming is not recommended. It was preserved in large part due to its significant archaeological record, with at least nine thousand years of human occupation recorded at campsites on hundreds of ridges near the beach. The vegetation is tundra, and permafrost, or year-round frozen ground, is found throughout the park. Noatak National Preserve was created to preserve the Noatak River; it includes tundra, forest, caribou, bears, and moose but no roads. Because it is a national preserve, hunting is allowed, and river rafting is also popular. Gates of the Arctic was named after a gap between the Frigid Crags and Boreal Mountain. These northern Alaskan parks are the remotest places in the national park system and ones where the few visitors must be the most self-reliant, as there are no facilities and many ways to die.

Yukon–Charley Rivers National Preserve lies between the state's two mountain ranges on the upper reaches of the Yukon River, one of North America's greatest rivers. The preserve includes evidence of considerable mining activity from the Klondike gold rush and is also open to hunting. Like others in Alaska, it can be reached only by boat or plane. Except for a few cabins for campers, there are no facilities. Any visitor to this remote park will be prepared for environmental extremes surpassing almost anywhere else in America; temperatures vary from -50°F in winter to almost 100°F during the short summers.

One aspect of Alaskan parks is often misleading when viewing a map of the state: their size. Nine of the ten largest park units are in Alaska, with Wrangell–St. Elias the largest of them all at 13,175,799 acres (a bit smaller than West Virginia). But even this is exceeded if contiguous parks are counted as one; the Western Arctic National Parklands, which comprise Gates of the Arctic, Kobuk, and Noatak, are even larger at 16,793,126 acres. Yet these parklands are still smaller than the Arctic National Wildlife Refuge, America's largest protected land area.

More than fifty years ago, Charles Hunt, a geologist and proponent of America's national park system, wrote that Alaska was the premier scenic attraction for Americans: "The fjords are equal to those of Norway. The glaciers of Switzerland are minor compared to those of Alaska; Mount McKinley [Denali today] is a mile higher than the Alps and closer to New York. And why go to Africa for big game?" (1967, 431). He felt that Alaska would and should become one of the world's leading tourist attractions, though on the negative side are "the fogs, mosquitoes, and charges for tourist facilities that are hardly commensurate with the services rendered" (431). A half century later, these complaints are still true.

Pacific Parks

The standard map of the national park system shows small insets of Hawaii, Guam, and American Samoa along the bottom of the map. But these small islands are in fact part of the largest geographic region within the system, the Pacific Ocean. This area includes America's westernmost and easternmost parks, as well as being home to America's only parks south of the equator (Fig. 4.2).

There are an estimated twenty-five thousand islands in the Pacific, most so small that they do not appear on the map. Geographers divide these islands into three groups: Melanesia, Micronesia, and Polynesia. Melanesia is the southernmost and includes Papua New Guinea and the Solomon Islands. Micronesia is home to the smallest islands and includes several island chains or archipelagos: the Marianas, Carolines, Marshalls, and Gilberts. Polynesia includes the Hawaiian Islands and many islands to the southeast, extending off the map to Easter Island. Samoa, Tuvalu, Fiji, New Zealand, and the Cook Islands are among them. Some islands are high islands with tall volcanoes, while others are heavily eroded low islands surrounded by a coral reef or sometimes just an atoll with the only dry land on the reef.

Figure 4.2 shows no land borders, but within the Pacific Ocean are many island countries scattered over vast distances. Within Micronesia, there are six countries: Palau, the Federated States of Micronesia, Republic of the Marshall Islands, Nauru, Kiribati, and the United States. The U.S. portion is made up of two territories, Guam and the Commonwealth of the Northern Mariana Islands, which are both parts of the Mariana archipelago. These are low islands

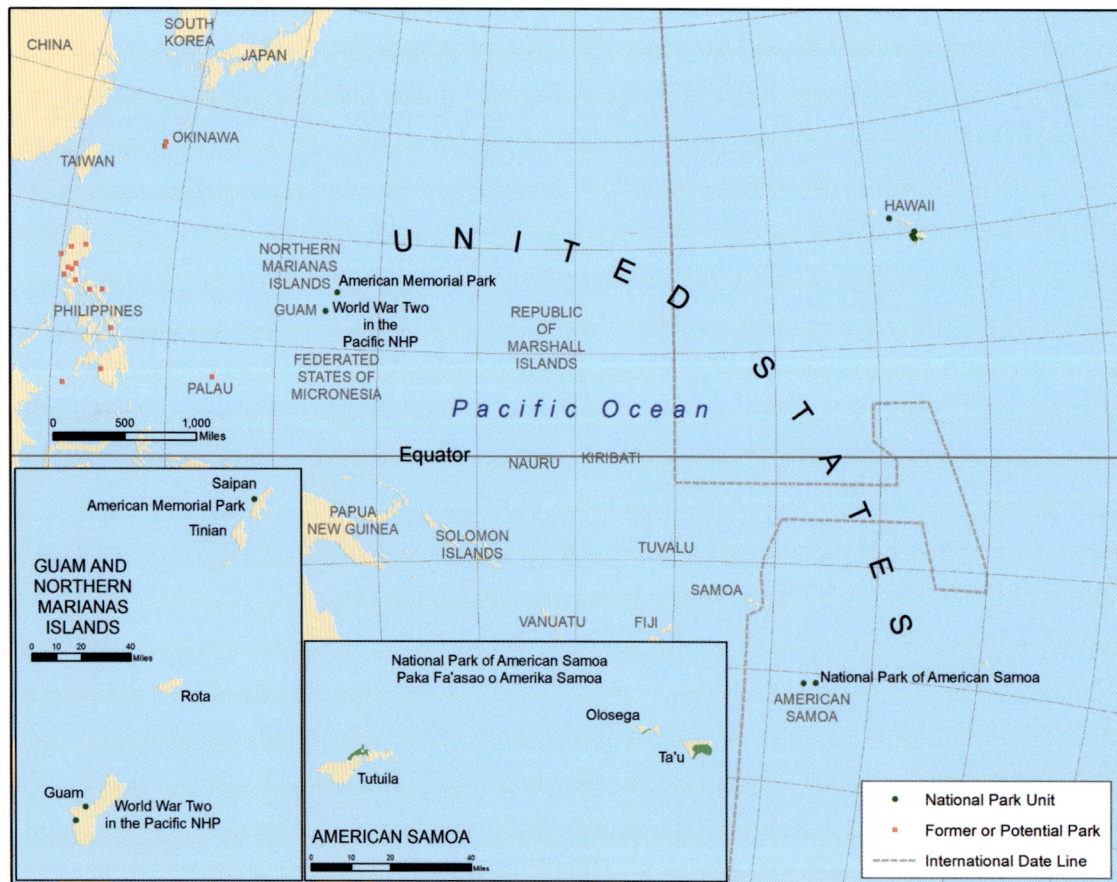

FIG. 4.2 Map of Pacific national park units in 2021. This is the largest region of the system, containing not just Hawaii but several territories and uninhabited islands.

with tropical rain forest climates. At 210 square miles, Guam is the largest island in Micronesia. Guam's War in the Pacific National Historical Park includes battlefield remains and the invasion beach where American forces landed in 1944 to recapture the island. The NPS also helps manage the American Memorial Park on the island of Saipan within the Northern Mariana Islands. This park was established in 1978 only months after the islands became part of the United States, the fastest any new territory received a park unit.

The Polynesian parks better fit the expectations for tropical national park units, with tall mountains and active volcanoes. Hawaii is home to two mountain parks, Haleakalā and Hawai'i Volcanoes (Fig. 4.3). Hawai'i Volcanoes is undoubtedly one of the most famous parks and appears on the evening news whenever fresh lava appears; this is a regular event since the Kīlauea volcano began its current eruption in 1983. But it is far from being the only active volcano in the park system. California's Lassen erupted in 1916, Mount Rainier is merely dormant, and most volcanoes in parks are in Alaska. Hawaii's volcanoes are gentle and well-behaved compared to the explosive, ash-spewing mountains on the mainland (Hunt 1967). Hawaii is also home to many cultural parks; Native Hawaiians are almost alone among living cultures in their representation within the system.

The National Park of American Samoa is perhaps the most directly and appropriately named of all the park units. It preserves rain forests and coral reefs on parts of three islands. The section on Ta'u includes half of a volcano, one side of which collapsed into the sea. The remainder of Lata Mountain stands 3,054 feet above sea level, the tallest peak in the territory and one of the more spectacular volcanic features in the park system. The park also serves as a cultural park showcasing native Samoan culture and has several unusual features due to that culture. Many signs are in both English and Samoan, including those with the park's name: Paka Fa'asao o Amerika Samoa. Samoa has a long tradition of communal landownership; land for the national park is leased from several villages. This unusual arrangement has parallels elsewhere with Arizona's Canyon de Chelly National Monument, owned by the Navajo Nation.

The parks in Guam, Northern Mariana Islands, and American Samoa are not easy to get to. Guam is reached most easily from the rest of the United States via Japan and Korea (90 percent of the territory's tourists are from Japan). American Samoa can best be reached via

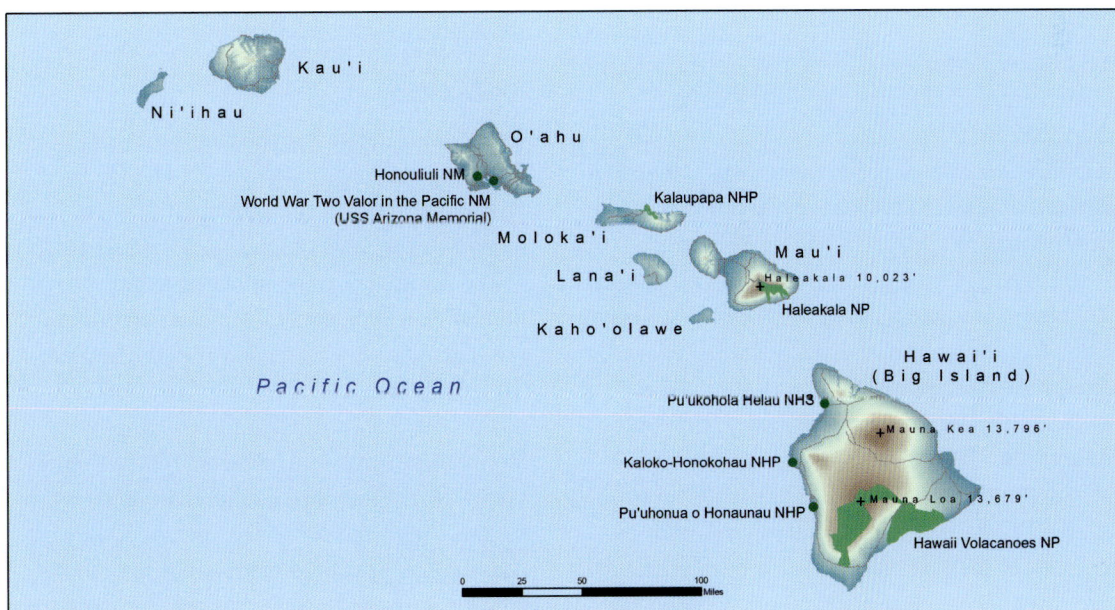

FIG. 4.3 Map of Hawaii national park units in 2021. While everyone is familiar with Hawai'i Volcanoes National Park on the big island, the state is also home to a variety of cultural and historic parks.

FIG. 4.4 Map of parks and physiographic provinces in 2021. This map shows the distribution of park units relative to the major landforms of the country.

a twice-weekly flight from Honolulu. Guam and the Northern Marianas are also unique in being west of the International Date Line and so are the only U.S. park units in the Eastern Hemisphere. They are also one day ahead of the rest of the national park system and the first places in the park system to welcome in a new year.

The Pacific region is unique in that here the U.S. national park system has overlapped with those of other countries (Fig. 4.2). The Philippines was a U.S. territory from 1898 until 1946, and thirteen national parks were created there during that time (Karnow 1990; United Nations Environment World Conservation Monitoring Centre 2020). These parks include mountains, canyons, caves, and historic sites, but they were never counted as part of the U.S. national park system. Another former U.S. territory is Okinawa, administered by the U.S. Navy from 1945 to 1972. Two Japanese "quasi-national parks" were created in Okinawa during that time. Palau, the Federated States of Micronesia, and the Republic of the Marshall Islands were U.S. territories until 1986. During this time, the NPS established seventy-three national register sites

and five national historic landmarks in these three countries, all still overseen by the NPS. The nation of Palau has the richest coral reefs in the Pacific, as well as World War II battlefields, and agency employees have worked with the government of that country to create a U.S. national historical park there, though so far without success (National Park Service 1991).

Western Parks

The American West is the historic hearth of America's national park system, and for most visitors the West is the sentimental favorite, where the most famous parks are located. This is an area of tremendous diversity for the parks, with spectacular mountains, volcanoes, glaciers, canyons, caves, amazing geological features, Indian ruins, historic sites, and lakes. Like that of Alaska, the geography of the western states is dominated by their mountain ranges, which can be divided into the coast ranges, the towering Sierra Nevada, the Cascade Range, scattered mountain ranges in the interior, and the Rocky Mountains farther east (Fig. 4.4). These mountains are all part of the same mountain chain as the Alaska and Brooks Range and continue southward into Mexico and Central America. In South America they are known as the Andes and continue into Antarctica.

The Cascades are volcanic in origin and still contain several active volcanoes. The largest of these is Mount Rainier, southeast of Seattle. California's Lassen Volcanic National Park last erupted in 1916; it became a national monument soon after. Crater Lake erupted 7,700 years ago before the mountain collapsed into itself to form a 1,949-foot-deep lake, the deepest in the country. The Sierra Nevada is the tallest mountain range in the contiguous forty-eight states, and Sequoia, Kings Canyon, and Yosemite contain large portions of the range's high country, an unmatched high-altitude wilderness. The highest point is 14,505-foot-tall Mount Whitney, located along the eastern boundary of Sequoia; it is best viewed from the east outside the park. To the east several parks are found within the scattered sections of the Rocky Mountains. Yellowstone was created on a high plateau within the Rockies, with most of the park above seven thousand feet elevation and several peaks over ten thousand feet. Other parks in the Rockies are Grand Teton and Rocky Mountain National Parks. Though much is made of the Continental Divide in the Rockies, this watershed boundary passes through only three parks (Glacier, Yellowstone, and Rocky Mountain) in this region and does not make up the boundary of any park, as it does in Alaska.

One feature of many western mountain parks is glaciers (Williams and Ferrigno 2008). Glacier National Park is home to twenty-five glaciers, but Mount Rainier is the most heavily glaciated peak in the forty-eight contiguous states. Other Cascade volcanoes and the Olympic Mountains sport numerous glaciers, and Yosemite, Sequoia, and Kings Canyon also have several dozen small glaciers high in the Sierra Nevada, the southernmost glaciers in the country (Guyton 1998). In the Rocky Mountains, Yellowstone, Grand Teton, and Rocky Mountain National Parks also contain glaciers.

Between the Sierra Nevada and the Rockies are several distinct areas: the Columbia Basin, the Colorado Plateau, and the Basin and Range country (Fenneman 1916; Hunt 1967). The Columbia Basin has relatively few national park units, the largest being Craters of the Moon, which contains a vast lava field. There are many deep canyons, but none, except the one flooded behind Grand Coulee Dam, has been included in the system.

The Colorado Plateau is named after the Colorado River, which runs through it; this region is centered on the Four Corners states of Colorado, New Mexico, Arizona, and Utah. Elevations are above five thousand feet and often over seven thousand feet, with mountains reaching nearly thirteen thousand feet. The plateau contains seventeen national park units, a collection unmatched anywhere but the Washington, D.C., area. If there is any place outside of Alaska that should be singled out as the heart of the national park system, it is this wonderland (Fig. 4.5).

Many of the scenic parks here were created around the plateau's many canyons; except for the Grand Canyon of the Yellowstone, almost every significant or well-known canyon in the park system is in this region. Foremost among them is the Grand Canyon, one of the planet's most spectacular natural features. This canyon was first protected as a national monument in 1909 and later expanded as a national park. For many, it remains one of the most recognizable national parks. As visitors to the canyon will learn, much of the Colorado Plateau is made up of different layers of colorful sandstone rocks. Deep canyons have cut through these layers at Zion and Canyon de Chelly and provided the name for Canyonlands. Colorado National Monument contains more sandstone canyons; it is in the state of Colorado but was named after the Colorado Plateau.

Sandstone also erodes into a variety of unusual formations, including arches sculpted by wind (as at Arches National Park) and natural bridges created by water erosion (in Natural Bridges and Rainbow Bridge) (Barnes 1987). Where erosion has not cut deep enough to create one of these, the result may be a massive cave in the side of a cliff. Many of these were used by ancient Native Americans for shel-

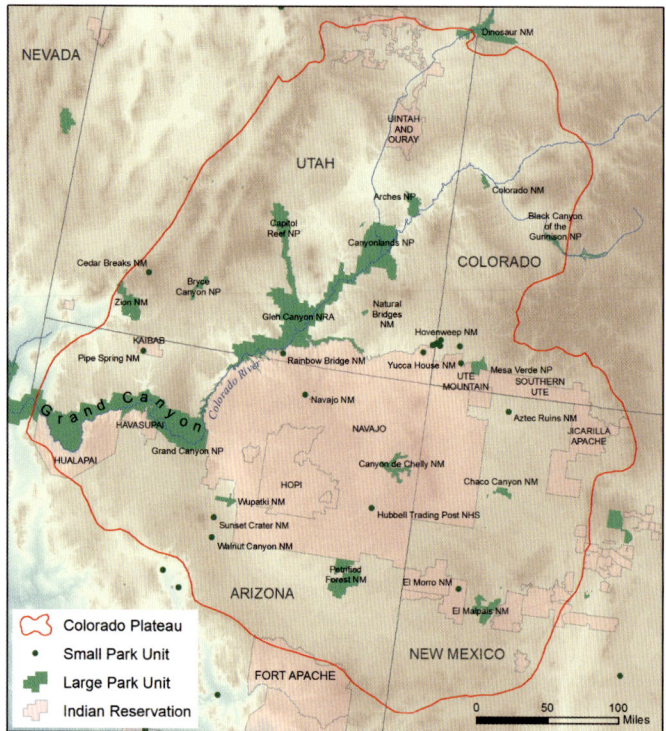

FIG. 4.5 Map of parks of the Colorado Plateau in 2021. These parks comprise the scenic heart of the system and most of its spectacular canyons.

ter, and the ruins they left behind can be seen at Mesa Verde, Canyon de Chelly, Navajo, Montezuma Castle, and several other small units. Capitol Reef also showcases colorful sandstone shapes.

Colorado Plateau parks are also home to most of the system's highest elevations—not the highest peaks; those are in Alaska, the Rockies, the Sierra Nevada, or the Cascades. Instead, the plateau's parks have the highest roads and visitor centers. While Rocky Mountain National Park has the highest visitor center, at 11,796 feet, Cedar Breaks in Utah is second, at 10,380 feet. The North Rim of the Grand Canyon, Great Sand Dunes, and Black Canyon of the Gunnison all have visitor centers over eight thousand feet, and thirteen more are over five thousand feet. These altitudes can make for surprising weather: the average overnight low temperature at Bryce Canyon is in the forties even in the summer. Visitors planning to stay for sunrise or sunset (when the light is best) are advised to dress warmly for the eight-thousand-foot elevations. It is surely one of the coldest year-round parks in the forty-eight contiguous states. (Hawai'i Volcanoes is another surprisingly cold park.)

The Colorado Plateau and nearby portions of the Southwest are also home to the largest collection of archaeological units in the system, featuring the ruins of Native American civilizations (Rohn and Ferguson 2006). The Anasazi or Ancestral Puebloans, Sinagua, Salado, Hohokam, and others built towns throughout the plateau and in the deserts to the south. Mesa Verde and Navajo have the most spectacular ruins among those surviving in cliffs, but they can also be found at Tonto, Montezuma Castle, Walnut Canyon, Bandelier, Canyon de Chelly, and Gila Cliff Dwellings. Chaco Culture contains entire intact towns; Wupatki contains smaller settlements abandoned after nearby Sunset Crater erupted. Casa Grande Ruins gets its name from a massive three-story adobe structure.

Other park units are scattered around the Colorado Plateau. Bryce Canyon contains spectacular and colorful limestone formations, while Petrified Forest contains the namesake petrified logs, as well as the colorful formations of the Painted Desert. Black Canyon of the Gunnison is another spectacular canyon, though not of sandstone.

Despite the stunning scenery, the Colorado

Plateau does not always impress visitors, especially those from eastern states who are put off by the absence of trees and confused by the vast distance they can see. Those too timid to venture off I-40 will not see any of the region's canyons, its Indian ruins, or its volcanoes, caves, arches, and natural bridges. The spectacular scenery and Native American ruins exist in part because of the dryness that produced the barrenness so distressing to some visitors. According to Charles Hunt, with

> a humid climate the plateau would have lost most or all its spectacular features. Grand Canyon and the other canyons along the Colorado River would become as drab as Hells Canyon of the Snake River—still vast, without the impressively distinctive sculpturing and without the color. The uplands would be grassed over and perhaps have trees and scenery not much different from that of the Appalachian Plateau. The painted deserts would be gone, and Monument Valley would be a collection of rounded hillocks. There would be no Rainbow Bridge or other natural bridges, arches, or alcoves. There would be more agriculture, more industry, and the land would be as crowded with people as is the rest of the country. (1967, 306–7)

We can only hope this fate never befalls this lovely place.

One of the largest regions of the West is the Basin and Range country, running from Oregon to Mexico and characterized by isolated mountain ranges stretching in a north–south direction with deep valley basins between them (Hunt 1967). Much of this landscape is within the Great Basin, an area including most of Nevada, as well as parts of neighboring states in which rivers do not flow to the sea but to dry lakes or basins. Great Basin National Park protects the Snake Range in Nevada, a typical mountain range except for containing Nevada's only glacier, on the summit of Wheeler Peak. Chiricahua is located on a southern Arizona range and contains a maze of hoodoos, or natural rock pillars, as well as an ecology that is a mix of American and Mexican flora and fauna.

Much of this area is desert. Deserts exist throughout the world, typically around thirty to thirty-five degrees latitude, both north and south of the equator (Trimble 1989; Mares 1999; Warner 2004; Walker and Landau 2018). They exist here because this is a zone of high pressure, with warm, dry air that produces little rain. Deserts can also be formed if moisture-laden air is blocked by mountains. Because the Basin and Range area is downwind from the Sierra Nevada and the coast ranges, it is a dry landscape; most of the moisture carried by the wind falls on the mountains as the air cools and rises over them. Downwind of them the drier and warmer air produces desert conditions. Much of Nevada, Arizona, and Southern California is desert as a result. These areas can be divided into the Great Basin Desert in Nevada, the Mojave in California and southern Nevada, the Sonoran in Arizona and southeastern California, and the Chihuahuan in Texas and New Mexico. Except for the Great Basin, these deserts are well represented in the national park system.

Death Valley contains both the lowest and some of the highest elevations in the Mojave Desert; it reaches a low point of 282 feet below sea level, the lowest elevation in North America (Hunt 1975; Walker and Landau 2018). Death Valley's highest point is 11,028 feet, giving it a vertical relief of 11,310 feet (Fig. 4.6). The only place on dry land in North America with more relief is the north side of Denali. Death Valley is also one of the driest as well as being the hottest place on Earth, having recorded 134° F in the shade in 1913.

To the south, Joshua Tree protects the desert plant of that name, which grows in abundance here. The park has also become a major rock-climbing destination (Dilsaver 2017). Mojave Preserve protects more of the desert and has the largest and tallest stand of Joshua trees. In Arizona the Sonoran Desert is included in two parks, both representing varieties of cactus. Saguaro National Park is split into two sections east and west of Tucson and is named for the iconic tall and branching cactus that thrives around the city. Organ Pipe Cactus National Monument is to the west and is named for a different type of branching cactus. Big Bend includes a large part of the Chihuahuan Desert in West Texas, which also includes White Sands, the Guadalupe Mountains, and Carlsbad Caverns. This desert has few cacti.

FIG. 4.6   Death Valley National Park. Snowcapped mountains are a common sight in this California desert park, the hottest place in the world. NPS.

Those expecting sand dunes in the desert will be surprised by their rarity, though they are the central feature of White Sands, and several smaller dunes can be found in Death Valley and Mojave. The biggest collection of dunes is at Great Sand Dunes in Colorado's San Luis Valley, adjacent to the high peaks of the southern Rockies (Geary 2016). Sand dunes can also be found in Alaska at Kobuk Valley National Park and in many seashores.

Within the Intermountain West, a common park type is a reservoir-based national recreation area. While the NPS unsuccessfully waged a political battle to keep dams out of Yosemite, elsewhere it has welcomed dams and the reservoirs behind them. This first took place when the massive Hoover Dam was built on the Colorado River, impounding Lake Mead (Dodd 2007). The dam was built under the direction of the Bureau of Reclamation, which was aware of the recreation possibilities but had no wish to be involved in such an activity. The solution was to subcontract to the NPS and allow that agency to develop the reservoir for recreation and operate it as a park unit while the Bureau of Reclamation continued to operate the dam. This

relationship proved workable for both agencies and was followed in many more cases.

Another type of park found only in the West is the fossil park (National Park Service 2020c). Fossils are found in rocks formed in the Triassic (252–201 million years ago, visible at Petrified Forest) and Jurassic periods (201–145 million years ago, found at Dinosaur) and during the Eocene (56–34 million years ago, at Fossil Butte and Florissant Fossil Beds) and Miocene epochs (23–5 million years ago, seen at John Day Fossil Beds and Agate Fossil Beds). Rocks of these ages are best exposed in western states; older rocks in the Midwest were removed by glaciations, while those in the Southeast were buried under sediment. Florida was underwater and Hawaii did not yet exist when most of these rocks were laid down.

The high and rugged mountains in the West have produced many spectacular park roads. The Going-to-the-Sun Road in Glacier National Park is often considered the most spectacular, but Rocky Mountain National Park has the Trail Ridge and Fall River Roads, Zion has the Mount Carmel Road, and Yosemite has the Tioga Pass Road. These are not just scenic drives; instead, they are spectacularly engineered, often seeming to cling to cliffs. Travelers will experience several climates and a range of ecosystems within a short drive.

Eastern Parks

The eastern United States contains half the parks, but these are mostly small historic sites or battlefields, and they lack the scenic splendor of the West and Alaska. But there are exceptions: the Black Hills of South Dakota contain an amazing collection of national park units beyond Mount Rushmore. Here Jewel Cave is one of the biggest known cave systems in the world, while Wind Cave is known both for its underground sights and aboveground wildlife. To the east, Badlands National Park contains a serrated but beautiful landscape, and Minuteman Missile National Historic Site includes a former underground nuclear missile base. To the west of the Black Hills in Wyoming is Devils Tower, America's first national monument.

The Midwest probably does not rank as a scenic wonderland in many people's view for the simple reason that most of the scenery has been removed. This took place during the ice age, when a massive continental glacier scraped and ground its way southward from the Arctic (Benn and Evans 2010). Thousands of feet thick, these icy sheets leveled terrain, reshaped drainage systems, and removed fossil-bearing rocks. But islands of scenery remain. One is Isle Royale in Lake Superior, one of the few large scenic parks in the East and one of the most remote in the forty-eight contiguous states.

The Midwest is the land of great lakes and great rivers, though decades of industry and pollution have rendered many of them untouchable. Despite this, long stretches of the upper Mississippi, Missouri, and other rivers are protected in the park system. Four national lakeshores are found around Lakes Superior and Michigan. Other large recreation units can be found in Ohio (Cuyahoga Valley), New York City (Gateway), and Boston (Cape Cod and Boston Harbor Islands), along the Delaware River, and near Washington, D.C. Catoctin Mountain Park is a small recreation area in Maryland notable for including Camp David, the presidential retreat (an earlier presidential retreat can be found in Shenandoah) (Kirkconnell 1988). You won't find it on the park map, but it's on the north side of Park Central Road between the Greentop camp and the parking area for the Hog Rock nature trail.

The Midwest and Northeast include many historic sites devoted to the nation's birth and the lives of its founders and inventors, as well as to less happy memories. In central Pennsylvania, Johnstown Flood and Flight 93 National Memorials commemorate two terrible disasters that each claimed thousands of lives. The two memorials are a short distance apart and are among very few park units devoted to the loss of life outside of a battlefield.

The Northeast includes the northern end of the Appalachian Mountains. Katahdin Woods and Water National Monument was created in 2016 near Mount Katahdin and the northern end of the Appalachian Trail. Acadia is an old park created in 1916 as the first park unit east of the Mississippi River. It contains Mount Desert Island and several smaller rocky islands; the summit of Cadillac Mountain is the highest point on the U.S. Atlantic coast. The network of carriage roads is one of the park's most striking features.

The Southeast can be divided into the Appalachian Mountains and the Coastal Plain (Hunt 1967). The highest peaks in the Appalachian Mountains are at its southern end, in North Carolina and Tennessee. Though the Great Smoky Mountains are not even as tall as the Black Hills of South Dakota, they are impressive within the region and form the largest park unit east of the Mississippi. Shenandoah contains a long mountain ridge, while the Blue Ridge Parkway runs along other ridges between this ridge and the Smokies. With the addition of several smaller parks, national forests, wild rivers, caves, and canyons outside the park system, the southern Appalachians make up one of the country's great assemblages of protected lands.

The Coastal Plain is America's largest low-altitude region; the North Slope of Alaska is the closest equivalent (but much different in climate) (Hunt 1967). Much of the Gulf and south Atlantic beaches and barrier islands are within the park system in the form of Padre Island, Gulf Islands, Canaveral, Cumberland Island, Cape Lookout, Cape Hatteras, and Assateague Island National Seashores. These are among the most fragile park units, threatened by storms, rising sea levels, mainland development, and overcrowding. For southeasterners, they provide the relief from the summer heat that westerners find on mountains, and their lighthouses are widely loved icons. Assateague Island is also known for its herds of horses, among the largest wild animals found in parks east of the Mississippi River. The tropical Everglades in South Florida is the largest river protected in the national park system, with Everglades, Big Cypress, and nearby Biscayne Bay part of this vast expanse of water and grass. This is also a fragile environment; a massive restoration plan was put together to restore this ailing landscape but has largely been forgotten.

One southeastern national park stands out on maps of the system as a dot located in the Gulf of Mexico west of Florida. Dry Tortugas National Park really is out to sea, seventy miles beyond Key West, a series of small islands and reefs. The name comes from the lack of freshwater on these islands, as well as the presence of many turtles (tortugas), seen by Spanish explorers. It is a long boat ride or expensive floatplane trip from Key West, but it is worth the trip. The park includes a massive brick fort from the 1840s, one of dozens that were built along the nation's coasts. Castillo de San Marcos, Forts Pulaski, Sumter, Monroe, and McHenry, and even the Statue of Liberty (which stands on Fort Wood) preserve other examples.

The Blue Ridge Parkway is the best-known example of a parkway, a park type created only during a short period of time in the 1930s; the Southeast is home to this parkway, as well as the comparable Natchez Trace Parkway, Shenandoah (a parkway in all but name), and Colonial Parkway in Virginia. Perhaps Alaskan parks aren't always the biggest: the Natchez Trace Parkway and Blue Ridge Parkway stretch for 444 and 469 miles, respectively, though they are rarely more than one-quarter mile wide. They are the longest continuous national park units.

A southern pastime well preserved in the park system is conflict. The region is home to an enormous assortment of Revolution battle sites, Civil War battlefields, and even War of 1812 sites. The siege lines of Vicksburg, the locations of the "devil's own day" at Shiloh, the battle above the clouds at Chattanooga, and the end of the fighting at Appomattox Court House are all preserved, along with countless other battles and skirmishes. Horseshoe Bend National Military Park commemorates Andrew Jackson's victory over the Upper Creeks in 1814; he led the Tennessee, Kentucky, and Louisiana Militias in a victory over the British army at the Battle of New Orleans a short time later (preserved as the Chalmette Battlefield section of Jean Lafitte National Historical Park and Preserve).

The nation's civil rights struggles have become one of the newest sets of places to be commemorated and preserved. Many center on the struggle by African Americans for equal rights. Little Rock Central High School National Historic Site, and Birmingham Civil Rights and Freedom Riders National Monuments are among the newest park units to be added. They will surely be joined by more.

Many national recreation areas are centered around reservoirs in western states, but there are no similar units to be found in eastern states. There certainly are many reservoirs in the East, so why are these not national recreation areas? The answer is partly history and partly the presence of competing agencies. The national recreation area concept did not exist before the 1930s and was uncommon before the 1950s; by the time the NPS got involved with reservoirs in the West, other agencies had long since been doing the same in eastern states. The Tennessee Valley Authority manages many lakes for recreation on the Tennessee and Cumberland Rivers, while the Army Corps of Engineers does the same on the Mississippi, Missouri, and Ohio River systems. The boat ramps, picnic areas, and campgrounds built by these agencies often look indistinguishable from those of the NPS (Foresta 2013). But there are NPS recreation parks in the eastern national park system in the form of either beach parks, men-

FIG. 4.7   Map of Caribbean national park units in 2021. While Puerto Rico has only one unit, much of the Virgin Islands are within the park system.

tioned above, or rivers. New River Gorge National Park and Preserve in West Virginia, Gateway National Recreation Area in New York City, and Upper Delaware Scenic and Recreational River are examples of the latter.

Caribbean Parks

Several parks are found on islands dividing the Atlantic Ocean from the Caribbean Sea (Fig. 4.7). Like Hawaii, Puerto Rico is at the eastern end of a chain of islands, in this case, the Greater Antilles (Hunt 1967). The island is located farther south than Hawaii and has a similar tropical rain forest climate. But it has only one park unit, San Juan National Historic Site, containing a Spanish fort built in the 1530s.

The much smaller Virgin Islands are east of Puerto Rico at the western end of the Lesser Antilles islands. The Virgin Islands has the distinction of being one of the newest areas of the United States, as it was purchased from Denmark in 1917. One result is that the national park units on these islands are the only ones in which people drive on the left side of the road. But since, as in some Alaskan parks, many visitors arrive on cruise ships and do little driving, this is not much of a concern.

Most of the island of St. John is within Vir-

gin Islands National Park, with beaches, reefs, and hiking. Christiansted National Historic Site preserves Danish colonial history in the Virgin Islands dating to the early 1700s. Christopher Columbus came ashore at Salt River Bay in 1493, the only place within the United States he ever visited. It is also one of the few bioluminescent bays in the Caribbean.

Coral reefs are a type of resource that has appeared in the national park system recently (National Park Service 2020b). Reefs are found throughout the tropical waters of the United States in the Caribbean, the southern end of Florida, and the Pacific, but here they are easy to reach. Virgin Islands Coral Reef and Buck Island Reef are the only park units created specifically around reefs; the former is entirely underwater. There are four other park units in the Pacific with reefs (one each on Guam and American Samoa and two in Hawaii), as well as Dry Tortugas and Biscayne Bay in Florida. (Utah's Capitol Reef is named after a long sandstone ridge.)

These Caribbean parks are served by airports with large numbers of flights to the U.S. mainland. For those living in northeastern or midwestern cities with time and money to travel, they are among the easiest of America's park units to get to, and almost half of the visitors come from these areas (and 92 percent are from the continental United States). From a visitation perspective, they could be thought of as northeastern or midwestern parks.

Urban Parks

A recent geographic grouping of parks is those in cities. This would have been unimaginable in 1864, when less than 25 percent of the population lived in urban areas. A typical farmer of the nineteenth century would have had a difficult time comprehending or appreciating a place like Yellowstone or Yosemite (though such typical farmers would have been extremely unlikely to have ever visited these places).

Much of the urbanization of the national park system stems from decisions made in the 1930s when the NPS was given responsibilities for memorials and monuments within the District of Columbia. Many of the best-known sites are clustered on or near the National Mall, running from the Capitol to the Lincoln Memorial. These are monuments constructed to memorialize key figures in the history of the country and public buildings such as the White House. They attract visitors from all over the country and world and constitute a memorial landscape of significance to all Americans.

Other NPS sites are scattered throughout the city. These have much in common with other urban monuments and historic sites throughout the country. Ford's Theater and Clara Barton National Historic Sites are examples of these. They are small sites, often just individual buildings, scattered throughout the city and attracting a much smaller number of visitors than the National Mall.

Because of the unique political geography of the District of Columbia, the NPS has also been assigned duties here that elsewhere fall to local government. Since 1933 the agency has been responsible for many local parks in the city, including Anacostia and Rock Creek Parks, which have ball fields and golf courses, as well as countless small parks and squares throughout the city. The NPS is also responsible for maintaining parts of the Washington, D.C., highway and freeway system in the form of the George

Washington Memorial Parkway, the Rock Creek and Suitland Parkways, and the Baltimore–Washington Parkway.

In the 1930s the NPS also began administering many historic sites and parks, which were often located within eastern cities. These parks are typically quite small and have limited facilities, often with a narrow appeal. These sites may be famous, such as the Statue of Liberty, or might be almost unknown, such as William Howard Taft National Historic Site in Cincinnati.

Although they are not always visible on a map, there are urban park clusters in Boston, New York, Philadelphia, Baltimore, San Francisco, and especially the National Capital Region (Fig. 4.8). These are day trip parks or even ones visited on school field trips. Aside from the national capital parks, few, if any, would serve as vacation destinations. Their managers don't worry about bears or bison but whether owners pick up after their doggies and if Frisbees are being thrown safely. The park system will become more urban over time if for no other reason than continued urban growth.

There are many more parks within urban areas than you think. The creation of an industrial America in the nineteenth and twentieth centuries was accompanied by urbanization on a scale never before possible; by 1920 over 50 percent of Americans lived in cities, increasing to more than 80 percent by the end of the century. Since 1950 the Census Bureau has identified Metropolitan Statistical Areas (MSAs), which contain large cities (with more than fifty thousand people) and their suburbs. There are now 383 of them, ranging from New York City at over twenty million people down to Carson City, Nevada, with fifty-four thousand people. Most people have some idea where the country's big cities are located, but MSAs can be tricky; their boundaries are defined by counties or equivalent units, and MSAs in western states can be enormous. A third to half of most eastern states are included in an MSA; the entire states of New Jersey, Delaware, and Rhode Island are in one. There are 335 parks that are within or that overlap an MSA. Most of the Black Hills are within the Rapid City MSA; the Grand Canyon is

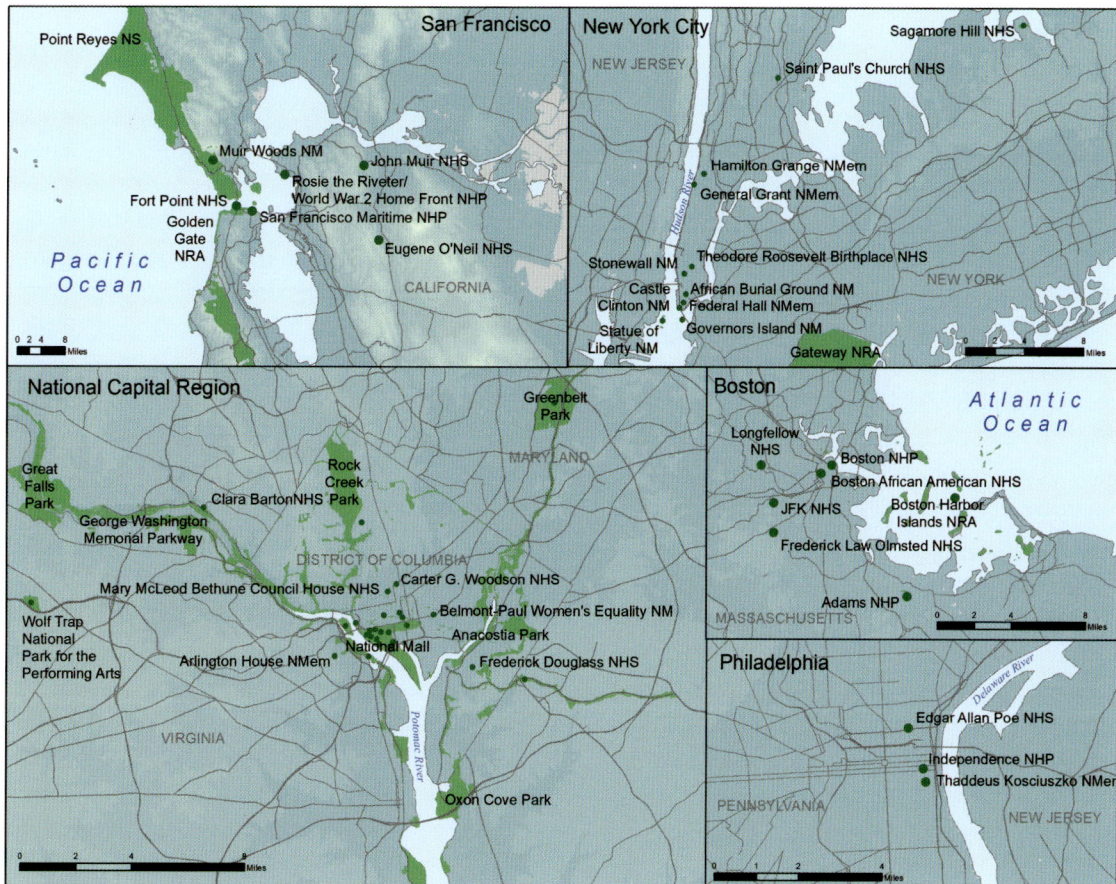

FIG. 4.8   Map of urban national park units in Washington, D.C., New York, Boston, Philadelphia, and San Francisco in 2021. These cities contain most of the nation's urban units.

within the Flagstaff and Lake Havasu–Kingman MSAs. Death Valley is partly within the Las Vegas MSA and partly in Los Angeles. All Hawaiian parks are metropolitan. Only in Alaska are parks nonmetropolitan, though even here Denali partly overlaps the vast Anchorage MSA.

Beyond the Nation's Boundaries

The boundaries of the United States are not necessarily the limit for the national park system. But while the NPS does not (yet) have any sites in another country, it does not need to: national memorials do not have to be located where the event commemorated took place. The National Mall has several NPS war memorials dedicated to events that took place in France, Belgium, Luxembourg, Germany, Japan, the Philippines, Korea, and Vietnam. This list will likely grow in the future as later wars receive memorials.

Park Names

This tour of the nation's park system has listed many names, some that are familiar, some that are not, and others that are strange. What's in a park name? Names can be important to advertising a location to potential visitors. Names are a quite important part of a park's identity and can affect its survival. They have also often proven politically contentious.

Most parks are named for the features they contain or the event that happened there, as included in proclamations or legislation. George Washington Carver National Monument is a site devoted to this man, Lake Meredith National Recreation Area contains exactly what the name says it does, and Rosie the Riveter / World War II Home Front National Historical Park is also reasonably clear, at least to older visitors. The name of Amistad National Recreation Area along the Texas-Mexico border is an exception; it was agreed upon by President Eisenhower and Mexican president Adolfo Mateos in 1959.

But some park names do not communicate the park purpose very well: What is Ninety Six National Historic Site about? Mesa Verde National Park is located on a mountain called Mesa Verde, but the park exists to preserve ancient Native American ruins. Fort Bowie National Historic Site was created in large part because of several important events at nearby Apache Pass, which were left out of the park's name (Gomez 1984). Some names are misleading: Montezuma Castle and Aztec Ruins have nothing to do with these peoples but were named by nineteenth-century settlers and explorers who associated the spectacular Native American ruins they found to the powerful Aztec civilization in Mexico (Protas 2002). There are undoubtedly many people who are disappointed that Petrified Forest does not have any standing petrified trees; would they feel that way if it were called Petrified Logs National Park? Navajo's name is especially confusing, as the monument is not related to the Navajo natives (though it is located within their reservation) and often mixed up with the Monument Valley Navajo tribal park.

Dinosaur National Monument was created around a small fossil quarry where dinosaur bones were being excavated; when the monument was expanded to include the nearby canyons of the Green and Yampa Rivers the name was not changed. This created some confusion, as the name did not encompass the canyons, nor did the canyons contain dinosaur fossils. And when the Bureau of Reclamation drew up plans to build a dam and reservoir in the canyons the name worked against efforts to protect the monument. The dinosaur quarries would not be affected by the dam, so the purpose of the park would not be threatened (M. Harvey 1994).

Two park units are named after the devil, a theme continued abundantly within many units, perhaps especially Death Valley, which has had a Devils Hole, Golf Course, Cornfield, Speedway, and Throne. These are relics of a time when explorers gave names based on perspectives different from those of modern visitors seeking a wilderness experience. These names appear old-fashioned or silly and are often offensive to Native Americans, whose cultures view them very differently. Devils Tower was known to Cheyenne, Lakota, Sioux, Kiowa, and Crow by a variety of names, with Bear's Lodge one of the most common. This refers to a story in which children were chased by giant bears; a rock the children took shelter on rose out of the ground while the bears clawed at the rising column, leaving the distinctive striations in the rock. Not only does this story not refer to the devil, but these peoples see the tower as sacred.

Names change over time, and those for parks are no exception (Table 4.1). Custer Battlefield National Monument became Little Bighorn Battlefield National Monument to better reflect what happened there in 1876, as well as reflecting varied perspectives on the battle and its meaning. Mound City Group became Hopewell Culture, replacing a descriptive label about a group of Native American mounds with a title that reflects the culture that created them. Some renamings are an attempt to return to an original or more authentic name, as when Fort Marion was renamed Castillo de San Marcos, the name given by the Spanish soldiers who built it. A national monument given the Paiute name Mukuntuweap was later changed to Zion to make the name easier to pronounce and more appealing to visitors. Several parks have sought proposed name changes that never happened, including a 1939 attempt to rename Mesa Verde as Cliff Dwellings National Park to better communicate why the park exists (D. Smith 2002). A similar effort sought to rename Organ Pipe Cactus into Arizona Desert or Sonoran Desert National Monument (Lissoway 2004).

Place-names can be controversial. The name Denali, meaning "great one," was used for centuries by Native Alaskans to refer to the highest peak in the land. In 1896 it was given the name Mount McKinley after the then-presidential candidate from Ohio (Monmonier 2006). Many preferred the original name, and in 1975 the state of Alaska began to refer to it as Denali again. The NPS agreed, and when Mount McKinley National Park was expanded in 1980 it was renamed Denali National Park. But the

TABLE 4.1 Park name changes, 2021

Original name	Later name(s)
National Cemetery of Custer's Battlefield Reservation	Custer Battlefield NM, 1946; Little Bighorn Battlefield NM, 1991
General Grant NP, California	Kings Canyon NP, 1940
Cinder Cone NM, California	Lassen Volcanic NP, 1916
Lassen Peak NM, California	
Mount Olympus NM, Washington	Olympic NP, 1938
Mukuntuweap NM, Utah	Zion NM, 1918; Zion NP, 1919
Gran Quivira NM, New Mexico	Salinas NM, 1980; Salinas Pueblo Missions NM, 1988
Sieur de Monts NM, Maine	Lafayette NP, 1919; Acadia NP, 1929
Mount McKinley NP, Alaska	Denali NP and Denali Npres by ANILCA, 1980
Lehman Caves NM, Nevada	Great Basin NP, 1986
Mound City Group NM, Ohio	Hopewell Culture NHP, 1992
Bryce Canyon NM, Utah	Utah NP, 1924, Bryce Canyon NP, 1928
Fort Marion NM, Florida	Castillo de San Marcos NM, 1942
Kill Devil Hill Monument, North Carolina	Wright Brothers Nmem, 1953
Fort Jefferson NM, Florida	Dry Tortugas NP, 1992
Hopewell Village NHS, Pennsylvania	Hopewell Furnace NHS, 1985
Virgin Islands NHS, Virgin Islands	Christiansted NHS, 1961
City of Refuge NHP, Hawaii	Pu'uhonua o Honaunau NHP, 1978
Fort Clatsop Nmem, Oregon	Lewis and Clark NHP, 2004

mountain's name remained McKinley on federal maps; the Ohio congressional delegation interfered to block any attempt at renaming the mountain. Only in 2015 was their disgraceful political manipulation overcome, and the mountain was restored to its original name (Jule Davis 2015; Schuppe 2015). Similar controversies have arisen over attempts to rename Mount Rainier and Devils Tower. A much more minor controversy exists in the name of Harry S Truman National Historic Site; the NPS omits the period after the S, while others include it (Harvey and Harvey 2017). Harry Truman started the confusion himself, perhaps as a joke.

Hawaiian parks are unique in the prevalence of Native Hawaiian names; this is a relatively recent phenomenon. Since the 1960s interest in preserving the Hawaiian language and restoring Native names became a major movement in the state (Monmonier 2006). An important step was a decision by the U.S. Board on Geographic Names to allow the diacritical mark ' to be used on federal maps to represent the 'okina, or glottal stop, often used in Hawaiian. This opened up possibilities for renaming many Hawaiian park units and even the way the names of islands such as O'ahu and Hawai'i are written (but not the name of the state, which remains Hawaii). But tourists have trouble with these names,

and several parks are known by other names: Pu'uhonua o Hōnaunau is also known as the City of Refuge, while Kalaupapa National Historical Park is the leper colony. But don't call it this if you visit; the NPS strongly disapproves of this term. Mahalo!

You might have noticed that many parks have long names: President William Jefferson Clinton Birthplace Home National Historic Site is the longest. It may seem like park names are getting longer since the days of Zion and Glacier to today's Rosie the Riveter / World War II Home Front National Historical Park, but there is almost no statistical correlation between the length of a park's name and its age. For the record, the longest ever park name was Alibates Flint Quarries and Texas Panhandle Pueblo Culture National Monument, created in 1965 but shortened to Alibates Flint Quarries National Monument in 1978. Despite the big name, it is the smallest park unit in Texas. It is, however, bigger than Maryland's Greenbelt Park, the park unit with the shortest name.

Many names are easily confused: Hopewell Culture and Hopewell Furnace, Minute Man and Minuteman Missile, San Juan and San Juan Island, Martin Luther King, Jr. National Historical Park and the Martin Luther King, Jr. Memorial, or Pu'uhonua o Hōnaunau and Pu'ukoholā

Heiauor. There are two Saint Croix Islands in the park system, one in Maine and the other in the Caribbean; the latter is home to Buck Island Coral Reef National Monument and Christiansted National Historic Site. Christiansted was originally Virgin Islands National Historic Site but was renamed in 1961 because of confusion with the nearby Virgin Islands National Park.

Within parks, place-names also often show a lack of variety. The name Death Valley was bestowed by members of a lost part of immigrants heading to California; that name inspired later visitors to continue the theme with names such as Hells Gate, Dantes View, Funeral Mountains, and Coffin Peak (Lingenfelter 1988). Death Valley is not the only park with a common theme for many place-names. Many prominent features within the Grand Canyon were named by geologist Clarence Dutton in 1882, inspired by ancient mythology or Hindu deities, and many of the names in Gates of the Arctic were given by Bob Marshall in 1929 (Dutton 1882; Cole 1992). These were literate and imaginative individuals, though sometimes perhaps too imaginative.

Park names are also valuable. Delaware North, a large corporation that operates lodges in several parks, sued the NPS in 2015, claiming that it owned several names in Yosemite, including the name Yosemite National Park and Half Dome. In 2016 the NPS changed several names in the park (Ault 2017). The Ahwahnee became the Majestic Yosemite Hotel, Curry Village became Half Dome Village, Yosemite Lodge became Yosemite Valley Lodge, Wawona became Big Trees Lodge, and merchandise bearing the label Yosemite National Park was removed from NPS-owned gift shops. Both the new and former names now appear on park maps.

Similar cases have appeared in other parks. The Xanterra Travel Collection, a large corporation, tried to trademark several Grand Canyon names. In 2011 the NPS sued the Hot Springs Advertising and Promotion Commission to prevent its use of a logo that contained the phrase "Hot Springs National Park Arkansas." The NPS claimed to be the only organization with the right to use that phrase, but it lost in court (Brantley 2011). Imagine the consequences if the NPS loses control over other park names.

Conclusions

This brief survey of the national park system named only a few of the 424 special places in it and described even fewer in detail. It is very likely that your favorite park was not mentioned, especially if you live in the eastern states or territories, but perhaps you will have some ideas about visiting new ones.

Perhaps the most easily overlooked fact about the park system is its size: it is enormous. It stretches almost ninety-five hundred miles from Guam to the U.S. Virgin Islands, from above the Arctic Circle to south of the equator. It extends across 82 degrees of latitude and 152 degrees of longitude, reaching almost halfway around the world. Much of this extent is in the Pacific Ocean, and the center of the park system is a point about halfway between Honolulu and Los Angeles. But being near the center of the park system doesn't mean that Los Angeles is near all the parks: the War in the Pacific National Historical Park on Guam is closer to Moscow than Los Angeles. Virgin Islands National Park in turn is closer to the West African city of Dakar, Senegal, or to Rio de Janeiro, Brazil, than to Los Angeles.

Another way to think of their geographic extent is that parks span nine time zones, from the Atlantic to the Chamorro, and there are parks in all of them. Virgin Islands National Park is one of several in the Atlantic time zone. The National Park of American Samoa is seven hours behind Virgin Islands National Park, but thanks to its location west of the International Date Line, Guam's War in the Pacific National Historical Park is fourteen hours ahead (and already tomorrow).

Some parks are big enough to span multiple time zones; the Manhattan Project National Historic Site is in three (with a fourth located between them). Time zones can be tricky when moving around within some parks. Theodore Roosevelt National Park in North Dakota is split into two time zones: the north unit is on Central time, and the south unit is on Mountain time. Lake Mead National Recreation Area is split between Nevada in the Pacific time zone and Arizona in the Mountain zone. The situation here is even more confusing because most of Arizona does not observe daylight saving time (DST), so during the summer all of Lake Mead will have the same time. Glen Canyon National Recreation Area is split between Arizona and Utah; the headquarters and main marina are in Arizona and do not observe DST, while marinas in Utah do.

Parks in northeastern Arizona are famously confusing because while the state does not observe daylight saving time, the section of the Navajo Nation that occupies the northeastern part of the state does. To make matters more confusing, the Hopi Indian Reservation, which lies within the Navajo Nation, does not. Allow plenty of time when visiting Canyon de Chelly, Navajo, or Hubbell Trading Post: not only are they much farther away than you think, but you'll never be sure what time it is when you get there. Parks in Guam, the Northern Mariana Islands, Hawaii, Puerto Rico, the Virgin Islands, and American Samoa also do not observe DST; these are tropical islands where the hours of sunlight hardly change and DST serves no purpose. At the other extreme is Alaska's Noatak and Gates of the Arctic, which extend north of the Arctic Circle. Visitors there will enjoy twenty-four-hour sunshine for a few months in the summer, making the state's participation in daylight saving time irrelevant.

Many visitors to a park will never bother with the time except to note the next sunrise or sunset. These can be magical times: Saguaro has the most spectacular sunsets to be found anywhere, especially at Gates Pass in the West Unit. The best views of the colorful pinnacles at Bryce Canyon are at either sunrise or sunset; no visit to the park is complete unless you've seen

at least one (but come prepared—either will be cold at any time of the year). Acadia in Maine claims the first sunrise in the country, though this is true only in the winter: in June and July the sun never sets on Gates of the Arctic (and, for a few days in late December, never rises). Sunsets and sunrises can require some advance planning. If you are visiting Acadia in June, you must be on top of Cadillac Mountain as early as 3:00 a.m. to find parking in order to view the 4:46 a.m. sunrise. Death Valley and Manzanar have unusually early sunsets because of towering mountain ranges to the west. Watching the sunrise from the 10,023-foot-high top of Hawaii's Haleakalā volcano has become such an attraction that the park now requires reservations up to sixty days in advance to experience it.

While watching sunsets or sunrises is a memorable park activity today, one national park unit exists solely because of the view during a single sunrise. It's not on a high mountain peak, in a geological wonderland, or in a serene desert but in downtown Baltimore, Maryland. The American flag visible by the dawn's early light over Fort McHenry on the morning of September 14, 1814, signaled to an invading British fleet that their continued attack was useless. One witness to this sunrise was Francis Scott Key, who was inspired by the sight to write what would become the national anthem of the United States. This in turn led to the fort becoming a national park unit in 1925. The flag that yet waves twenty-four hours a day over Fort McHenry National Monument and Historic Shrine is a replica; the original is in the Smithsonian Museum of American History, protected in a dark room. It will never again see the light of day. Unfortunately, Fort McHenry doesn't open until 9:00 a.m., so, as in 1814, you will have to watch the sunrise from elsewhere.

FIVE

The Political Geography of the National Parks

Think about the last large national park you visited. You probably remember the moment you entered the park, thanks to a large park sign, perhaps built with logs and stones, next to a pullout (Fig. 5.1). You might have even stopped for a picture. Entering a park can be memorable, but beyond that, we don't usually give boundaries much thought. But they are an essential element of every unit, and, as with states or countries, there are different boundaries and customs on either side of the boundary. We take the size and shape of park boundaries for granted and might assume they were natural, but they have been created by bureaucrats in Washington, D.C., and reflect a range of competing interests and concerns.

The geography of park boundaries is a vital aspect of parks. How big should a park be? There have been several points of view on size, which remains a contentious issue. Where should the boundaries be located? Those who drew boundaries relied on straight lines, survey systems, terrain, and other ways to draw them. Once created, park boundaries are frequently subject to change, may contain private land that must be acquired, and may include noncontiguous areas, making for difficult management. The status of many parks has been changed, as when a national monument becomes a national park, and some parks have been eliminated from the system. Parks overlap a range of political jurisdictions, including counties, states, and, most importantly, congressional districts.

Specifying Boundaries

The Size of Parks

Perhaps the most basic question about a new park unit is how big it should be. Big enough to include the protected resource, but what does that mean? Should it include just the resource or the mountains the make up its backdrop, the watershed it lies in, and land for campgrounds and a visitor center? Should there be room enough for wilderness hiking? As discussed in chapter 3, differences arose from the beginning. When Yosemite was set aside in 1864 the reservation included only the spectacular valley, which was then and now the primary attraction. The later Yellowstone park was much larger than Yosemite canyon, as the Yellowstone re-

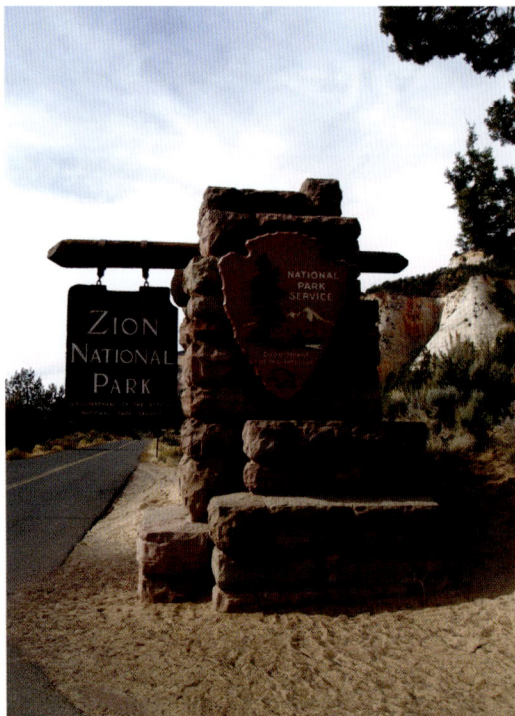

FIG. 5.1 Zion National Park entrance sign. One of the most eye-catching entrance signs to be found in the system, this rustic stone and timber work of "parkitecture" marks the eastern entrance to Zion. NPS.

gion remained little explored and mapped at the time. The boundaries had to be made a large rectangle because there was little idea exactly what was there and where it was located (Ise 1961; Mackintosh 1991; Runte 2010).

These two cases show several choices about national park boundaries that remain important today. A big park proved useful when Katmai was created in 1918 to preserve a volcanic eruption in a little-known part of Alaska (Norris 1996b). The monument was centered on the Mount Katmai volcano, but decades later scientists determined that the eruption had occurred seven miles away at a previously undiscovered volcano. The monument, then the largest park unit in the country, was expansive enough to encompass both volcanoes.

In later years, bigger parks were thought useful because they provided buffers against outside development and provided more opportunities for tourism development and recreation. This debate was fierce after the Antiquities Act of 1906 allowed the creation of national monuments (Rothman 1994a, 1994b; Harmon, McManamon, and Pitcaithley 2006). The act stated that these should be as small as practicable, which for a geological feature or an Indian ruin might be as few as forty acres. But as the Antiquities Act was applied to more places, the notion of "as small as practicable" expanded: for the Grand Canyon a large monument was the smallest practicable, though even then only a small portion of the canyon was protected. Specifying the size of monuments has been an unending source of controversy. Saguaro was created in 1933 to protect a spectacular cactus forest growing along the lower slopes of the Rincon Mountains in southern Arizona, but park advocates wanted the monument to encompass the entire mountain (Burtner 2011). The advocates saw the opportunity to create a monument that spanned desert valley bottom to forested mountain top, displaying a range of vegetation not found in any other park. Not all were in favor of this, and opponents sought to reduce the size to the cactus forest. The large monument survived.

The army contended with the same issues when it established its system of battlefield parks, transferred to the NPS in 1933. This system included Chickamauga and Chattanooga, Antietam, Shiloh, Gettysburg, and Vicksburg (Sellars 2005, 2007). Most were called national battlefield parks and included the entire battlefield. Antietam was an exception; due to cost considerations, only a few key battle sites were purchased from the landowners, and it was known as a national battlefield site instead of park (Snell and Brown 1986). These two approaches have come to be known as the "Chickamauga plan" and the "Antietam plan." The former required the entire battlefield, while the latter attempted to preserve and mark a battlefield with limited resources. The distinction between these two approaches has been lost for battlefields, as all but four of these units are now larger than they were when they were created, but "park" and "site" still denote size differences for historic sites. National historical parks are generally larger and have more varied historical resources than do national historic sites.

Drawing Boundaries

How and where are park boundaries to be drawn? Legislation or proclamations may describe the boundaries of a new park unit, or they may be notated on a separate map. Perhaps the most basic distinction in the way boundaries are drawn is between western and eastern parks. The boundaries of western parks were usually based on public lands that had been surveyed and could be transferred to the administration of the National Park Service, but when parks moved east there was no public domain; the land had to be acquired for a new national park unit. The process of drawing boundaries was tied up with the task of obtaining the land.

Western Parks

The boundaries of large western parks are defined in a variety of ways. Yellowstone National Park was defined by straight lines between specified points. The park was in a remote and largely unexplored area, so straight lines provided a simple and quick way of enclosing the area. This precedent was not followed for later parks, because land surveyors had reached the West.

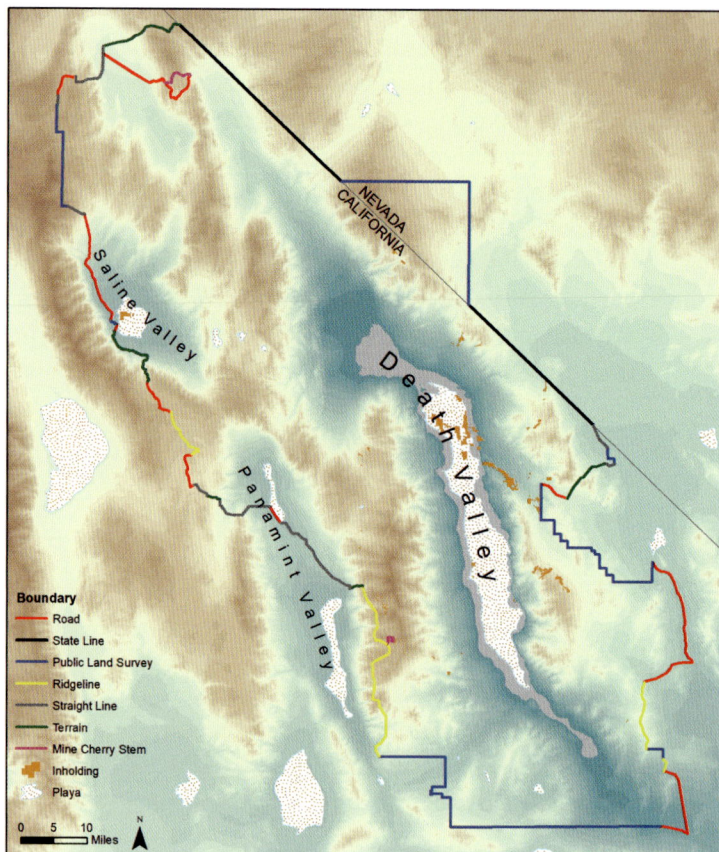

FIG. 5.2 Map of Death Valley National Park boundaries in 2018. This vast park has boundaries that are based on political boundaries, survey lines, mountain crests, roads, and other features.

The Public Land Survey System provides the framework for most western park boundaries. This system was originally devised in 1785 in order to survey the western frontier ahead of settlement (Hubbard 2009). Surveyors divided up the land into township squares six miles to a side, each of which was subdivided into thirty-six one-square-mile sections that could be further subdivided as needed for sale or homesteading. The boundaries of any area could be quickly specified using this system.

Ridgelines sometimes make up park boundaries in rugged western parks. The western boundary of Grand Teton; the eastern boundary of Sequoia–Kings Canyon along the crest of the Sierra Nevada; the eastern boundary of Organ Pipe Cactus National Monument; and portions of the eastern and northern boundaries of Yosemite, Mount Rainier, and the current Yellowstone boundary also run along ridgelines.

The Continental Divide is a popular attraction in parks such as Yellowstone and Rocky Mountain, but it passes through only a few

units and forms the boundary only of Noatak in Alaska. A more meaningful watershed boundary is between the Pacific slope and the Great Basin: Lassen, Yosemite, Sequoia, and Kings Canyon have boundaries that run along this divide. The boundary of Idaho's Craters of the Moon is the closest match to the terrain of any park unit: it follows the edges of the jagged lava flows the monument protects. Other lava-flow-based parks exist, but none have such straightforward boundaries.

There are many other possibilities for drawing park boundaries, including following state or tribal boundaries, though parks frequently cross these. The Great Smoky Mountains National Park crosses the Tennessee–North Carolina line, but the southern boundary of the park includes a large indentation for the Qualla Boundary, the last major remnant of Cherokee lands in the eastern United States. In newer parks, roads can serve as a useful boundary, as at Mojave National Preserve, bounded on the north and south by freeways. Parks may use a combination of strategies. Death Valley's boundaries are partly based on the Public Land Survey but in places follow the California-Nevada state line or roads, or they are drawn along ridges as straight lines between peaks (Fig. 5.2).

Occasionally, roads lead to a mining claim or private land the NPS wants to leave outside the park; the park boundary may run along one side of the road, around this intruding land, and back out along the other side of the road. This kind of road boundary is called a "cherry stem," with the private land being the cherry and the road the stem. Perhaps the greatest cherry stem to ever have existed in the park system was at the northern end of Death Valley due to a mine left out of the park (Fig. 5.2). The mine closed after the boundary was drawn, and in 2019 the park filled in the cherry stem. Not all cherry stems impose on park boundaries: Gila Cliff Dwellings National Monument lies near the center of the vast Gila Wilderness of southwestern New Mexico. The wilderness boundary excludes the tiny monument and its long access road, which together make up a cherry stem protruding inside the wilderness.

Needless to say, there are no reasons to expect park boundaries to have any environmental or ecological significance (Newmark 1985; Wagner et al. 1995; Landres et al. 1998). They very rarely contain entire watersheds or ecosystems of any size; such considerations have seldom been important in drawing park boundaries.

Small Monuments

The one-square-mile (640-acre) sections created by the Public Land Survey System can be subdivided into half sections of 320 acres, quarter sections of 160 acres, or quarter-quarter sections of 40 acres, the minimum size allowed for sale to farmers (Webster and Leib 2011). These subdivisions are also very useful when drawing boundaries for small western monuments, for which a quarter-quarter section was also the smallest size. Most have gotten bigger over time, but Pipe Spring in northern Arizona remains this size today.

Nowhere is the reliance on the Public Land

Survey more appropriate than at Homestead National Historical Park. This Nebraska site commemorates the Homestead Act, under which settlers could claim land and take title to it after several years of farming. The first homestead claim was by Daniel Freeman on May 20, 1862, for 160 acres. His claims were all in Section 26 of Township 4 North, Range 5 East, in the sixth Principal Meridian. Within this section, he claimed the south half of the northwest quarter, the northeast quarter of the northwest quarter, and the southeast quarter of the northeast quarter. When Congress created the monument in 1936 it specified the boundaries using the same description (Fig. 5.3), though more land has since been added.

Many small western monuments have had their boundaries specified in a similar fashion, using subdivisions of sections to select the land needed to preserve a scenic or historic feature. A few are simple squares, while others were made of combinations of sections and subdivisions down to quarter-quarter sections, sometimes with stair-stepping boundaries where a diagonal line is needed. The use of quarter-quarter sections as the smallest size means their acreage is often evenly divisible by forty, a telltale sign of a Public Land Survey monument. Tonto National Monument was proclaimed in 1907 as Section 34 of Township 4 North, Range 12 East in Arizona (Fig. 5.4). In 1937 the monument was expanded with three quarter sections to the northeast to give the NPS control over the entrance road (Dallett 2008). The straight boundaries do not match the rugged terrain but contain the two ruins and the monument's limited visitor facilities. The trail to the upper ruin zigzags steeply up the canyon slope just inside the boundary, one of the best examples in the park system of how these monument boundaries were not defined with visitors in mind.

The difference between terrain and the reliance on squares and rectangles may be nowhere as apparent as at Arizona's Navajo National Monument (Fig. 5.5). The monument consists of two 160-acre squares, each enclosing a separate ruin, along with a 40-acre square many miles to the west around a smaller ruin. These squares are superimposed on a rugged

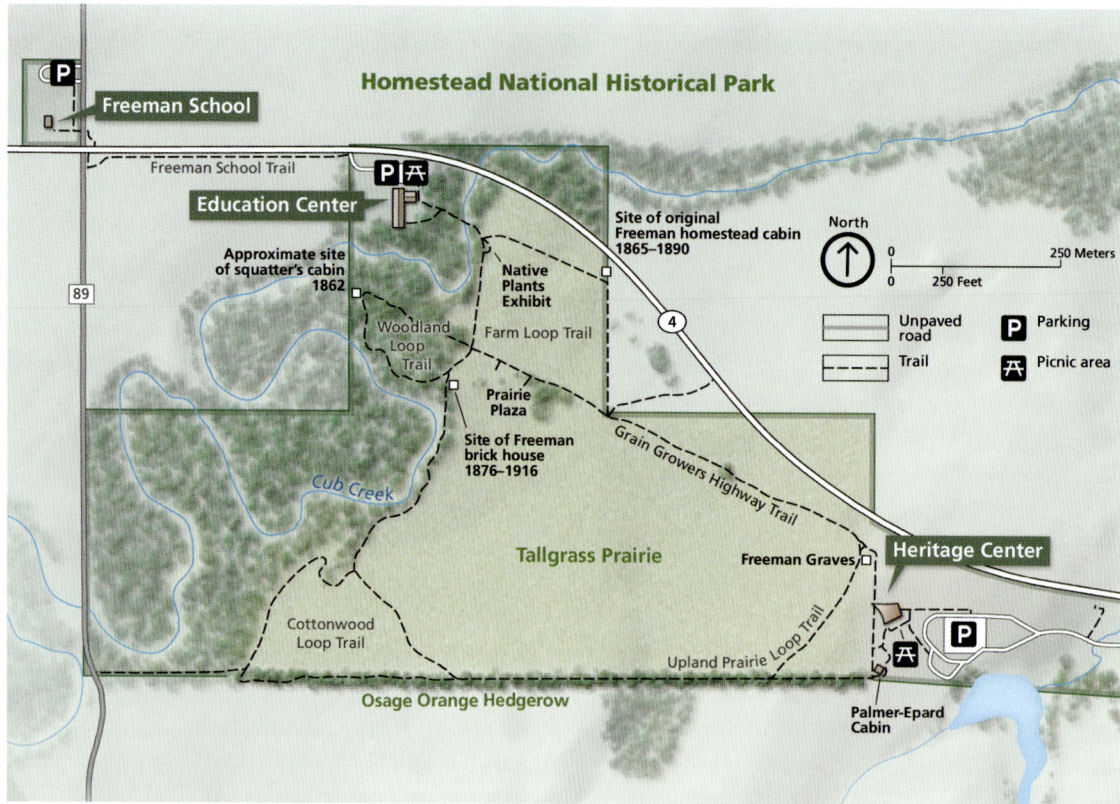

FIG. 5.3   Map of Homestead National Historical Park, 2021. This small Nebraska unit commemorates the first place claimed under the Homestead Act. NPS.

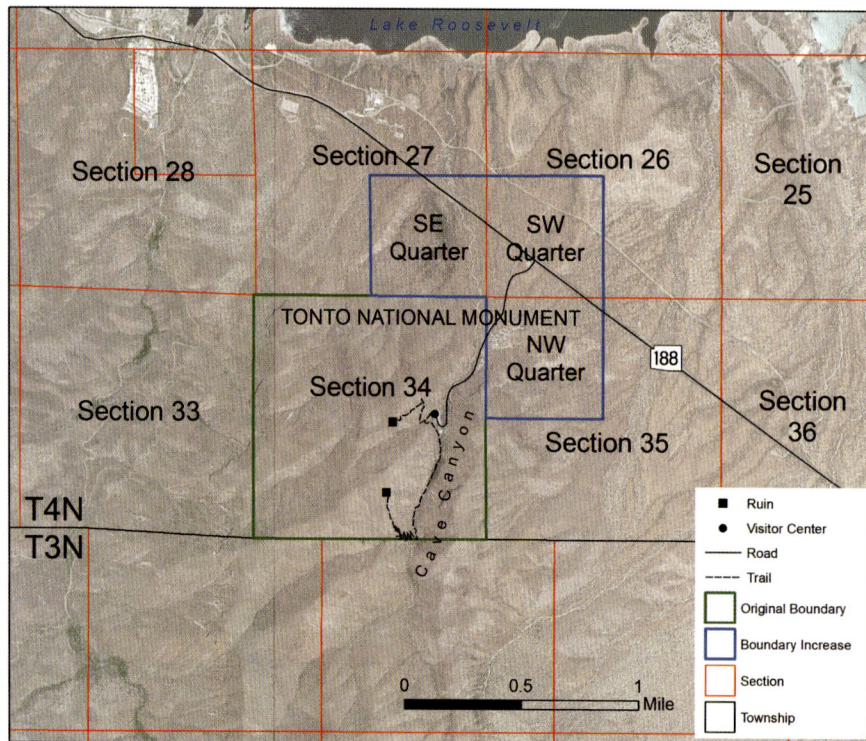

FIG. 5.4   Map of Tonto National Monument boundaries, 2021. The small Arizona site preserves Native American ruins sheltered under cliff overhangs. Like those of many western units, its boundaries are based entirely on the squares of the Public Land Survey System.

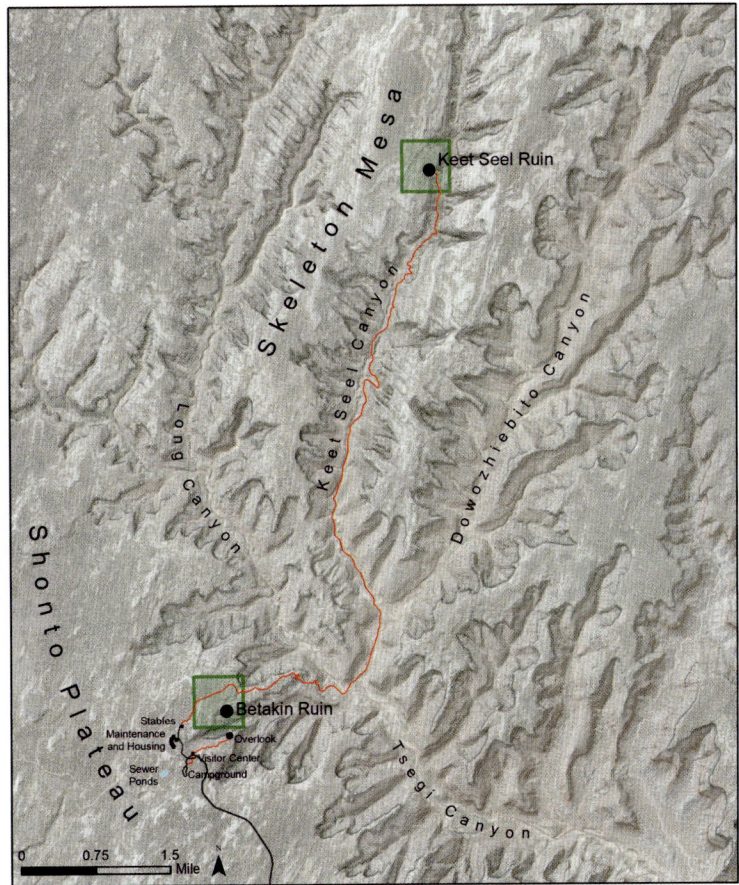

FIG. 5.5   Map of Navajo National Monument boundaries, 2021. This remote Arizona site preserves the largest Ancestral Pueblo ruins in the country, as well as offering an example of how poorly square boundaries fit the canyon and mesa landscape of the Colorado Plateau.

topography of deep canyons incised into the horizontal layers of the Colorado Plateau. The boundaries neither conform to the canyons nor allow room for a visitor center, campground, and other needed park facilities.

Most eastern units were not within the area of the Public Land Survey and had boundaries based on the older metes-and-bounds system. This system made use of lengthy descriptions of landmarks such as trees or posts along a boundary, together with distances and directions between them. The difference in the boundary descriptions could be striking. Acadia National Park was created as Sieur de Monts National Monument in 1919. The proclamation ran to over 4,300 words, of which about 3,970 are taken up with a seven-page metes-and-bounds specification for the boundaries. In contrast, the much larger Zion National Park was created in 1909 with a 286-word proclamation whose location was expressed in public land description using only 80 words. The Public Land Survey System offered the NPS an easy tool to

rapidly mark out new park units, though perhaps it was sometimes too easy.

Eastern Parks

The first parks were created from the public domain, limiting them to the interior West. An important precedent for the expansion of the system was set in 1907 when a landowner near San Francisco donated a grove of redwood trees to the Department of the Interior to become Muir Woods National Monument (Dilsaver 1994b; Rothman 1994a, 1994b). This was the first park unit to incorporate private land, and the donation provided a precedent that allowed the NPS to expand eastward out of the public domain and later into cities, becoming a truly national park system.

But in most cases, the land had to be purchased first. Acadia, Great Smoky Mountains, Mammoth Cave, and Shenandoah were established within lands that had been privately owned; the land was either donated directly to the NPS or purchased by state governments and donated for the parks. The process was smoother when the land was owned by large and willing sellers. For example, the Little River Lumber Company owned most of the north side of the future Great Smoky Mountains park from Clingmans Dome to Cades Cove, and it was happy to sell the land after logging the mountain bare (Schmidt and Hooks 1994). Despite this, other lumber companies strongly resisted the new park (Tooman 1995).

It was during this period that the NPS developed one of its most important partnerships, with John D. Rockefeller Jr. (1839–1937). This billionaire, the richest man ever, had a strong urge to protect America's natural treasures. He bought and donated much of the land that became Acadia National Park and then went on to buy and donate land for the Great Smoky Mountains. (The Rockefeller Memorial in Newfound Gap honors his efforts.) He went on to fund major land purchases that helped create Grand Teton and Redwoods National Parks. The John D. Rockefeller, Jr. Memorial Parkway, which links Grand Teton and Yellowstone, is named for him. His son Laurance (1910–2004) continued his legacy by donating the land for Virgin Islands National Park and Puʻukoholā Heiau National Historic Site to the government along with land for expanding Haleakalā (Winks 1997). He also donated what is now Marsh-Billings-Rockefeller National Historical Park in Vermont.

These partnerships were rare, and raising money from the public was slow. In 1951 the NPS was allowed by Congress to purchase land directly for a new park for the first time at George Washington Carver National Monument (L. Watt 2017). This transaction involved a single small parcel of land with a willing seller. The purchase set a precedent, and in 1959 the NPS again purchased land, this time for Minute Man National Historical Park in Massachusetts (Hemmat 1986), a process repeated on a much larger scale in 1961 for Cape Cod National Seashore (Runte 2010). From this point onward, land purchases by the NPS became an accepted way of acquiring land, limited largely by the budget Congress sets for land acquisition. This process was formalized in 1965 when the Land and Water Conservation Fund was created to purchase private land for conservation purposes, including national parks. The fund has a revenue source in oil and gas drilling royalties, but much of this money is regularly diverted for other purposes, so the money available varies from year to year.

Some western parks were also created on purchased and donated land. The struggle to create Redwoods took fifty years, with the final purchase in the 1960s costing $92 million, a record at the time (National Park Service 2010). The final park included four separate parcels, with a long thin section along Redwood Creek called "the worm." Although quite different from the simple rectangle of Yellowstone or the relatively compact Yosemite, Redwoods is indicative of the forces at work when creating new parks. Its borders can be considered gerrymandered, as are voting districts, to preserve as many stands of redwoods within the limits of land acquisition and the willingness of logging companies to part with future profits.

Donations are still common, and private organizations raise money to purchase land from willing sellers and donate it to the NPS. In 2004 the Conservation Fund purchased two ranches outside Petrified Forest National Park to allow the park to expand. The Mojave Land and Desert Trust has bought and donated land for Death Valley and Joshua Tree National Parks and Mojave National Preserve. In 2019 the trust donated 680 acres of land within the boundaries of Death Valley to eliminate inholdings (Mojave Desert Land Trust 2019).

Not all landowners are willing to sell or donate their land. The government has the right of eminent domain, allowing it to take private property for the public good, with appropriate compensation. This was first heavily used by the NPS when Cape Cod National Seashore was being developed. Almost 20 percent of the land in the park was obtained this way (Hemmat 1986). The creation of San Juan Island National Historical Park in Puget Sound required substantial land purchases, but since the creation of the park came at a time when rural lots on the island were rapidly appreciating in value as second homes, few landowners were willing to sell (Cannon 1997).

Perhaps the most extreme example of land

acquisition was for Delaware Water Gap National Recreation Area in Pennsylvania and New Jersey. The land was acquired by the Army Corps of Engineers for a reservoir on the Delaware River, after which the NPS would develop it as a recreation area (Albert 1987). The NPS frequently used its powers of eminent domain to take the land with just compensation. Before the reservoir project was canceled in 1978, the NPS had purchased 7,344 properties containing as many as 2,600 homes. Eminent domain remains a useful but controversial method for land acquisition, but it is not always possible; the enabling legislation that created a park may prevent its use.

One solution to land acquisition problems is to not own the land. This has occurred with several units in parts of the United States where land is owned communally instead of individually. Parks on Indian reservations and in American Samoa are cooperatively managed or on land leased by the NPS. Another solution is not to have any land at all: 99 percent of Buck Island Reef National Monument is water, as is 95 percent of Biscayne National Park. The coastal waters that make up most of these units are under federal control, so they are part of the public domain.

To create a park in an area where private ownership predominates, a two-step process must be followed. A park is authorized, and then the NPS must go through a lengthy process of land acquisition. Once enough land has been acquired, the park is established. The boundaries are dependent on the final pattern of land purchases rather than specified in the original legislation. These parks will have two dates for their creation, which can be confusing.

Shenandoah National Park is one of the best examples of this. It was authorized in 1926 to consist of at least 250,000 acres and not more than 521,000 acres (Whisnant, Whisnant, and Silver 2011). At the time, the NPS could not buy land directly, so all this acreage would need to be donated to the federal government, a process made more difficult because the state of Virginia initially refused to use any state money to buy land. The NPS never had enough money, and so it gave priority to land along the scenic crest of the mountain and the upper slopes.

There was no possibility of purchasing entire watersheds or the lower slopes of the mountains. Eventually, the money ran out, and the park was established in 1935. Because the farms the NPS bought had irregular sizes and shapes, the park ended up with a jagged "sawtooth" boundary. The sawtooth remains today, a legacy of nineteenth-century settlement patterns (Fig. 5.6).

Big Bend grew out of efforts to create a national park in Texas in the 1930s (Jameson 1996). Agency staff studied six sites around the state, including the Guadalupe Mountains, Palo Duro Canyon, the Davis Mountains, McKittrick Canyon, and Frio Canyon. Big Bend was selected and authorized in 1935. Then the state had to purchase the land for the park for donation to the federal government. Raising money to buy land was a slow process during the Depression and war years, and it took until 1944 for the NPS to assemble enough land. The planned boundaries of the park were adjusted continuously as land purchases were carried out or rejected, but there were always four main features planned for the park: the Chisos Mountains and Boquillas, Mariscal, and Santa Elena Canyons. Even after the park was established, private land remained in the park, and it took until 1972 for the NPS to buy out these last properties.

Island and Shoreline Parks

Several parks are made up entirely or partly of islands. For some, such as Boston Harbor Islands, the shoreline provides a natural park boundary. Others have boundaries based on a buffer extending outward from the shore. Apostle Islands includes everything within a quarter mile from the shore, the Channel Islands extend out to one nautical mile (Fig. 5.7), and for Isle Royale, the boundary is four and a half miles from shore. What if there is no shore? Dry Tortugas, Virgin Islands, Virgin Islands Coral Reef, Biscayne Bay, and Buck Islands Reef have ocean boundaries defined by latitude and longitude coordinates, which can only be found with a GPS navigation system.

Many inland parks have shorelines along lakes, but these are rarely used for park boundaries. This is a good thing, because many are

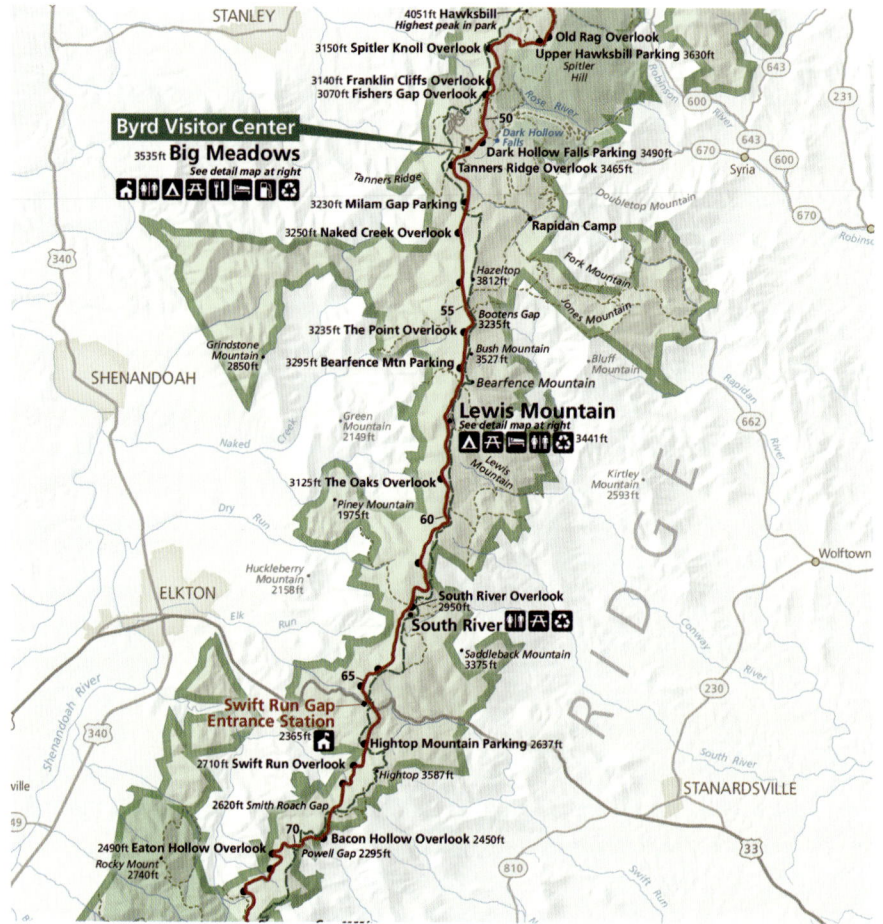

FIG. 5.6   A portion of Shenandoah National Park's "sawtooth" boundary, 2021. The NPS bought land from farmers first along the top of the mountain and then worked its way down the sides. When the money ran out the boundary was complete. NPS.

FIG. 5.7   Channel Islands National Park, 2021. This is one of several island units whose boundaries are defined by distance from the shoreline, in this case, one nautical mile (the dark green line). The outer blue line denotes a national marine sanctuary that extends out to six nautical miles from shore. NPS.

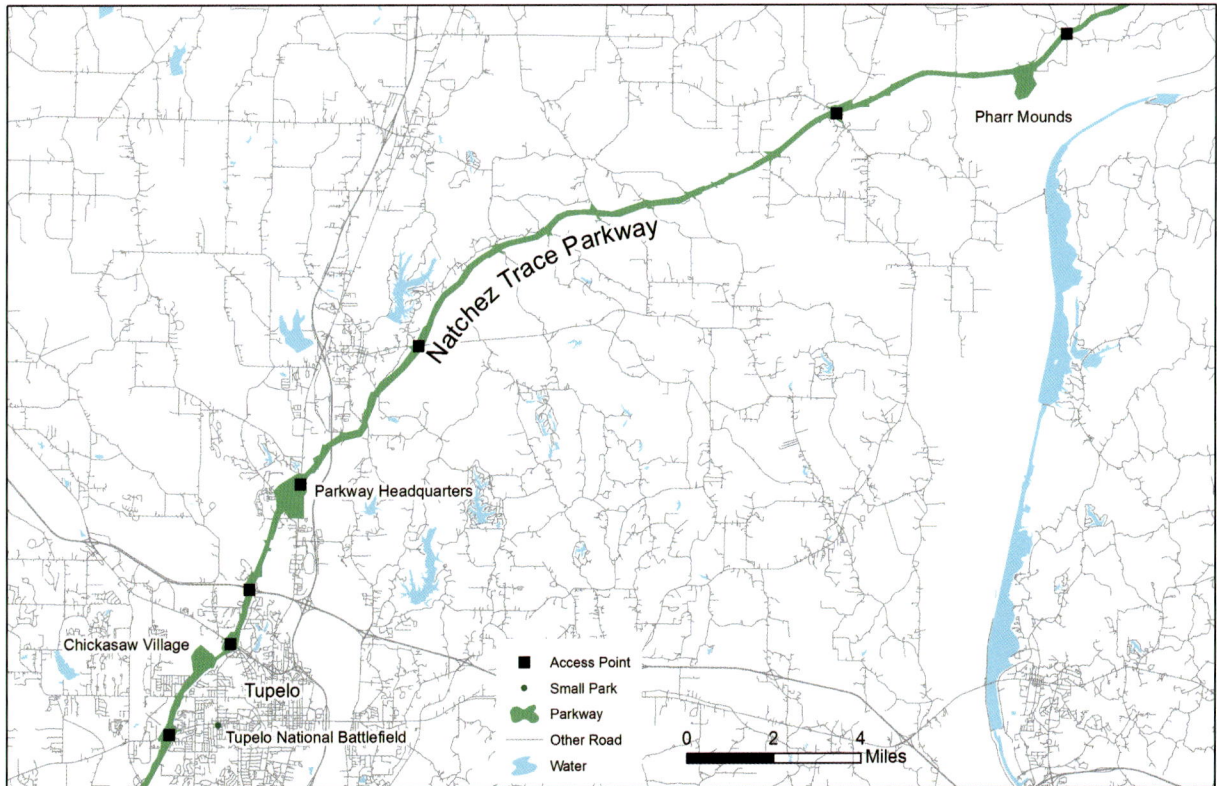

FIG. 5.8 A portion of the Natchez Trace Parkway boundary in 2019. The park is a 444-mile road with wider areas every so often where drivers can park and visit historic sites or a visitor center.

reservoirs with constantly fluctuating water levels. Boundary lines may run right through a lake, as in Amistad National Recreation Area in Texas, half of which is in Mexico. Lighted buoys mark the international border, which is defined by the former Rio Grande channel. The north end of Ross Lake is in Canada, though few boaters on the American side ever make it that far. Those that do can easily see the park's northern border as a wide swath cut through the forest by surveyors establishing the international boundary.

Linear Parks

Some parks are linear, presenting unusual boundary problems. The Natchez Trace Parkway runs for 444 miles and the Blue Ridge Parkway for 469. They are also very narrow parks, with most of the Blue Ridge less than a quarter mile wide and the Natchez Trace usually eight hundred feet wide or less, though this varies to provide for recreational facilities and protect views from the road (Fig. 5.8).

River units are also linear but don't use shorelines as boundaries. Eastern rivers, such as Obed in Tennessee and Gauley in West Virginia, show the sawtooth pattern of private land purchases, while those farther west, such as Buffalo in Arkansas and Ozark in Missouri, have boundaries based on the Public Land Survey System, with many straight lines (Fig. 5.9). Those in the Midwest tend to have boundaries that fit the river much more closely, as with Saint Croix National Scenic Riverway in Wisconsin and Minnesota. Rivers also differ from parkways in often having several noncontiguous sections protecting separate sections of the river. Flowing rivers are continuous features, but the parks that protect them don't have to be.

Urban Parks

Many national historic sites are found within cities, created from downtown buildings or public spaces. The size and shape of these park units are determined by each city's street and lot patterns. Many cities are laid out in the form of a grid pattern, with streets crossing at right angles and square or rectangular blocks subdivided

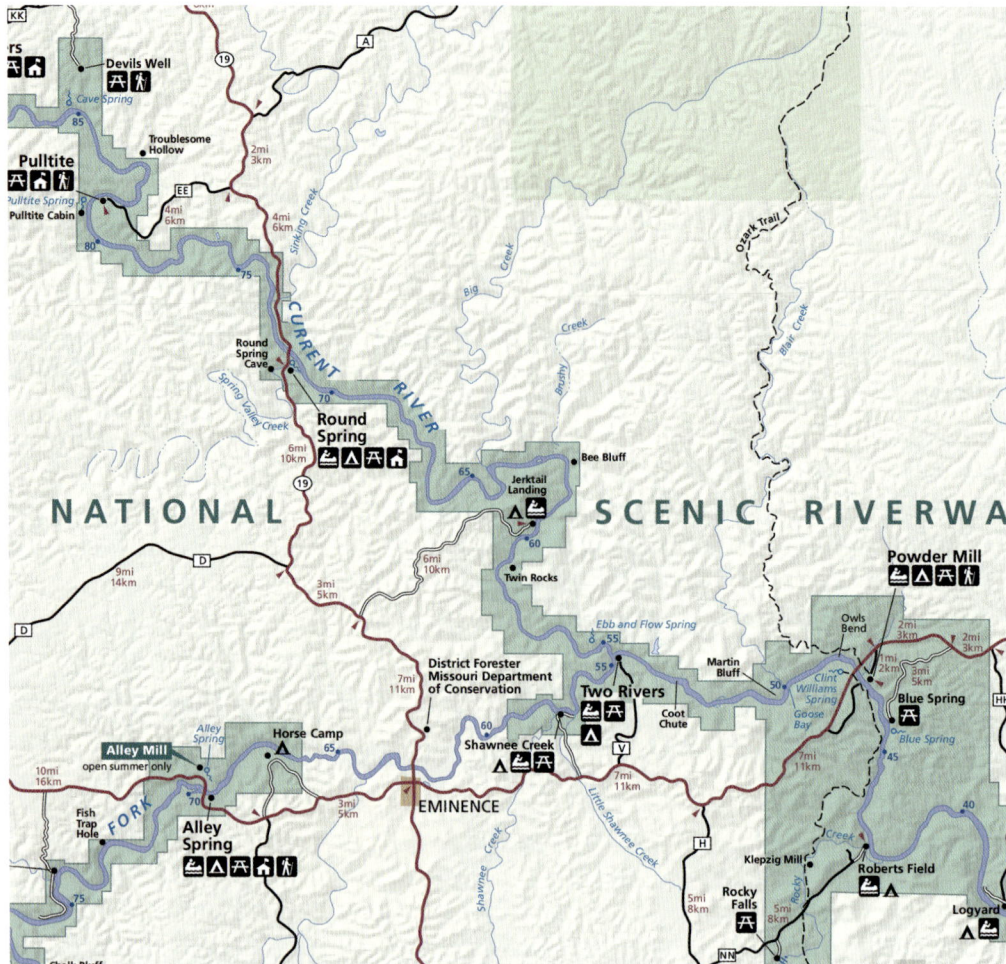

FIG. 5.9　Map of a portion of the Ozark National Scenic Riverways, 2021. While the river curves, the boundaries are based on the straight lines of the Public Land Survey. NPS.

into lots of regular size (Stanislawski 1946, 1947; Reps 1965, 1979; Rose-Redwood 2008).

Philadelphia was one of the first cities in America to be given a grid street pattern when it was founded in 1682, with rectangular blocks, public squares, and wide streets between the Schuylkill and Delaware Rivers (Reps 1965). The city grew into this grid, beginning along the Delaware River waterfront. In 1753 Independence Hall was built as the capitol of the colony of Pennsylvania. It served as the site where the Declaration of Independence and the Constitution were debated and signed and was once home to the Liberty Bell. In 1812 the state capital was transferred to Harrisburg, and Philadelphia's business district moved west, leaving Independence Hall in a declining area. In 1956 Independence National Historical Park was created to preserve not just the hall but also newly

developed parks when the city demolished rundown buildings to create Independence Mall and other grassy plazas. The sprawling park encompasses six entire city blocks, Washington Square, and numerous other properties in the area. It commemorates not only the founding of the country but also one of the country's first urban street grids.

Grids and blocks vary widely. Klondike Gold Rush National Historical Park has two urban units, one within the original 1853 Seattle townsite (now the Pioneer Square historic district) (Fig. 5.10) and the other in Skagway, Alaska, in a small town. The Seattle unit is within the former Cadillac Hotel, a three-story brick building dating from 1889. It sits on a 0.15-acre lot within a grid aligned with the cardinal directions, so streets run north–south or east–west. In Skagway the grid was aligned to a nar-

FIG. 5.10 Klondike Gold Rush National Historical Park in Seattle, 2021. This unit, housed in a former hotel, has boundaries that represent the street grid of early Seattle. NPS.

row valley between towering mountain ranges. Here the NPS owns several wooden buildings on Broadway Avenue, the town's main street. Although newer than those in Seattle, these buildings are smaller and tend to have false fronts with business names prominently displayed. This park unit, like the one in Seattle or Independence Hall, preserves not just historical events but also a glimpse of what American cities once looked like.

Because urban parks are based on individual buildings, they tend to be much smaller than small western monuments based on the Public Land Survey. Thaddeus Kosciuszko National Memorial, the former home of the American Revolution hero, is at the corner of Pine and 3rd Streets in downtown Philadelphia. This is the smallest unit in the park system, with a land area of 0.02 acre, or 871 square feet (twice the size of a residential two-car garage).

In these cities, national park units are just small elements within downtowns. But in Washington, D.C., park units were built into the city's original plan. This plan was created by Pierre "Peter" Charles L'Enfant in 1791 in his master plan for the district (Berg 2007). He included what we know as the National Mall as a broad boulevard extending west from the Capitol. On

a hill in front of the President's House (now the White House), a statue of George Washington was to be erected; the Washington Monument was built here instead. The city plan also included diagonal streets with circles or squares at major intersections. These provided opportunities for statues or memorials, a common feature in the National Capital Parks East.

The National Mall has perhaps the strangest boundaries in all the park system: nobody seems to know exactly what they are. The NPS has conflicting definitions of what the mall includes, and much of what the public considers part of the mall belongs to other agencies, among them the Smithsonian Institution (Benton-Short 2016). Four other government agencies control parts of the mall: the National Gallery of Art, the General Services Administration, the Architect of the Capitol, and the District of Columbia. In addition, any proposed building or alteration must run a gauntlet of agencies and commissions for approval. The NPS has little freedom to act within this unit.

Park units that encompass not just a building but also part of an urban street grid have become more common, as with Martin Luther King, Jr. National Historical Park in Atlanta, Birmingham Civil Rights National Mon-

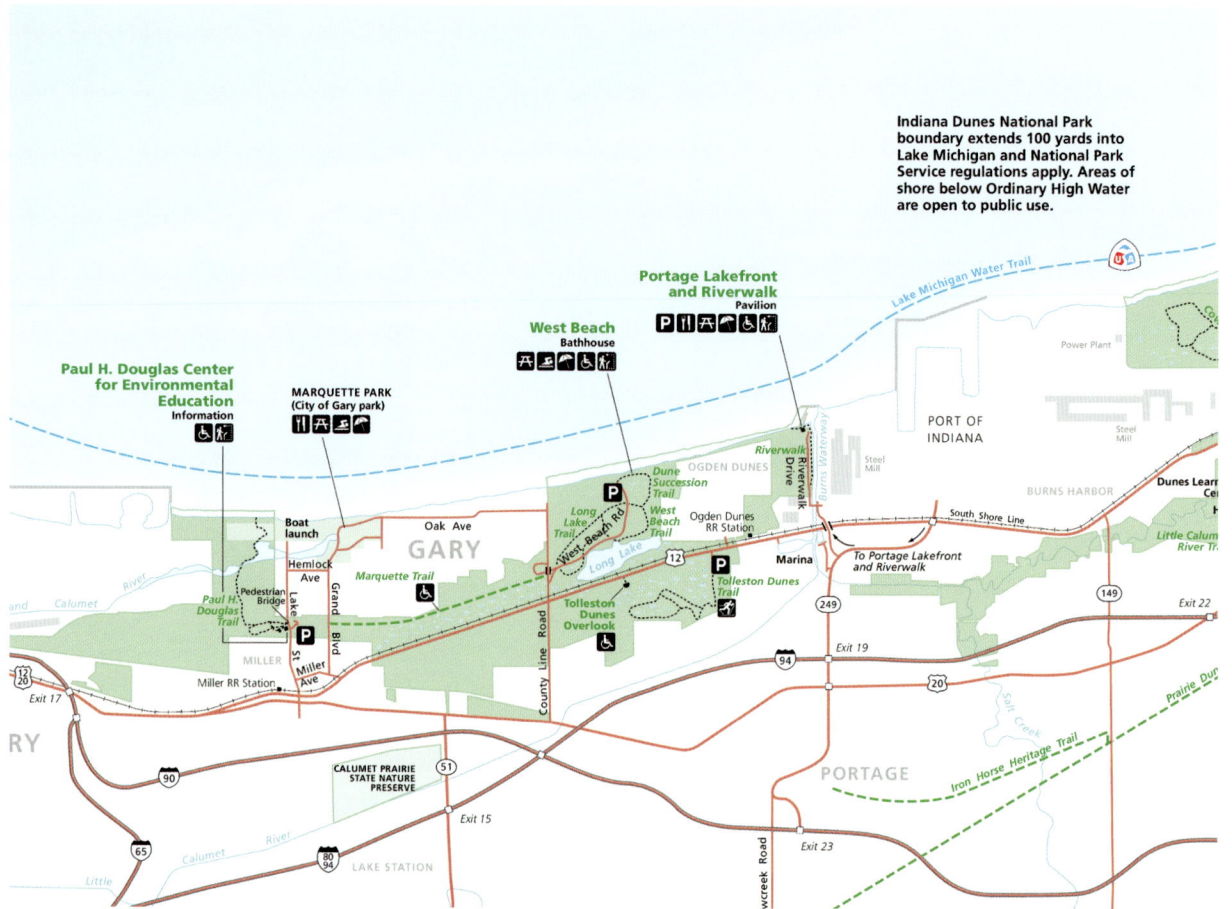

Indiana Dunes National Park boundary extends 100 yards into Lake Michigan and National Park Service regulations apply. Areas of shore below Ordinary High Water are open to public use.

FIG. 5.11   Map of Indiana Dunes National Park, Indiana, 2021. The park has boundaries that encompass different sections of the lakeshore while avoiding steel mills and a port. Several sections of the park are separate islands of nature amid an industrial landscape. NPS.

ument, and Stonewall National Monument in New York City; the latter is unusual because it includes only streets and not the surrounding buildings. In the last two cases, these monuments commemorate not just buildings but also protests that spread out into the streets and public parks. These streets are still in use, not closed to traffic and preserved, as at Appomattox Court House, and little of these monuments is owned by the NPS.

Parks with Multiple Units

The popular perception of a national park unit is a contiguous area that contains only federal land and that is managed by its own superintendent. Yosemite and Big Bend are examples. However, about 37 percent of park units consist of multiple sections. Some have small outlier sections, such as Death Valley's forty-acre Devils Hole, many miles from the rest of the park.

Other parks are more balanced with their separate units, such as Saguaro, which is divided into East and West Units. Each unit has a visitor center, roads, picnic areas, and trails to hike. Other examples include Navajo, San Antonio Missions, Hovenweep, Salinas Pueblo, Independence, Klondike Gold Rush, and Chickamauga and Chattanooga. Indiana Dunes consists of ten parcels, only two of which are on the lake, and some are nine miles from the beach (Fig. 5.11). War in the Pacific National Historical Park has six different units on Guam, plus a separate visitor center. The record for greatest spacing is held by Manhattan Project National Historical Park, with three units in Tennessee, New Mexico, and Washington with the closest nearly a thousand miles apart. But for the greatest number of separate units, Nez Perce National Historical Park is unmatched, with thirty-eight locations in four states.

Creating multiple-unit parks allows the NPS to protect a resource without the need for a vast park, a sort of extreme version of the Antietam plan. It recognizes that many related events took place at widely scattered locations, something a single-unit park would not be able to accommodate.

Separate park units have been merged together. Grand Canyon began as a national monument but was expanded into a national park in 1919 and was later enlarged by absorbing Marble Canyon National Monument and a second Grand Canyon National Monument. Unrelated units could also be combined: in 1961 the Natchez Trace Parkway (created in 1938) absorbed Meriwether Lewis and Ackia Battlefield National Monuments, both established in the 1930s. These commemorated the home of one of the Lewis and Clark expedition leaders and a pivotal battle between the French and Chickasaw tribes in 1736, respectively. Both sites are now just two of many historic sites along the Natchez Trace Parkway. Many parks are also administered jointly while being treated as separate entities. The Natchez Trace Parkway is also responsible for operating the nearby Tupelo National Battlefield and Brices Cross Roads National Battlefield Site, which are both a short distance off the parkway. The Natchez Trace National Scenic Trail parallels parts of the parkway and is administered by the parkway as well.

It is also possible for a park to be split into separate units. Hawai‘i National Park was split into separate Hawai‘i Volcanoes and Haleakalā National Park on the big island and Maui, respectively, in 1960, and World War II Valor in the Pacific National Monument was split into three separate sites, only two of them administered by the NPS, in 2019.

Inholdings

Regardless of their location, parks often contain some non-NPS-owned land, called inholdings. Park maps usually do not show these, and the extent to which they are common in national park units can be quite surprising (Fig. 5.12). Only 171 park units are 100 percent federally owned, and these tend to be smaller parks. Most parks have a mix of federal and nonfederal lands, and larger parks are more likely to have inholdings. In thirty parks, less than half the land is owned by the NPS, including Canyon de Chelly National Monument in Arizona, owned by the Navajo tribe, and the tiny Hohokam Pima National Monument in Arizona, owned by the Gila River tribe. Only 14 percent of the land within the Santa Monica Mountains is owned by the NPS and only 0.05 percent of Upper Delaware Scenic and National River. (The park map does not show this and gives the impression that the park is nothing more than the Delaware River.) Even Yellowstone still contains 1.58 acres of private land within its boundaries.

The degree to which the NPS owns its parks varies by type of unit. National parks are the most likely to be NPS-owned, followed by many smaller units such as battlefields and monuments. National historic sites and recreation areas have more private land within them.

Inholdings are seen as a problem best eliminated. Some units have managed to purchase, trade for, or receive ownership of inholdings, but for many others, the private lands remain. Land purchases by the federal government have been limited and controversial, and there is never enough money to go around. In addition to the private land, there are mining claims, water rights, and grazing leases, which are legal rights to use land that can be even more difficult to remove than ownership.

Lake Mead contains the community of Meadview, one of the few places you can live within a park without working there. The community had 1,224 residents in the 2010 census. This community is far off main roads, but much more visible is the Hoover Dam Lodge, a sprawling casino with a fifteen-floor hotel that sits on an inholding near the park visitor center. The park has been working for years to buy the lodge and tear it down. In 2003 Congress gave the NPS $20 million to buy it, but the deal fell through. In contrast, when in 1994 the NBA Cleveland Cavaliers team left the Richfield Coliseum basketball arena between Cleveland and Akron, the land was eventually sold to the NPS, which then demolished the coliseum and replaced it with a prairie. In 2000 the land was added to Cuyahoga Valley National Park.

Mojave National Preserve is one of the most

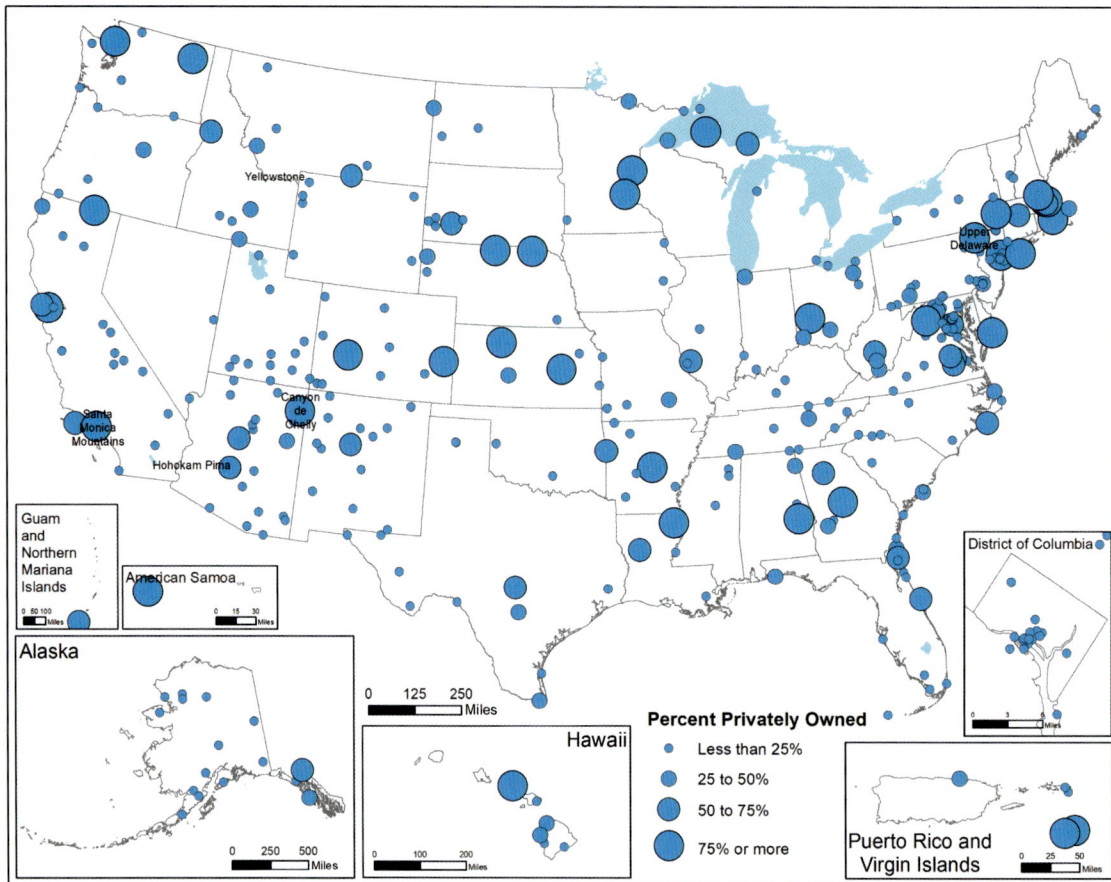

FIG. 5.12   Map of percent of park unit acreage privately owned in 2018. The ideal of a national park containing no private land is rarely met; some parks, such as Arizona's Canyon de Chelly, contain no federal land at all.

extreme examples. At its creation in 1994 Mojave National Preserve contained 240,000 acres of nonfederal land (15 percent of the total park area), 2,447 active mining claims, 12 grazing permits, and 125 rights-of-way for roads, railroads, pipelines, power lines, and communications lines (Dilsaver and Wyckoff 2005). Most of these remain within the preserve.

Western park inholdings are often specified using the Public Land Survey System and show up as squares, as in Mojave National Preserve, where a checkerboard pattern of private land exists in Lanfair Valley, relics of a short-lived farming community. Mining claims are different; they are customarily a maximum of fifteen hundred feet long and six hundred feet wide and can be oriented in any direction. Some may have been purchased by claimants and are now private lands. These rectangles show up as private lands on several park maps and are even visible in several park boundaries, such as Great Basin and Death Valley.

Boundary Changes

On August 17, 1909, a 160-acre quarter-section square was surveyed around Rainbow Bridge, a great sandstone arch standing 290 feet tall and spanning a 234-foot-wide dry creek in southern Utah (Hassell 1999; Sproul 2001) (Fig. 5.13). This square was used as the boundary of a national monument created the following year. The monument is the oldest park unit to retain its original boundaries. Just as with states (Stein 2008), boundary changes of parks have been common (Fig. 5.14). Most of the larger and older units have had their boundaries adjusted numerous times, and many have been greatly expanded. Rocky Mountain National Park holds the record with fifteen boundary changes, followed by Acadia with fourteen, the Great Smoky Mountains with thirteen, Colonial with twelve, Hot Springs with eleven, and Yosemite with ten. Thirty-nine units have had five or more changes, and 209 units have had

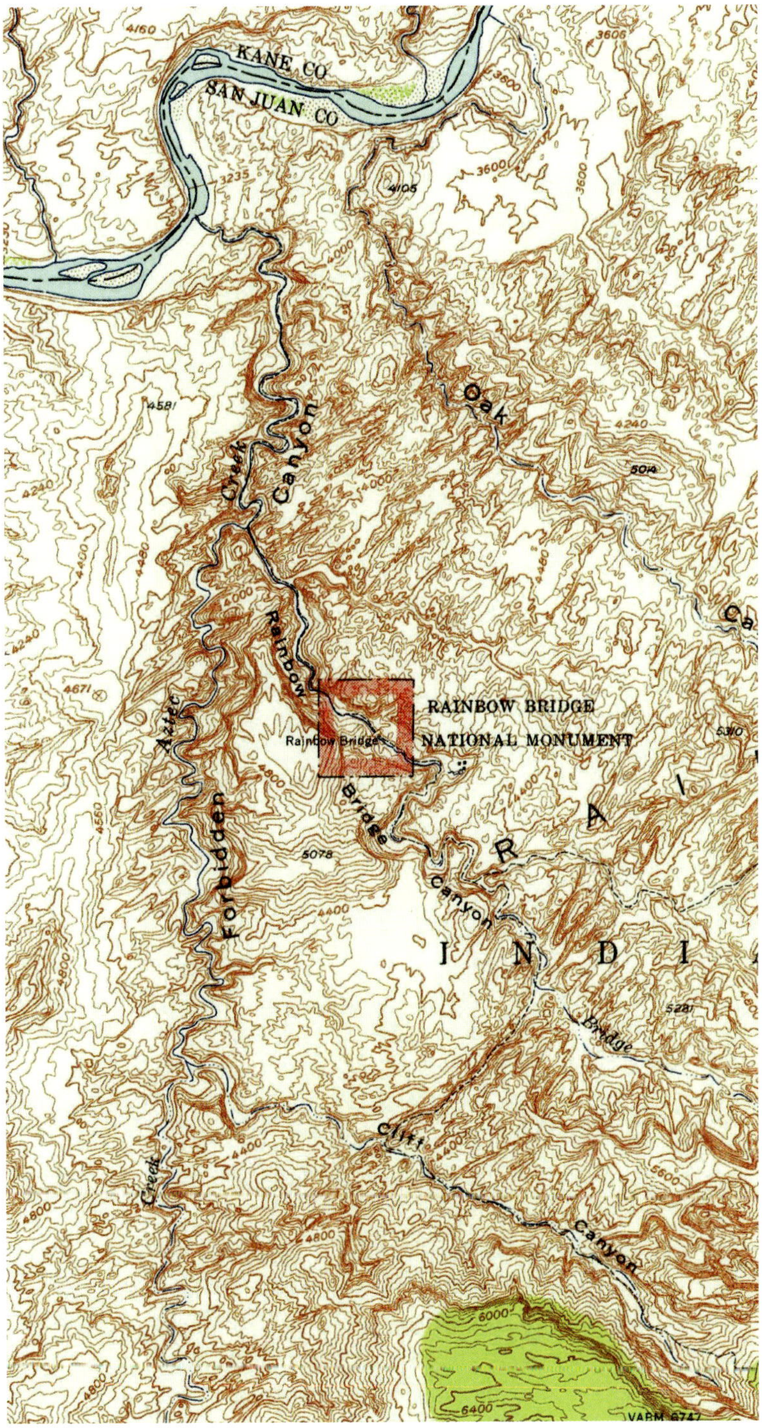

FIG. 5.13   Rainbow Bridge National Monument, seen on the 1953 Navajo Mountain, Utah, United States Geological Survey topographical map. This simple square is the oldest unaltered boundary in the national park system. Its relative location has greatly changed since Lake Powell was constructed on the Colorado River.

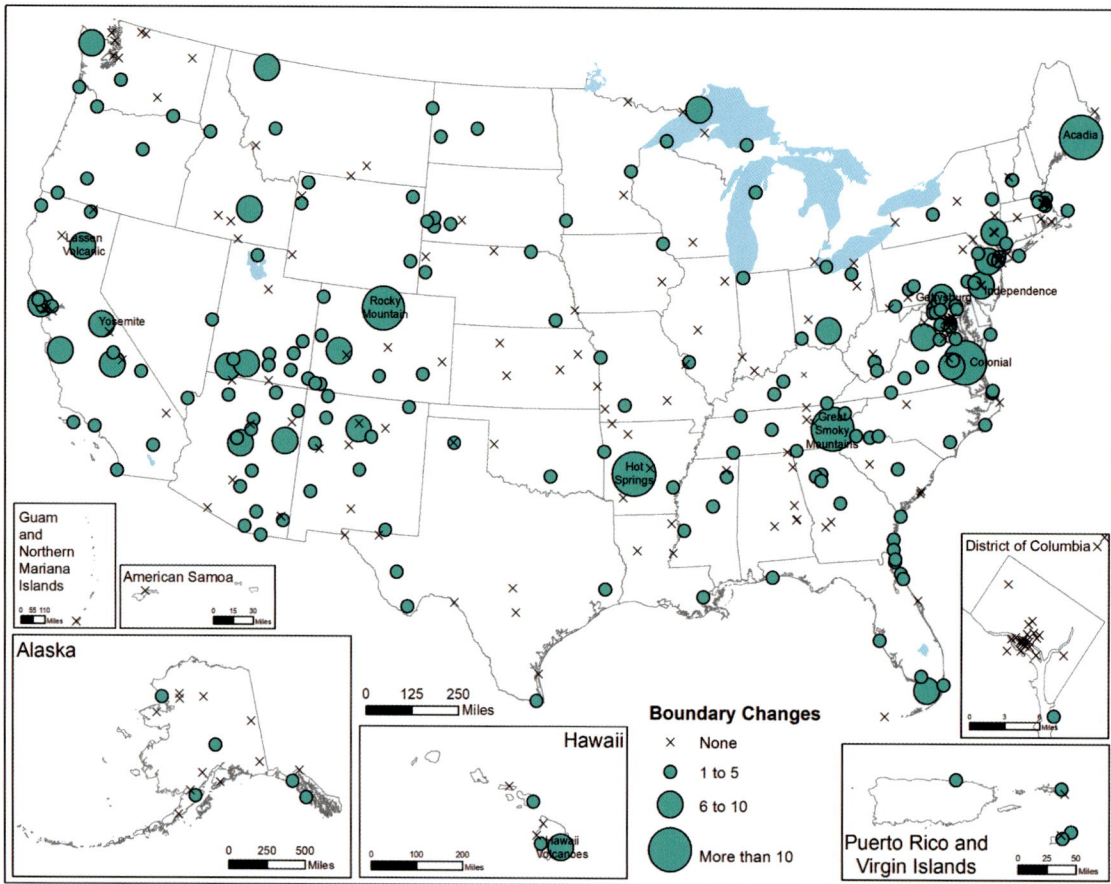

FIG. 5.14 Map of the number of boundary changes by park unit in 2018. Rocky Mountain has had fifteen, more than any other unit. Those without any changes are usually newer and smaller.

at least one change. For 361 units outside of the national capital parks, the average is 1.7 boundary changes.

Why do park boundaries change? Current NPS policy provides criteria for boundary changes with five considerations for additions and deletions. The opportunity to include significant features related to the individual park purpose is the first.

Another important reason for boundary changes is to add additional resources. Death Valley National Monument was created in 1933, but several areas containing water resources were added in 1937 (Fig. 5.15). In 1953 an outlier unit was added to protect the nearly extinct species of desert pupfish, a tiny fish living in only one underwater cave. In 1994 the park was designated a national park and greatly expanded. Arches started as two small scenic units around the Windows and Devils Gardens sandstone outcrops, both units based on

the Public Land Survey System and with stairstepping diagonal boundaries. In 1938 these were combined within a larger monument and then expanded further when it became a national park. The original Grand Canyon was a much smaller park than it is today and grew in large part by absorbing adjacent national monuments added later and even much of what had been Lake Mead National Recreation Area. It also lost land to the Havasupai tribe in the 1970s. Despite this, the park does not contain the entire Grand Canyon; the NPS only has a small part of the canyon, with two tribes, the USFS, and the Bureau of Reclamation all having some control over its land and waters. Rocky Mountain, Redwoods, and especially the Everglades are other examples in which the NPS has only a small part of the resource after which the park is named.

Adding a resource accidentally left out of the initial legislation or proclamation is a re-

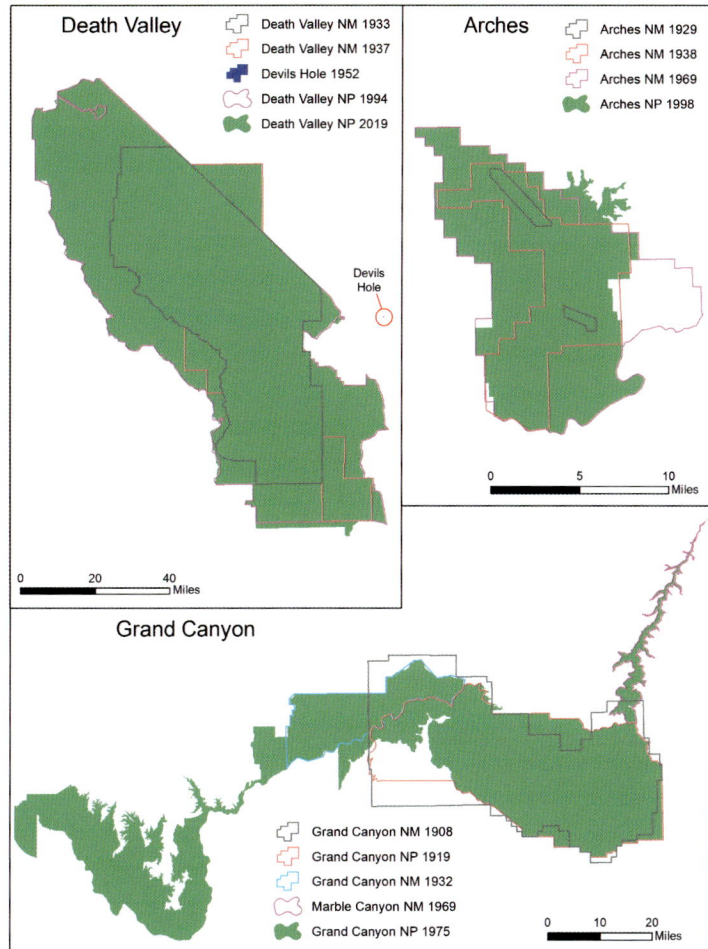

FIG. 5.15   Map of examples of boundary changes over time. These three parks have been altered substantially after being created as small monuments. Arches has also lost considerable land.

lated cause for revision and has been common; the median number of years before the first boundary change is only fourteen. The ease of specifying boundaries in the Public Land Survey System could cause problems, as in Navajo National Monument and Mesa Verde National Park, when the initial boundaries missed important ruins that the parks were designed to protect (Rothman 1991; D. Smith 2002), and in Great Sand Dunes National Monument, where several township sections specified in the original proclamation were later found not to exist when the area was surveyed (Coary 2016).

Second, boundaries may be revised for reasons of improved management. The boundaries for many early units were specified without any regard to future visitation or facilities, and more land may be necessary for visitor facilities, as in Navajo National Monument when additional land was leased from the tribe for a visitor center. One common change was to include the main entrance road to prevent incompatible roadside development. This happened at White Sands National Monument in 1934 (Schneider-Hector 1993), as well as at Crater Lake, Allegheny Portage Railroad, Dinosaur, Tonto, and Montezuma Castle. These are the opposite of the cherry-stem roads discussed earlier.

Third, lands may be added to prevent harm to resources within the park, though adding land as a buffer zone is specifically discouraged by NPS policies. The remaining considerations are that any changes must be feasible regarding cost and operations, and changes are only recommended if there are no feasible alternatives to protect resources and operate the park.

Some expansions are so large as to fundamentally change the nature of a unit. Dinosaur National Monument was created in 1915 as a small eighty-acre site surrounding an active fossil quarry. In 1938 it was expanded into a two-hundred-thousand-acre park containing the

Green and Yampa Rivers. Some of the large expansions were accompanied by name changes: General Grant National Monument was enlarged as Kings Canyon National Park in 1940, and the tiny Lehman Caves National Monument in Nevada was enlarged and renamed Great Basin National Park in 1986.

A similar situation existed at Great Sand Dunes (Geary 2016). The original monument encompassed the enormous dune field, which was managed as an isolated element of the San Luis Valley's landscape in Colorado. By the 1980s it was apparent that this was only part of a much larger eolian system that included the watershed along the western side of the Sangre de Cristo Mountains, the dunes, a sand sheet to the west, and salt flats. When the monument was redesignated and expanded in 2004 the new boundaries encompassed most of this eolian system, providing one of the best examples of revised park boundaries encompassing an intact environmental region.

Even Yellowstone's boundaries were changed. In 1917 the idea of a Greater Yellowstone ecosystem arose, leading to calls for expansion for winter range for elk and other animals (Haines 1996). In 1929 the park's boundaries were revised, with 81 square miles added to the park but 159 square miles of the park transferred to adjacent national forests. With this adjustment, Yellowstone now contained summer range for bison in the Lamar Valley, and the eastern boundary followed a watershed. In 1932 7,600 acres were added in the northwestern section.

Parks have generally gotten larger over time, but are bigger parks better? Larger boundaries may provide a buffer against development impacts, but given issues such as air pollution, nighttime lights, and global warming, creating a buffer may no longer be feasible. Another size issue is the cost of running the park. While park budgets are often related to visitation (Rettie 1995), the cost of operating a park may depend more directly on the size of the park (Rothman and Miller 2013). Sequoia National Park has four times the budget of Death Valley but is one-quarter the size (and has fewer visitors).

Expanding a park can cause problems. In the 1940s the NPS sought to obtain inholdings at Lava Beds National Monument (Brown 2011) and by 1946 had acquired all the private land in the park. In 1951 the boundaries were expanded, requiring more acquisitions, completed in 1956. This experience, as well as similar stories at Petrified Forest and Big Bend, shows that expanding an existing park will almost certainly increase the amount of non-NPS land within the boundaries. Since inholdings are undesirable and can be sources of problems, expanding boundaries may be a mixed blessing.

Several examples exist of parks being substantially reduced to exclude areas desired for other economic uses or when resources are lost. Yosemite and Joshua Tree lost large areas to mining, Casa Grande gave up some area for irrigation canals, and Natural Bridges lost several caves whose archaeological resources had been destroyed.

Hamilton Grange National Memorial provides one of the most unusual examples of boundary change: the entire park was put on wheels and driven to a new location (Dunlap 2008). The memorial is a house built by Alexander Hamilton in 1802 on the north side of Manhattan Island. It was a victim of changing relative location, as the once-rural house became encompassed by urban growth. In 2006 the house was closed to the public and moved several blocks to a less crowded location, reopening in 2011.

Parks and Other Boundaries

While park units have legal boundaries, they also exist within and across many political boundaries and jurisdictions within the United States. Twenty-seven units cross state lines or are present in multiple states, and many others extend across county or municipal boundaries. Nez Perce holds the record, with units in six counties in Idaho, Montana, Oregon, and Washington.

These overlaps complicate planning and sometimes make for strange situations. Yellowstone occupies part of three states, and that portion of the park in Montana once had its own county until 1978, when it was split between Gallatin and Park Counties. (There is still

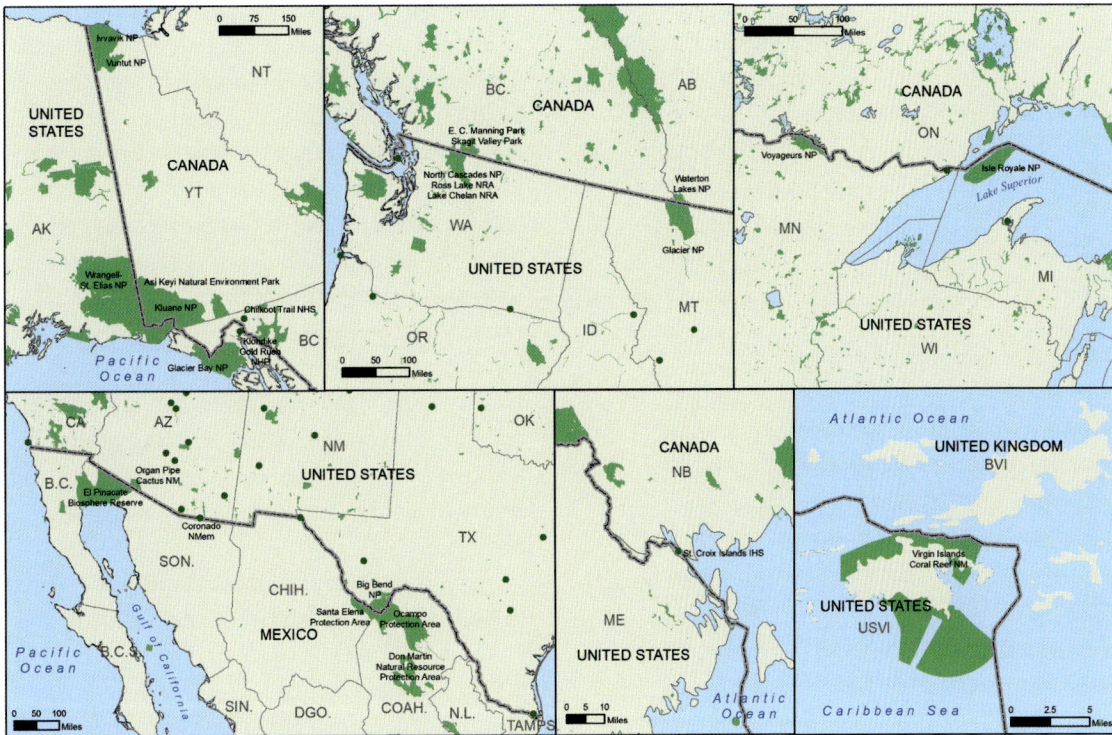

FIG. 5.16   Map of park units along the national border, 2021. America's national parks commonly share a boundary with Canada, Mexico, and even the United Kingdom.

a Yellowstone County in Montana, but it is not near the national park.) The Statue of Liberty has perhaps the most unusual situation: the Supreme Court held that Liberty Island is in New York, even though it is on New Jersey's side of New York Harbor. Since 1965 the monument has also included Ellis Island, which is split between New York and New Jersey. Most of the island is in New Jersey and completely encloses the New York portion, which was the original island before it was expanded.

The degree to which parklands overlap county boundaries is often a source of concern for local governments. Because parks are government land, they do not pay property tax. Creating or expanding parks may severely cut into a county's tax base, though the loss may be offset by increased expenditures by more visitors to the park. Efforts to expand Fort Raleigh National Historic Site in North Carolina were opposed by Dare County, as 75 percent of the county's land area was already within national park units (Cape Hatteras National Seashore, Wright Brothers National Memorial, and Fort Raleigh), and the county feared further loss of its tax base (Binkley and Davis 2003). The federal government has long dealt with this problem in the rural West and may provide "payments in lieu of taxes" to these counties.

Parks on the Border

Today there are fourteen park units along the boundaries of the United States (Fig. 5.16). Most of these parks border Canada; not surprisingly, the largest extent of these parks is in Alaska. Wrangell–St. Elias has the longest border with Canada (about 250 miles); on the other side lies Kluane National Park and Reserve. These parks, combined with Glacier Bay and a Canadian provincial park, make up a World Heritage site. Yukon–Charley Rivers, Glacier Bay, and Klondike Gold Rush National Historical Park share shorter stretches of border, but only at Klondike can one cross the border. The Chilkoot Trail was a gold rush era route to the goldfields and can still be hiked thirty-three miles from the sea to Bennett Station in Canada. Chilkoot Trail National Historic Site contains the Canadian section of the trail; hikers must register and have their border-crossing paperwork checked at the Parks Canada office in Skagway, Alaska, four blocks up from the

NPS visitor center. Since 1998 the two parks are sometimes referred to as being part of Klondike Gold Rush International Historical Park, though they remain legally separate.

In the contiguous states, North Cascades, Ross Lake, Glacier, Voyageurs, and Isle Royale also lie along the Canadian border, though the last has only a water boundary. Organ Pipe Cactus, Coronado, Big Bend, and Amistad lie along the border with Mexico. Amistad National Recreation Area contains the northern side of Amistad reservoir, formed by a dam on the Rio Grande. Boaters entering the lake on the U.S. side may waterski or fish on the Mexican side of the lake but are not allowed to land there without prior arrangement. Finally, Virgin Islands Coral Reef National Monument also contains an international boundary, that between the United States and the United Kingdom. The park's water boundaries, made up of straight lines between geographic coordinates, include those between the U.S. and British Virgin Island territories, a zigzagging line midway between the islands of St. John and Tortola.

Two park units commemorate boundary disputes between the United States and its neighbors. Chamizal National Memorial in El Paso, Texas, is a legacy of an 1852 change in the Rio Grande separating the United States and Mexico. The change in the river created the question of whether the boundary should be changed as well. This was not resolved until 1964, with the United States and Mexico splitting the disputed area into two parks. Chamizal is no longer directly on the border and is divided from it by a freeway. It and the nearby Parque Público Federal El Chamizal on the other side of the Rio Grande's concrete channel are peaceful refuges along a troubled boundary.

San Juan Island National Historical Park in Washington's Puget Sound commemorates a boundary dispute between the United States and Great Britain during the 1859–72 Pig War (Cannon 1997). While the 1846 Oregon Treaty specified a U.S.-Canada boundary line, it was vague about maritime boundaries between British Vancouver Island and American territories. The Strait of Juan de Fuca separated the two south of Vancouver Island, but the border to the east was disputed, with both sides claiming the same islands. Both countries attempted to settle San Juan Island, and after a brief skirmish over a dead pig they agreed to jointly occupy the island. This lasted until the German kaiser Wilhelm I arbitrated the border dispute and awarded the island to the United States. The British consulate in Seattle took an interest in the creation of the new unit and assisted with the interpretation of the British view of events. The consulate also worked to gain permission for park employees to fly the British Union Jack over the former British settlement, one of the few places where this flag is flown by a U.S. government agency. (Perry's Victory and International Peace Memorial in Lake Erie is another.)

And then there is Saint Croix Island International Historic Site, a small island (also known as Dochet Island) on the Saint Croix River, which separates Maine from New Brunswick, Canada. It is unique in the park system in having the word "international" in its title and is often said to lie on the U.S.-Canada border, which runs down the Saint Croix River. But the boundary lies a quarter mile to the east of the island, which was assigned to the United States in a 1797 treaty. Its significance is that it was the site of a 1604 settlement by the French, one of their first in North America.

The island became a national monument in 1949 and an international historic site in 1984. Due to the significance of the site to Canadian history, Parks Canada established its own Saint Croix Island National Historic Site in 1968, consisting of a museum and overlook on the Canadian side of the river. The island is considered important to both countries but may be visited by citizens of neither; the NPS has closed it to the public to protect its archaeological resources.

Law Enforcement Jurisdictions

Federal law enforcement agencies have territorial (geographic) and offense jurisdictions (National Park Service 2015). Territorial jurisdiction is based on the area that an officer has powers in; offense jurisdiction is based on the type of crime. The U.S. Fish and Wildlife Service has territorial jurisdictions on its preserves and reserves. FWS agents also have offense-based jurisdiction throughout the country for wildlife laws such as violations of the Endangered Species Act and the transport of endangered or exotic animals or animal parts such as rhino horns, eagle feathers, and elephant tusks.

Law enforcement rangers in national parks have territorial jurisdictions that apply not just within parks but to any NPS-owned land outside a park (National Park Service 2015). These jurisdictions vary depending on where a park is. In some parks, rangers can enforce all state and federal laws, as can local law enforcement; this is known as concurrent jurisdiction. In other parks, rangers can enforce all state and federal laws, but local law enforcement has no jurisdiction. This is exclusive jurisdiction. Elsewhere, they can enforce only federal laws relating to parks, called proprietary jurisdiction. However, in these parks, rangers may be deputized by state or local law enforcement. Since parks can cross state and county lines, different jurisdictions may exist in different sections of the park.

Yellowstone has exclusive jurisdiction because Wyoming, Montana, and Idaho were territories when it was created (Hampton 1971). The park has a federal courthouse where all crimes committed in the park are tried. This courthouse, part of the District Court for the District of Wyoming, also has jurisdiction over those parts of the park in Montana and Idaho. A legal scholar speculated that a crime committed in the fifty square miles of the park in Idaho cannot be prosecuted in the Yellowstone courthouse (Kalt 2005). The law requires that a trial take place in the jurisdiction where the crime occurred with a jury selected from residents, but nobody lives in the Idaho portion of the park. Can a trial take place for a crime committed there?

Identifying the jurisdictions of other parks is not easy, the NPS does not publicize this information. In 2015 the NPS took exclusive jurisdiction over Mammoth Cave, followed by three more Kentucky parks in 2016 (National Park Service 2016a). Parks in the District of Columbia, as well as in Shenandoah, are also exclusive, while all parks in Colorado and South Carolina are concurrent.

Mention should also be made of the U.S.

Park Police. This organization traces its history back to 1791 as an early law enforcement organization in the District of Columbia. In 1933 it was transferred to the National Park Service during the reorganization that also brought many monuments and battlefields into the park system. The jurisdiction of the Park Police has also been extended to the Statue of Liberty and Gateway National Recreation Area in New York and Golden Gate National Recreation Area in San Francisco. Officers do not wear the typical park ranger uniform, do not display the NPS arrowhead symbol, and are usually indistinguishable from local law enforcement.

Congressional Districts

Perhaps the most important jurisdictions are the nation's congressional districts. Every state has two senators, and the 435 members of the House of Representatives have individual districts, redrawn every ten years after each census. There are parks in all fifty states, so every senator represents at least one park (Fig. 5.17). Of the 435 House of Representative districts, 221 have a park unit within them (Fig. 5.18). The dozens of park units in the District of Columbia include some of the most recognizable and visited places in the system, but along with those of Puerto Rico, the Virgin Islands, Guam and the Northern Mariana Islands, and American Samoa, they have no voting representation in Congress.

The National Parks Conservation Association ranks members of Congress by their voting record and has designated those with a positive record toward the nation's treasures as Friends of the National Parks (National Parks Conservation Association 2017). Fifty-two senators and 176 representatives were counted as friends in 2017. Parks along the West Coast and in the Southwest, Northeast (excluding the District of Columbia), and upper Midwest are most likely to lie in a friend's district. In seventeen states both senators are Friends of the National Parks, but most large western parks such as Yellowstone are not in one of these states. States and districts with the most to gain by promoting and protecting parks are not always those with Friends of the National Parks, and vice versa. Needless to say, parks in the District of Colum-

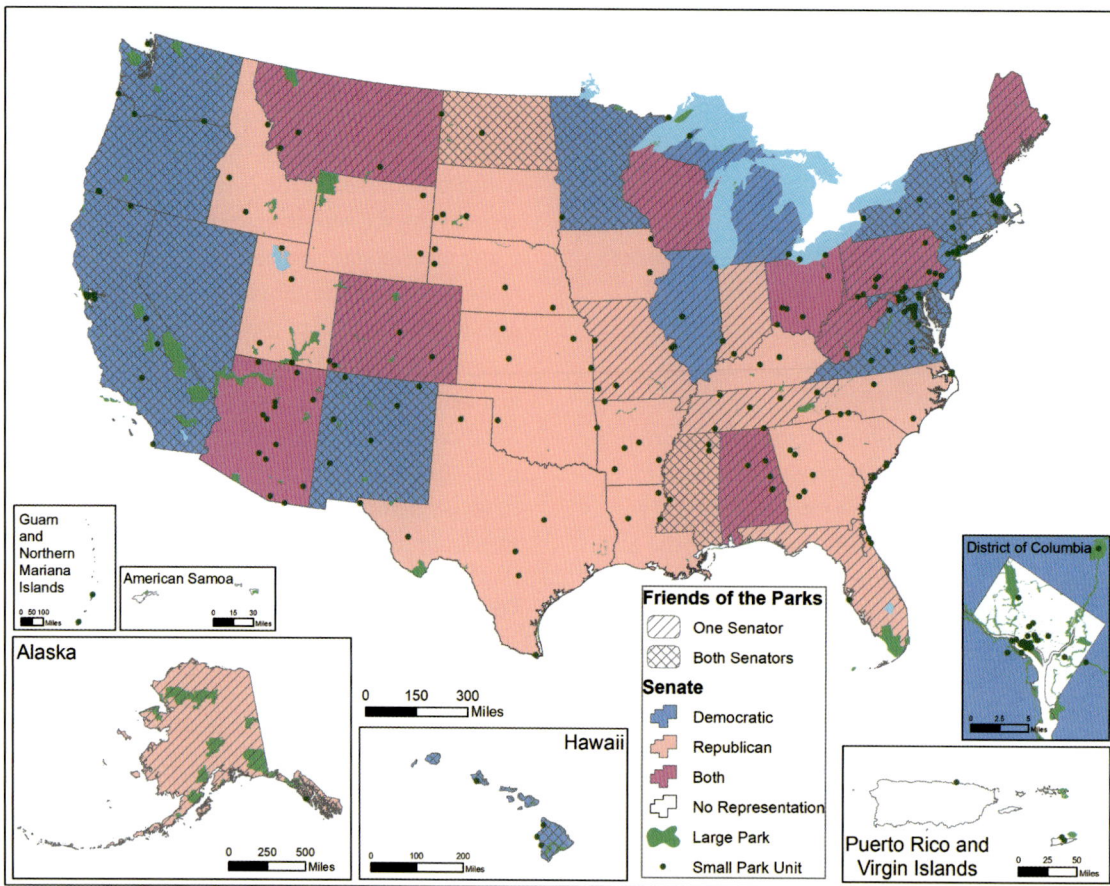

FIG. 5.17   Map of Senate districts and Friends of the National Parks in 2019. Though the electoral map can change rapidly, it is evident that Democratic senators are more likely to support parks than Republican senators.

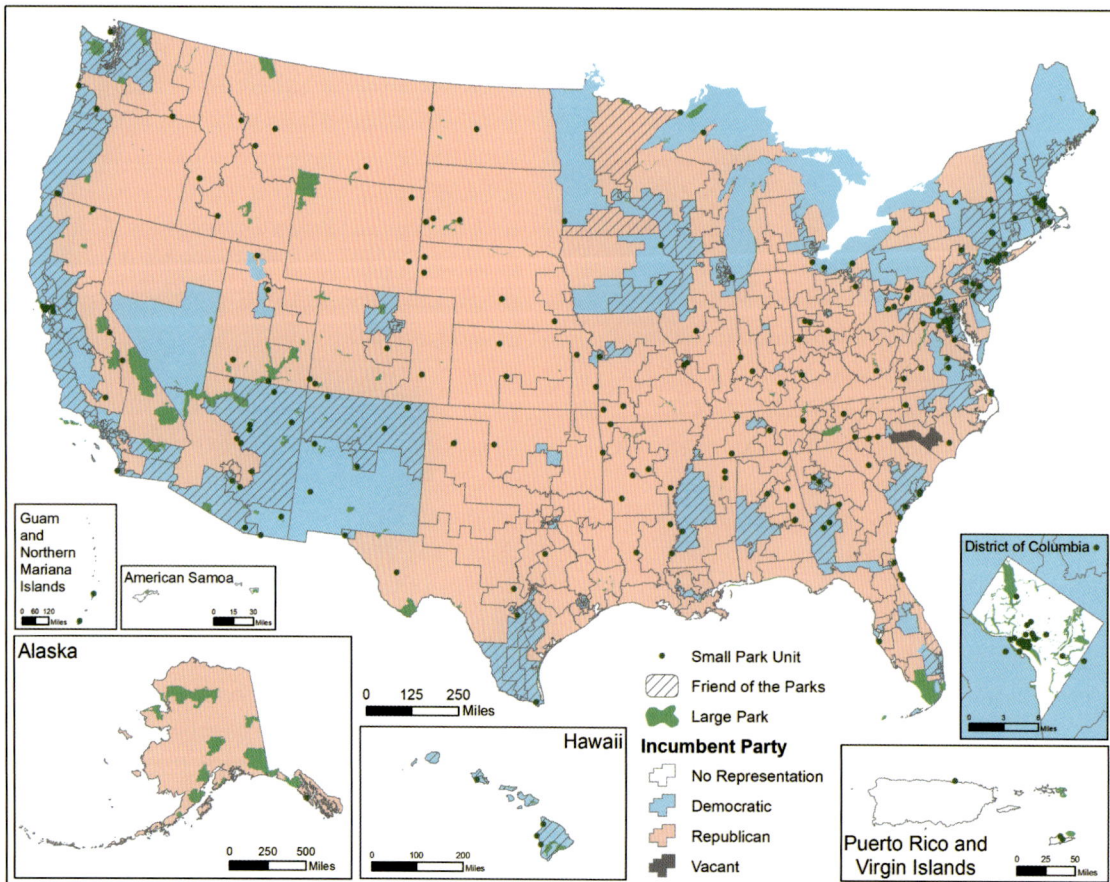

FIG. 5.18   Map of House districts and Friends of the National Parks in 2019. It is an unfortunate fact that most friends are Democrats, and a striking number of America's most famous park units lie within the districts of congresspeople who are not supporters of the national park system.

bia, Puerto Rico, and other territories have no Friends of the National Parks in Congress.

Most of these Friends of the National Parks are Democrats; it is an unpleasant but undeniable fact that since the 1970s party affiliation overwhelmingly predicts not only presidential use of the Antiquities Act but also the voting behavior of Congress regarding the park system. The voting record shows that Democrats are far more likely to vote to protect parks and create new ones, while Republicans are far more likely to undermine and attack America's national park system. The 114th Congress (2015–16) was the most antiparks yet, due largely to the actions of twenty members of Congress from Arizona, Alaska, California, Idaho, Montana, New Mexico, Nevada, Oregon, South Carolina, Tennessee, Texas, Utah, and Wyoming. All twenty were Republicans facing election contests against even more conservative Republicans, and attacking the nation's public lands treasures has unfortunately emerged as a strategy to gain votes among conservatives.

Political affiliation does not always determine attitudes toward parks. West Virginia's New River Gorge was created as a national river in 1978 but was redesignated as a national park in 2020. The change was supported by both senators of the state, one Republican and the other Democrat, with the modification of keeping part of the park open for hunting.

Throughout the history of the national park system there are several members of Congress who have been particularly strong allies of the system. Alan Bible, a Nevada senator from 1954 to 1974, was one of those allies and was commemorated with a visitor center in Lake Mead National Recreation Area named after him (Elliott 1994). Another is Dianne Feinstein, a senator for California since 1992. She introduced the bill that became the California Desert Protection Act of 1994, the largest expansion of the park system in the contiguous forty-eight states, as well as the California Desert Protection and Recreation Act of 2019, one of the most significant park laws since 1994. Stewart Udall served as a congressman from Arizona and then eight years as secretary of the interior. During his years in office, he saw more new park units added to the system than during the tenure of

any other secretary (Einberger 2018). The national parks need more friends like these.

The Dual Hierarchy

The NPS has a well-developed management structure, with a park superintendent reporting to a regional office, the regional director reporting to the deputy director of operations, who reports to the director of the National Park Service, who reports to the secretary of the interior.

Those outside the NPS but living near a park often take a different approach, contacting their congressperson directly. The congressperson may take the issue up with the secretary of the interior or the director of the NPS and perhaps obtain a response or change of policy before the issue has even risen through the NPS hierarchy. With park issues, a member of Congress may become directly involved in a seemingly minor local issue. Park superintendents must therefore work not only within their own hierarchy but also with members of Congress. Superintendents may find themselves transferred if they give priority to the NPS's environmental policies over the wishes of greedy business people just outside the park boundaries (Wagner et al. 1995).

This is the "dual hierarchy," a concern for rural western parks and perhaps most noticeable in Wyoming, whose population is only five hundred thousand, giving the average person far greater access to a senator than in any other state (Yochim 2009). Yellowstone's recent snowmobile crisis was due entirely to this dual hierarchy, with local business leaders using their congressman to force the NPS to accommodate their escalating demands.

Park Redesignation

Eighty-seven park units have been redesignated over the years, meaning they were changed from one type of park to another. The most common action was to promote monuments into parks or other specialized units (representing fifty-seven reclassifications). Many national monuments created in the 1910s and 1920s became national parks. Monuments were essentially undeveloped parks, a stepping stone to full "parkhood." This was true even of the larger monuments: Katmai National Monument was created in 1918 and became the biggest unit in the system in 1931, yet not until 1950 did the NPS assign an employee to it. Reclassification to a national park has often been pursued to increase the visibility and visitation of a unit (Comay 2013). Such a redesignation does in fact tend to increase visitation.

But not all monuments were equal in their chances of being promoted to a different type. There are two groups of national monuments: large ones very much like parks but proclaimed as monuments until they could be promoted, and small sites much in keeping with the spirit of the Antiquities Act. Of the forty-four large monuments (those over sixty-four hundred acres), twenty-eight (64 percent) were redesignated, usually with substantial increases in size. Fourteen (32 percent) remain as national monuments, the largest being Cape Krusenstern in Alaska (created in 1978) and Organ Pipe Cactus in Arizona (from 1937), and one was abolished. The most recent monument redesignation was White Sands in 2019, but efforts to redesignate Colorado National Monument as Rim Rock Canyons National Park and Cedar Breaks National Monument as a national park are ongoing (Lofholm 2014). Efforts to redesignate a park can be as difficult and lengthy as the effort to create one in the first place (Geary 2016).

Of the 101 small monuments proclaimed, only 25 were redesignated, 10 were abolished, and 65 remain as monuments. The oldest of these is Devils Tower, America's oldest national monument. These small monuments have generally not been enlarged much, as they encompassed small historic sites or geological features. Moreover, about 10 percent of these small monuments were abolished and eliminated from the national park system.

The timing of a park's creation was important in determining its designation. Until 1933 the only alternative to a park was a monument, so it was used for a variety of features. After 1933 many other more specific designations were created, so monuments fell out of use (until they were revived by President Carter). If the national preserve designation had existed in 1933, Death Valley might have been one. Re-

designation from a monument to another kind of unit is one way in which Congress can provide a more accurate or specific designation that was not available when many national monuments were proclaimed (Comay 2013).

Delisting

Twenty-nine units have been removed from the national park system (Hogenauer 1991a, 1991b; Weber 2016). They were created by legislation, proclamation, or interagency memorandum; they existed for a number of years; and they were then eliminated from the system (Fig. 5.19; Table 5.1). These abolished or delisted units fall into four groups. Five of these were historic sites that were little more than roadside markers. These were small War Department memorials that were transferred to the NPS in 1933 but did not fit into the system. Most were increased in size, in several cases becoming large battlefield parks, while Tupelo and Brices Cross Roads remained the same size. The remaining five were dropped from the system.

Father Millet Cross National Monument was surely the very smallest national park unit ever created, consisting of an eighteen-foot-by-eighteen-foot square containing a cross. It remains within Fort Niagara State Park near Niagara Falls, New York. Three of these sites consisted of nothing but one or more roadside markers. The New Echota Marker National Memorial (1930–44) commemorated the short-lived capital of the Cherokee Nation that existed in Georgia from 1825 to the 1830s. In 1957 the site of the capital was excavated and reconstructed as New Echota State Historic Site. Atlanta Campaign National Historic Site (1944–50) consisted of five roadside markers along U.S. Highway 41 commemorating events of the Atlanta campaign in 1864. Camp Blount Tablets National Memorial in Fayetteville, Tennessee, and White Plains National Battlefield Site in New York were to be small markers commemorating military events. Neither was recorded as having any acreage, though Camp Blount Tablets was originally to include an old stone bridge (long since collapsed).

Ten of these sites were western monuments created between 1908 and 1929 that were never developed (Hogenauer 1991a). Fossil Cycad

FIG. 5.19   Map of former national park units, 2022.

TABLE 5.1   Delisted national park units, 2021

Name	Created	Delisted	Origin	Fate
SMALL MARKERS AND MEMORIALS				
Father Millet Cross	1925	1949	Presidential proclamation	State park
White Plains	1926	1956	Congress	Roadside marker
New Echota	1930	1950	Congress	State park
Camp Blount Tablets	1930	1944	Congress	Roadside marker
Atlanta Campaign	1944	1950	Interior secretary proclamation	Roadside markers
RESERVOIR-BASED NRAS				
Millerton Lake	1945	1957	Interagency agreement	State park
Shasta Lake	1945	1948	Interagency agreement	National forest
Lake Texoma	1946	1949	Interagency agreement	Army Corps / state park
Shadow Mountain	1952	1979	Interagency agreement	National forest
Flaming Gorge	1963	1968	Interagency agreement	National forest
NATIONAL MONUMENTS				
Lewis and Clark Caverns	1908	1937	Presidential proclamation	State park
Wheeler	1908	1950	Presidential proclamation	National forest
Shoshone Cavern	1909	1954	Presidential proclamation	BLM
Papago Saguaro	1914	1930	Presidential proclamation	City park
Old Kasaan	1916	1955	Presidential proclamation	National forest
Verendrye	1917	1956	Presidential proclamation	State park
Fossil Cycad	1922	1956	Presidential proclamation	BLM
Castle Pinkney	1924	1956	Presidential proclamation	Private
Holy Cross	1929	1950	Presidential proclamation	National forest
OTHER UNITS				
Mackinac NP	1875	1895	Congress	State park
Sullys Hill	1904	1931	Congress	FWS
Chattanooga National Cemetery	1933	1944	Congress	
St. Thomas NHS	1960	1975	Interior secretary proclamation	Local museum
National Visitor Center	1968	1981	Congress	
Mar-a-Lago NHS	1972	1980	Interior secretary proclamation	Private
JFK Center for Performing Arts	1972	1994	Congress	Foundation
Oklahoma City Nmem	1997	2007	Congress	Local

National Monument is perhaps the most well known of these. It was created in 1922 as one of many fossil sites being preserved as national monuments in the early twentieth century (Santucci and Ghist 2014). The fossil site was heavily excavated in 1935 to the point where there did not appear to be any fossil specimens left. In 1936 interest grew in developing the monument for tourists, including a museum to interpret the scientific value of the site to the public. Instead, the monument was abolished in 1957. Lewis and Clark Cavern and Shoshone Cavern National Monuments were caves located near Yellowstone (P. Roberts 2012). Lewis and Clark was developed by the CCC and is open today as a state park; Shoshone Cavern was never developed by the NPS and is now on BLM land. It has never been developed or fully explored.

Wheeler and Holy Cross were both high-elevation monuments in the Colorado Rockies, and both were abolished in 1950 (Szasz 1977). The Mount of the Holy Cross is a 14,005-foot-tall peak with a rock formation on its eastern face that when filled with snow resembles a cross. In 1919 the idea of an annual pilgrimage to the mountain became popular, creating support for a national monument, which was proclaimed in 1929. By the 1930s interest in the area had dwindled, in part because of light snowfall and erosion of the cross formation.

Old Kasaan National Monument was created to preserve a recently abandoned Native Alaskan village in the Alaska Panhandle (Norris 2000). The village had become a tourist attraction, and interest grew to preserve it so that it would remain one. In 1916 it became a national monument under the Forest Service. Unfortunately, the village had been damaged in a 1915 fire, by vandalism, and by natural decay of the wood artifacts. After several decades of occasional interest but usually inaction, the national monument was abolished in 1955.

Papago Saguaro National Monument was proclaimed in 1914 to preserve an example of Sonoran Desert vegetation near Phoenix, Arizona (Yard 1919). Rather than preserve the vegetation, locals wanted to see the monument developed, leading to friction with the NPS. In 1930 it was handed back to local control and can be seen today as Papago Park, complete with a zoo, golf courses, and a botanical garden.

Six delisted units were postwar western reservoir-based national recreation areas that were developed but then transferred to other agencies, often after only several years in the park system. The NPS had a mixed early record with dams and reservoirs before embracing them as Lake Mead filled. After the end of World War II, the agency became involved in several reservoir recreation projects involving Bureau of Reclamation dams. The NPS soon had second thoughts and transferred these areas to other agencies. Most are today state parks or Forest Service areas (Weber 2016). Shasta Lake, Millerton Lake, Lake Texoma, Shadow Mountain, and Flaming Gorge National Recreation Areas were created, developed, and handed over to other agencies within a few years. The NPS did not turn away from reservoirs entirely, as several more reservoir recreation areas were added in the years after World War II and remain a part of the system.

The final group consists of miscellaneous sites. Mackinac National Park was created in 1875, only three years after Yellowstone, but it lost its status in 1895 when it was transferred to the state of Michigan, which continues to manage it as a state park (Widder 1975). Sullys Hill Park was created in 1904 as a game preserve and seems to have ended up with the Department of the Interior and later the NPS by accident (Hogenauer 1991a). It was transferred to the Fish and Wildlife Service, which maintains it as a game preserve. Mar-a-Lago National Historic Site was an unusual addition that has drawn a fair amount of notoriety (Rettie 1995). Mar-a-Lago is a large mansion and estate in Palm Beach, Florida, built in the 1920s by Marjorie Post, the richest woman in America at the time. When she died in 1973 she left the mansion and grounds to the U.S. government, perhaps to be used as a presidential retreat. The property ended up with the NPS, which did not have the money or interest to do anything with it. In 1985 it was removed from the system and sold to Donald Trump (Hogenauer 1991a), who eventually did make it a presidential retreat.

In addition, there exist several park units that came close to being abolished. The most dramatic of these stories is a small historic site in Ohio. In the late eighteenth and early nineteenth centuries, the settlement of Ohio introduced Americans to the mound-building cultures of the interior (Cockrell 1999). Among these were the mounds referred to as the Mound City Group near Chillicothe, surveyed and discussed by Ephraim G. Squier and Edwin H. Davis (1848). These mounds remained prominent in the following decades and survived agricultural settlement and other disruptions. They did not survive World War I, as a training base, Camp Sherman, was established on the site and the mounds were removed. As Camp Sherman had been under the control of the War Department, Mound City Group was as well. Mound City Group was quite different from the War Department's other national monuments, which preserved fortifications and battlefields. The War Department had little use for the monument, and so it gave the Ohio Historical Society a license to operate the monument as a state park. The state of Ohio removed the remnants of Camp Sherman and rebuilt the mounds following the 1848 map. In 1929 the monument formally opened as an Ohio state park. Structures were put up by the Works Progress Administration and other New Deal programs, and signs along the main road referred to it as Mound City State Park (Cockrell 1999).

The NPS finally gained control of Mound City Group in 1933 but was not very interested in the park. In fact, the NPS actively sought to permanently transfer it to the state of Ohio. This process started in 1937, but completion was stopped by World War II. After the war, locals preferred NPS control, and the NPS formally reclaimed the monument from the state in 1946. While the NPS inherited some buildings from the state park, it also inherited the widespread view that the park was a picnic and recreation area. The NPS worked hard to reduce these uses and put the focus on the interpretation of the Hopewell culture and mounds (Cockrell 1999). However, the NPS still had little interest in the mounds, and in 1953 the mounds were noted as a second-class park unit and a candidate for elimination from the park system. In 1954 the NPS leadership approved abolishment, but local opposition prevented it. The NPS tried again in 1956, but this time it was blocked by internal opposition. An NPS archaeologist argued that the site was important as one of only four archaeological units not oriented toward southwestern ruins. It represented the only Hopewell culture site in the NPS, and as the Hopewell culture had only been studied for a few decades, this site was significant. It was also the location of Squier and Davis's 1848 work and so could represent the early history of American archaeology (Cockrell 1999). After this defense the monument was secure within the NPS, and a belated Mission 66 plan was put together, calling for a new visitor center (which opened in 1961) and a new name. The boundary was expanded in 1980 to include the Hopeton earthworks, but not until 1990 was the land purchased. In 1992 Mound City Group became Hopewell Culture National Historic Site, its place in the park system presumably secure.

There are many other examples of parks that came perilously close to elimination. The Gila cliff dwellings were discovered in 1878 and quickly looted (Russell 1992). In 1907 the area became a national monument, but in 1955 the NPS decided to abolish it. After the park custodian revealed that there were many adjacent ruins in the area, the NPS decided to keep the monument. Tiny Devils Postpile was transferred to Forest Service control in 1947, and the NPS wanted to be rid of it permanently (Johnson 2013). But the tiny monument became more popular after the war, and the NPS decided to keep it.

There are many other similar examples of parks nearly lost or targeted by members of Congress. There are in fact many more than we think, including parks that we would never expect to be threatened with elimination (Rettie 1995). Regardless of their merits, the abolished units of the national park system serve to remind us that national parks are not necessarily permanent. They require continued support and attention lest they disappear.

Conclusions

Parks are large and expensive public facilities whose existence depends on Congress and the

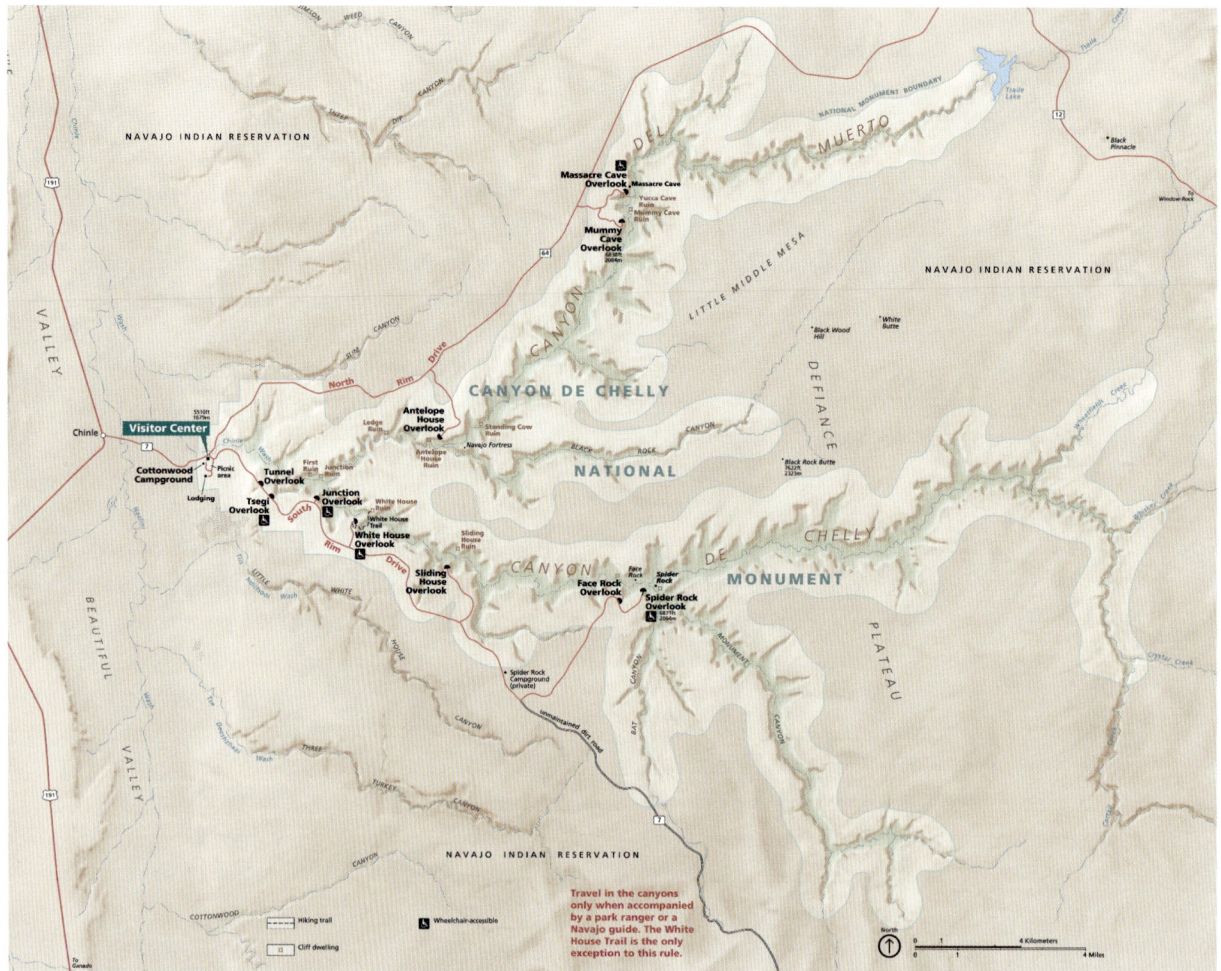

FIG. 5.20   Map of Canyon de Chelly National Monument in Arizona, 2021. The park boundary extends a half mile from the canyon rims. NPS.

president to create and fund them. This contentious political process is always subject to change, and it is impressive how many such places have been created and how few have been lost. Boundaries must be surveyed and are always subject to revision, and land must be purchased or received as donations. Changing conditions will likely lead to changed boundaries. A unit may become substantially bigger or be trimmed to exclude threats. While the sanctity of park boundaries may seem paramount to protecting these places, they are in fact quite malleable, and this should be thought of as an asset.

Each park was the outcome of a political process at a particular time and place, and reconstructing the reasons why each park had the boundaries it did (or why its boundaries have been changed) is very difficult. Administrative histories may provide some information but often lack detailed maps. There is no national park equivalent to Mark Stein's (2008) discussion of how and why each state's boundaries evolved. Stein's account shows that creating states and drawing their boundaries were long processes involving early colonial claims, tension between local interests and the federal government, a preference for straight lines and rectangular boundaries, the establishment of boundaries in areas not yet explored, the occasional use of rivers or ridgelines for boundaries, the proposal of boundaries vastly greater than those finally approved, and in one case (the northern boundary of Delaware) the use of a fifteen-mile circular buffer to establish a boundary. Indian lands played a role, and some boundaries preserve earlier national boundaries or other relics of political geography.

Much the same can be said for national park boundaries.

To one accustomed to working with digital maps with GIS software, the absence of circles or buffers in park boundaries is striking. Only a few island units and Canyon de Chelly, where everything within a half mile of the canyon rims is inside the monument (Brugge and Wilson 1976), have boundaries that use this approach (Fig. 5.20). Instead, the park system was largely based on a survey system and property boundaries defined by straight lines, squares, and rectangles.

National forests provide an interesting comparison with parks. Many small forests were created in the late nineteenth or early twentieth century. They differ from parks in that forests then went through a process of consolidation and renaming. Nevada's Toiyabe National Forest grew through the merger of five separate national forests and was then merged into the Nevada and later Humboldt National Forests, only to be separated again as a distinct forest (Thompson 2007; United States Forest Service 2012). Most national forests have gone through similar mergers, splits, and renamings. In contrast, park units have usually kept their historic (original) identities, at least as far as the public is concerned. Many parks have been merged and are operated as one unit in all but name.

The overlap of parks with counties, states, and reservations can make for strange situations. This can be seen with gay marriage, legal everywhere in the United States since a 2015 Supreme Court decision. Except that it's not: due to vagaries in federal law, the decision is not binding in American Samoa and on Indian reservations, and not all of these jurisdictions have accepted the practice. As of 2021, same-sex couples looking to tie the knot in a spectacular national park setting will not be allowed to do so in Arizona's Canyon de Chelly National Monument or in the National Park of American Samoa, among others.

One of the most important aspects of park geography is how parks intersect with congressional districts. If you live near a large interior park, you are likely represented by a member of Congress who is hostile to the park's existence and closely allied with extractive industries that are fundamentally opposed to the existence of protected places. It is sad that the phrase "I love national parks" has become a partisan statement likely to receive a rude response in some places.

SIX

The Internal Spatial Organization of the National Parks

Parks are carefully designed landscapes, often so well designed that it is hard to imagine they are not natural. This chapter examines the design and construction of visitor infrastructure such as roads, lodges, trails, campgrounds, restaurants, and visitor centers in parks. These cultural landscapes are important factors in making each national park unit a distinct place.

The era in which parks entered the system is crucial to the kind of development each park received. While many early parks had facilities built in the horse-and-wagon era, emphasizing guided tours and luxurious lodges, the coming of the automobile put increased pressure on the National Park Service to provide roads and inexpensive lodging for families. In the 1930s the Civilian Conservation Corps constructed roads, trails, campgrounds, housing, and other facilities in many parks using a rustic architectural style. Throughout these years the layout of these facilities in these parks was guided by landscape architects, who worked to showcase the parks' features. Their work created a common template for parks.

After wartime neglect, many parks were deteriorating and badly in need of expanded facilities to meet the influx of visitors arriving by automobile. Mission 66 was launched in 1956 as a ten-year program to rebuild parks and accommodate crowds. New roads, trails, campgrounds, lodging, and multipurpose visitor centers were the result. As before, these facilities were located carefully to structure the visitor experience in the park and channel movement through the park. Many parks owe their entire design to this project. Since the program ended,

development patterns have varied, with many parks receiving little or no development.

Building Parks

Even after a national park unit is created and its boundaries specified, much work must be done. Parks are created for people, and there must be facilities for these people. Exactly what these facilities should be, how common they should be, where they should be located, and who should pay for them have been topics of considerable discussion over the years and frequently contentious issues.

The Early Years

In June 1878 Congress appropriated $10,000 to be spent in Yellowstone, the first money spent on a national park (Hampton 1971). This was spent on roads and bridges, but the money did not last long. Spending on roads has continued ever since and has rarely kept up with demand. In the absence of more government money, the superintendents in charge of Yellowstone tried to use leases to obtain internal improvements; private companies were allowed to build lodges and roads. The park would get visitor facilities for free, with fees paid to the park by the companies helping pay for rangers. This scheme did not work out due to the superintendent's lack of authority, absence of any legal basis for enforcing regulations, and possibly corruption.

The solution was to bring in a disciplined group of uniformed professionals whose distinctive hats would become a symbol of national parks: the United States Army. The army was soon busy throughout the park; the soldiers established several patrol cabins and went out to search for poachers and to try to stop vandalism at thermal features (Haines 1996). There were sixteen patrol cabins in use when the army ended its work in Yellowstone, most of these still occupied by ranger stations today. The army built much of the park's basic road network (T. Davis 2016), fought fires, and set up an early interpretation program for visitors. Without the army's efforts to stop wildlife poaching, it is very likely the bison would have become extinct.

The army needed a base to operate out of and created Fort Yellowstone, an army post occupied from 1886 to 1918 (Hampton 1971; Haines 1996). The fort was laid out using a standard template used by the army throughout the West: a rectangular parade ground with officers' barracks along the sides, officers' houses at one end, and enlisted barracks, stables, and cookhouses at the other (also visible at other NPS sites such as Fort Bowie and Fort Union). Fort Yellowstone differed slightly due to its location on a small plateau; officers' quarters lined only one side of the parade ground, and barracks were behind them. In 1904 a new water system was installed with enough capacity to allow planting and watering grass on the parade ground as well as running a hydroelectric generator, bringing electricity to Yellowstone. Sidewalks and streetlights were put in, and substantial stone buildings replaced earlier ones.

In 1890 new Yosemite, Sequoia, and General Grant (later Kings Canyon) National Parks were established in California's Sierra Nevada (Myerson 2001). The Department of the Interior called on the army to patrol the new parks, and troops were sent out from the Presidio in San Francisco (which would itself become part of the park system a century later). Unlike Yellowstone, they did not build a permanent headquarters but set up temporary summer encampments each year, again with backcountry patrol cabins. Farther north the army also built the Nisqually road at Mount Rainier, which still serves as the park's main entrance (T. Davis 2016).

In 1918 the army turned Fort Yellowstone over to the NPS. The fort became the Mammoth Hot Springs area, still the park's headquarters.

Thirty-five of the fort's buildings survive in NPS use, as do many of the army's policies and even the campaign hat that was worn by the cavalry. This is familiar today as the ranger hat. Many soldiers with park experience later worked as civilian rangers for the new National Park Service, ensuring continuity as the NPS slowly gained experience.

The NPS Adopts the Rustic Style

When the Park Service was created, landscape architects were put in charge of developing the parks for visitors (Sellars 1997; McClelland 1998; Carr 1998). These architects had developed a set of plans for nineteenth-century country estates that included a common set of elements. Approach roads gave access to the estate and often included a gatehouse at the entrance (McClelland 1998). A curvilinear drive carried visitors through a carefully selected sequence of views, and trails allowed them to walk to other viewpoints. A bench or shelter would be constructed at these viewpoints to allow the viewer to rest and enjoy the view. Vegetation was manipulated as needed to obtain desired results: trees and shrubs could be planted to screen service buildings, while plants were cut back in other areas to open views. Using a variety of species and ages was preferred to create maximum scenic impact.

Many of these elements were adopted for national parks (Good [1938] 1999; McClelland 1998). Gatehouses were replaced by entrance arches in some early parks, succeeded by elaborate entrance signs, still common today. Roads and trails were laid out with careful attention to views, frequently employing loops to return travelers to their starting point without backtracking. The Carbon River Road at Mount Rainier was one of the first to be laid out this way, with a carefully planned sequence of views and experiences. Visitor roads, trails, and service roads were kept apart as well as invisible to each other.

The primary design principle was that preserving the natural landscape was the most important goal, and roads, trails, and buildings should be blended into that landscape. Buildings were to be designed in harmony with the

FIG. 6.1 A 1914 postcard view of Yellowstone's Old Faithful Inn. For many, this represents what a rustic park lodge should look like. Photo by F. Jay Haines. Wikipedia, Creative Commons Attribution–ShareAlike 2.0 Generic License.

land and should use natural materials from local sources, usually stone and timber (McClelland 1998). Lodges and camps in the Adirondack Mountains provided the model for later park lodges, with a rustic look, stone walls, log ceiling beams, and massive fireplaces (Kaiser 1997). Making buildings seem as though they had always been part of the landscape was desirable.

The NPS was eager to build new roads; in fact, Mather's strategy for boosting visitation and public support for parks required them. While many roads had been built by the army in Yellowstone and at Mount Rainier, with the arrival of the automobile these were now obsolete in their design (T. Davis 2016). The use of cars required longer turning radii, easier grades, more room to pass on narrow mountain roads, and a design that would allow higher speeds. Rustic tradition called for bridges to be built from tree trunks and branches, but these quickly decayed and were not suitable for ever-increasing vehicle traffic (McClelland 1998). Concrete arches quickly became favored both for their permanence and because they allowed unobstructed views.

Lodges replaced country mansions, with lesser lodges or permanent camps elsewhere in the park. In time, the idea of a stand-alone lodge evolved into that of a park village, including a lodge but also a ranger station, a store, a gas station, employee residences, and utilities. An early plan for General Grant (later Kings Canyon) National Park had buildings arranged around a square, a pattern used in later developments. Campgrounds were designed with one-way loops to ease movement, with individual lots for each campsite.

Other influences on the design of early parks include segregated facilities in southern parks (W. O'Brien 2015, 2018). Both Shenandoah and Prince William Forest Park had separate camping facilities for African Americans. Though separate, these campgrounds were never equal in terms of the quality of the facilities or the scenery. They are a regrettable phase of development in the park system.

Great Lodges

The rustic style became well known in part due to several large hotels built by private companies with contracts to operate concessions within parks. This began during the army era in Yellowstone but accelerated after the NPS was created. Much of the early development of parks was carried out by railroads or subsidiary companies; these adopted the rustic style of Adirondack camps to parks (Kaiser 1997).

Perhaps the most famous is the Old Faithful Inn, completed in 1904 by the Yellowstone Park Company (Fig. 6.1). It was built from logs cut within the park, with a fireplace made from

FIG. 6.2   The Desert View Watchtower at the Grand Canyon, 2017. This 1933 building was designed by Mary Colter to resemble an ancient Native American ruin. Photo by Mario Roberto Duran Ortiz. Wikipedia, Creative Commons Attribution–ShareAlike 4.0 International License.

rocks quarried nearby. The interior is seven stories tall and stunning. It functioned as part of a five-day counterclockwise tourist route, with travelers moving between different lodges each day (Barringer 2002). The replacement of stagecoaches by buses sped up the circuit and eliminated the need for several facilities, which were demolished. Old Faithful Lodge survived these changes and remains today many people's ideal of what a park lodge should be.

El Tovar (1905) at the Grand Canyon, Many Glacier Hotel (1915) at Glacier National Park, Paradise Inn (1917) at Mount Rainier, the Ahwahnee Hotel (1927) in Yosemite Valley, and Furnace Creek Inn (1927) in Death Valley are other outstanding examples of rustic lodges still in use.

After Mesa Verde was added to the park system, Native American architecture became another inspiration for rustic "parkitecture" design, with this park's buildings based on pueblo designs. Mary Colter's Grand Canyon buildings followed this example (Carr 1998; Berke 2002). Hopi House (1905) was designed to look like a Hopi pueblo building and housed a gift shop. Hermit's Rest (1914) was constructed like a mountain man's house (Grattan 1992). It used local stone and timbers and was deliberately made to look old. Phantom Ranch at the bottom of the canyon and the Bright Angel Lodge (1935) on the South Rim were also made of local stone in a rustic style. Colter's masterpiece was the Desert View Watchtower (1933), an overlook of the canyon built to resemble a Native American ruin (Fig. 6.2). The building is thoroughly modern, with a steel frame and concrete floors, but it so completely resembles a ruin that many visitors will not believe otherwise. All these buildings remain today and are considered masterpieces of the rustic style.

The era of the great lodges peaked surprisingly early; by the 1930s cabins or motels with

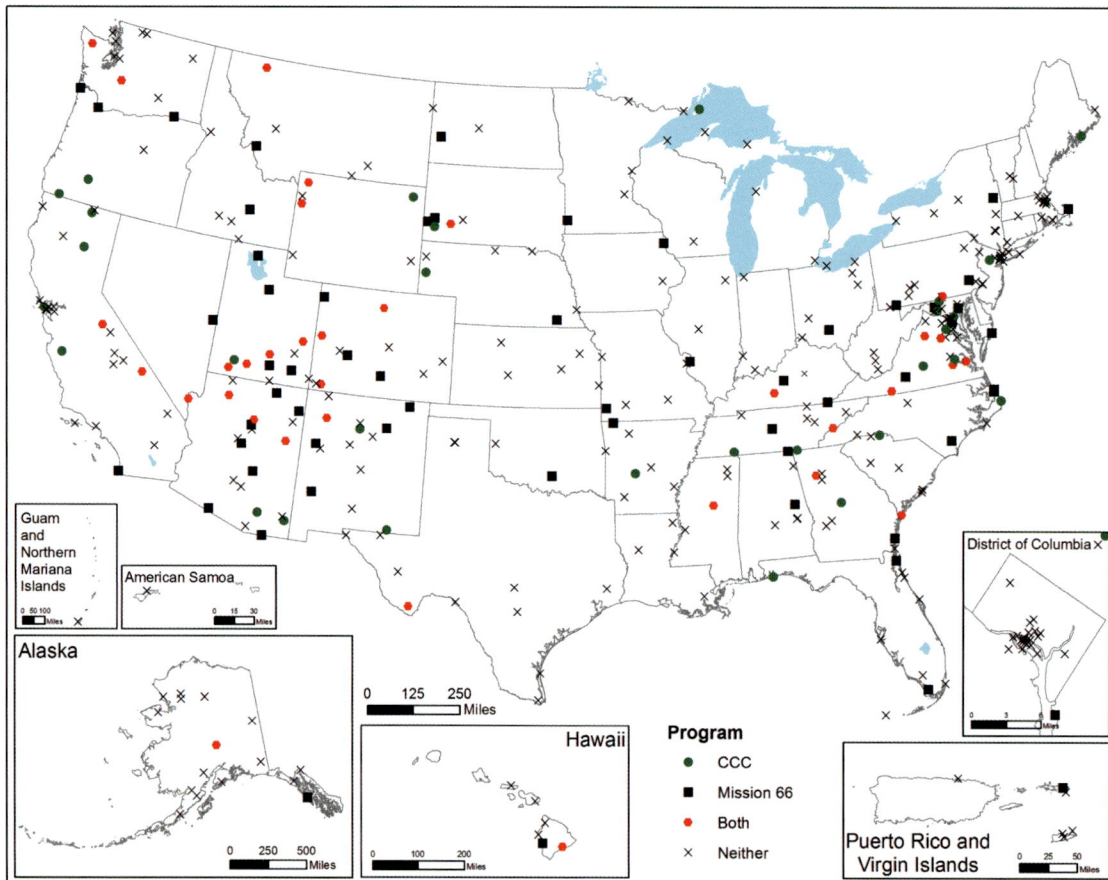

FIG. 6.3   Map of park units where CCC and Mission 66 work was done. Most older parks were developed during one or sometimes both programs.

parking for personal vehicles were becoming more popular (Barringer 2002). These also required gas stations, stores, cafeterias, and other amenities, accelerating the trend toward the park village approach.

The Civilian Conservation Corps

In the 1930s the Great Depression devastated the nation's economy and workers and resulted in many federal programs to put people back to work. Among these was the Civilian Conservation Corps (CCC), which put hundreds of thousands of young men to work on a variety of projects, often in the national park system (Fig. 6.3). CCC workers lived in spartan barracks and built roads, trails, campgrounds, bathrooms, picnic areas, water and sewage systems, and housing and maintenance facilities for the workers. As before, this work was done using a rustic style that emphasized local building materials and styles and was designed to

blend into the landscape. But the men did not build grand lodges.

For many small national monuments, the arrival of the CCC provided the first facilities of any kind and opened the parks up to visitors. At Chiricahua, the CCC built hiking trails into the spectacular rock hoodoos, a campground, and the park's visitor center (Pinto 2007). At Saguaro workers built roads, trails, picnic areas, check dams to prevent erosion, and an administration building, most of which remain in use (Clemensen 1987). They carried out similar projects at Death Valley (Fig. 6.4), White Sands, Crater Lake, Petrified Forest, Big Bend, Mesa Verde (Brown and Smith 2006), Devils Tower, Lava Beds, and many others.

Bandelier is one of the best surviving examples of the work of the CCC in the national park system. The men constructed the monument's entrance road as well as thirty-one buildings using pueblo-style architecture. They also built

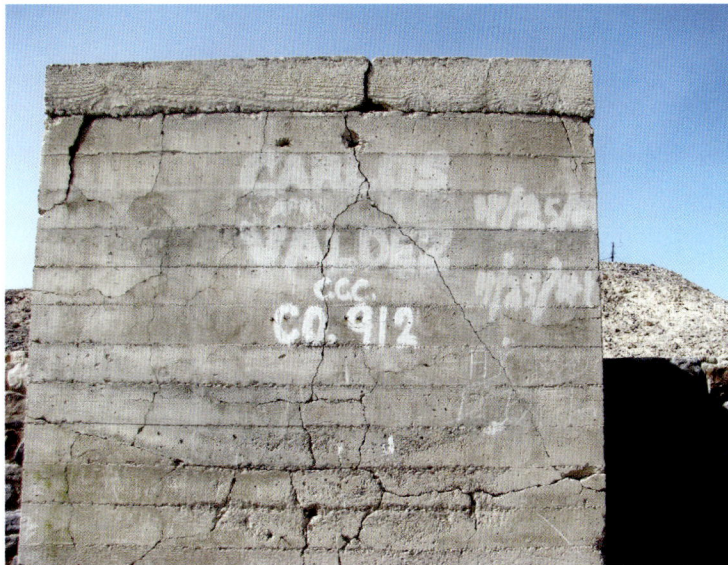

FIG. 6.4   CCC graffiti left on a garbage incinerator in Death Valley in 1941. Shortly after Carlos Valdez painted his name, the attack on Pearl Harbor ended his work. Company 912 was deactivated in 1942, and most members joined the military, substituting the tranquility of Death Valley for the horrors of war. Photo taken in 2009.

trails to ruins, along with the many ladders visitors must climb to visit some of them. These facilities are still in use today and still look much as they did in the 1930s.

Mission 66

During World War II park attendance plummeted (Rydell and Culpin 2006; Carr 2007). Yellowstone dropped from 581,761 visitors in 1941 to only 64,144 two years later. The coming of war also meant the end of the CCC program, and the parks were almost abandoned. Most services were shut down, and many agency employees left to join the war. After the end of the war in 1945, park attendance not only returned to prewar levels but also continued to increase year after year. In 1946 Yellowstone visitation reached 814,907, a new record, and it has continued to grow year after year.

The parks were in no shape for this growth, and discontent grew over the shabby condition of many as they were overrun by hordes of tourists (Barringer 2002). After ten years of trying to rebuild and keep up with the growth in visitation, NPS director Conrad Wirth decided to launch a bold program to rebuild and redesign the parks. The NPS forecast eighty million visitors across the park system in 1966, the fiftieth anniversary of the creation of the NPS, compared to forty-nine million in 1956. Ultimately, 127 million visitors arrived in 1966 (Carr

2007). A massive increase in new facilities was clearly needed. Wirth called it Mission 66, after the target year of 1966. Many parks needed new lodging, food services, campgrounds, roads, and trails, as well as employee housing.

Each park was responsible for coming up with its own Mission 66 plans; these were reviewed higher in the NPS bureaucracy and often modified or rejected (M. Bishop 1998). Geography was fundamental to the program. Tourists would be concentrated in fewer areas, and these areas would, in turn, be developed (or "hardened") to handle the crush of crowds without damaging natural resources. In Yellowstone this process began before Mission 66 when wooden walkways were installed around hot springs to prevent people from trampling large areas (Rydell and Culpin 2006; Carr 2007). Channeling people has remained a common NPS strategy.

A second and related geographic strategy was to move facilities (and heavy use) away from sensitive resources; these strategies are apparent at Yellowstone, where the West Thumb area was abandoned in favor of the new Grant Village, Canyon Village replaced facilities adjacent to the Grand Canyon of the Yellowstone, and the new Firehole Village was planned (but never built) to replace development around Old Faithful (Barringer 2002; Rydell and Culpin 2006). Some Mission 66 plans called for lodging or campgrounds to be moved out of the park altogether, as was unsuccessfully planned for Mount Rainier and Everglades (Carr 2007). Due to the distance from towns and the size of parks, some form of overnight accommodations was clearly necessary at many parks, but they did not have to be next to a major natural feature. New park facilities at Yosemite were located outside the park in a nearby town. A strategy that was not adopted was to protect resources by limiting visitor numbers (Carr 2007).

Mission 66 is often perceived as having been a road-building program, and 2,767 miles of road were built during its run (T. Davis 2016). However, many of these miles replaced older, substandard roads that were then abandoned, and only 1,197 miles were new. Among these were substantial additions to the Natchez Trace Parkway. The construction of a new Tioga road in Yosemite was a major controversy at the time; it was completed, but the improvement of the Denali park road was not. Road work had been proceeding into the park, and when work was halted in 1965 only the first portion of the road had been paved, the next section had been improved but was still dirt, and the remaining miles were unimproved. The road remains this way today.

The Visitor Center

One innovation of the Mission 66 program was the visitor center, a multipurpose facility that included ranger station, museum, bathrooms, auditorium, gift shop, and sometimes restaurant and park administrative offices (Allaback 2000; Rydell and Culpin 2006; Carr 2007). A large parking lot was provided for visitors arriving by car, which by 1950 was 99 percent of them. These buildings replaced many different buildings that would have been found in a park village in earlier eras. A visitor center was carefully located based on a park's master plan. It might be located near the entrance road, where all incoming visitors would be introduced to the park and its features; at a central location within the park's road network, where it would serve as a hub for the visitor experience; or adjacent to the park's major feature. If it was built near

a feature, then it would likely be sited in a location to allow visitors to view that feature. Visitor centers were sometimes criticized for being too close or too intrusive in the historical landscape, an example being Gettysburg.

Carlsbad Caverns and the Grand Canyon received similar buildings in the early 1950s, but these were called "public use buildings"; the name "visitor center" was chosen by the NPS director despite concerns that these centers might be confused with shopping centers, another innovation of the 1950s (Carr 2007). Both kinds of centers featured ample parking in front of one or more related buildings and often featured bold new architectural styles designed to attract motorists (Hine 1990; Hess 2004a).

These styles were based on modernist architecture, a sharp break from the rustic architectural tradition of park buildings from previous years (Allaback 2000; Rydell and Culpin 2006). This came about largely because of cost considerations. The earlier rustic buildings found in many parks had been built either many decades earlier by private companies or in the 1930s by the CCC; labor was now much more expensive, and hand-made buildings assembled out of stone and logs were unaffordable. Using stone and trees found within the park was also now unacceptable, and shipping them in was economically out of the question.

The NPS used architectural firms to design buildings and hired construction companies to build standardized designs (Allaback 2000). Because modernist styles were in vogue, the firms employed them when designing new visitor centers and other buildings. These new designs used concrete, steel, and glass as their building materials, all much cheaper than massive logs, hand-cut timbers, and hand-placed stones. These buildings often featured large interior spaces, large windows, and natural lighting. In the postwar era, the older rustic designs had also begun to seem out of date or even decrepit. A bold new look was appropriate for the parks. Not all visitors approved of this; many preferred the older rustic look (and still do).

Over a hundred visitor centers were built during the Mission 66 era. Although a good solution to the problem of constructing many build-

FIG. 6.5 The Quarry Visitor Center at Dinosaur National Monument, one of the best-known examples of Mission 66 structures built in a modernist architectural style. Photo by Burley Packwood, 2018. Wikipedia, Creative Commons Attribution–ShareAlike 4.0 International License.

ings throughout the park system, the modernist architecture of Mission 66 visitor centers has not always aged well (Allaback 2000). Rustic buildings built of stone and timbers improve with age as those materials age; if the clean concrete, steel, and glass of modernist buildings show fading or weathering, they can quickly look shabby. Many buildings have had to be modified over the years to improve their usability, but the unity of design of modernist buildings makes any alterations look glaringly out of place.

Dinosaur has what has become one of the most famous Mission 66 structures, the Quarry Visitor Center (Allaback 2000) (Fig. 6.5). This center was built to showcase excavations of dinosaur fossils on a cliff, which formed one wall of the structure. Unlike many Mission 66 visitor centers, this one was praised at the time, perhaps because it did not replace a rustic building. Unfortunately, it was built on unstable soil, a problem in many western parks, and foundation cracks appeared, requiring an eventual renovation.

Not all visitor centers were so loved. The circular Jackson Visitor Center at Paradise within Mount Rainier National Park was likened to a flying saucer and considered out of place high on the slope of the mountain. The roof also could not support the heavy snowfall of the alpine location; legend has it this was due to it having been designed by an architectural firm from tropical Hawaii. This visitor center was closed and demolished in 2008 and replaced by a more traditional rustic-style building.

But not all visitor centers are so elaborate; many small units have much simpler buildings. Some are even temporary, or so visitors may hope. North Dakota's Theodore Roosevelt National Park has a single-wide trailer for a visitor center (Fig. 6.6), though it does at least have a wheelchair ramp. It does not continue the tradition of rustic parkitecture; instead, it follows the Mission 66 style by imitating the kinds of buildings in widespread use outside the parks. It is an efficient solution to providing facilities with limited budgets, but it is surely not what visitors expected to be welcomed by after driving hundreds of miles.

FIG. 6.6   Theodore Roosevelt National Park (North Unit) Visitor Center, 2021. The building is a single-wide mobile home, one of the smaller visitor centers in the national park system.

Housing and Lodging

An important element of Mission 66 was new housing for park employees. Due to the remoteness and size of many parks, no housing was available for rent or purchase; accommodation had to be provided for employees within the park. But whatever housing existed was often old and substandard. In 1953 10 percent of all park workers lived in tents, and the frame buildings that existed were largely built in the 1920s or 1930s (Carr 2007). Many lacked heating, insulation, or even electric lighting. Replacing thousands of these substandard units was a priority of Mission 66, and a set of several standard house and apartment plans was drawn up by the NPS. These could be quickly and inexpensively built by local construction companies, and a total of 746 houses and 496 apartments were constructed during this program (Carr 2007).

In the 1950s the ranch-style house was the most commonly built house in the United States, and the NPS made heavy use of it. These houses were single-story and asymmetrical, with a low-pitched roof and wide eaves (Hess 2004b). They were adaptable to a wide variety of climates and building materials. The housing in many parks is still made up of this style, matching suburban neighborhoods built during this same period. These homes were not extravagant; three-bedroom houses had 1,260 square feet, while those with two bedrooms were no bigger than 1,080 square feet (slightly less than the size of the average American home in 1960 but half the size of a typical house in 2020). They were also built in areas out of sight of visitors, often hidden away near administrative or maintenance facilities. Despite living in a national park, residents of these homes did not usually enjoy much of a view.

New lodging was also built in some parks. Like the visitor centers and houses, lodging was based on new architectural styles. Fancy lodges were out of date; visitors of the 1950s preferred motels, where they could park next to their room. This form of lodging emerged along America's roadsides in the 1930s in response to the ever-increasing numbers of families vacationing by auto (Jakle, Sculle, and Rogers 1996). Motels were less pretentious than hotels and catered to motorists by putting parking right outside the guest rooms. They had a simple, often plain design, though some were based on themed architecture. A pool was often prominent out front, by the roadside, or next to an eye-catching sign. In the 1950s motels spread rapidly across America, and park planners looking to replace dilapidated lodging with modern facilities (minus the eye-catching sign and pool out front) appreciated their simple design. Jackson Lake Lodge (1955) at Grand Teton National Park was an early example and one in which the move from the rustic style was not widely appreciated. Adopting modern architecture for a lodge was quicker and less expensive; park planners also pointed out that while rustic styles might be relevant in a forested mountainous area with a strong rustic architectural tradition, they were not so in many other areas of the country that the park system had spread to by the 1950s (Carr 2007).

Changing Values

Mission 66 spawned a substantial backlash against park development, and many plans were dropped (Foresta 1984; Carr 2007). Yellowstone was one of the parks most affected by Mission 66 (Rydell and Culpin 2006; Carr 2007). Grant and Canyon Villages were built during this program to replace older areas too close to sensitive natural features. The public and many within the NPS were not enthusiastic about these new developments, whether due to their replacement of much-loved lodging or their new architectural styles. Canyon Village was infamous at the time as a symbol of all that was considered wrong about Mission 66. The village was centered on a large parking lot flanked by restaurants and stores, much like a suburban shopping center. The layout was a distant echo of the park village square concept pioneered decades earlier in the rustic era (McClelland 1998).

Small units, such as Devils Postpile, also saw their Mission 66 plans cut (Johnson 2013). These plans were sometimes substantial; tiny Rainbow Bridge National Monument in Utah was to be expanded and receive a visitor center, campground, trails, and residential and maintenance facilities, along with two dams and a drainage tunnel to keep the rising Lake Powell out of the monument, but none of this came about (Sproul 2001). The only facility that was ever built in the monument was a floating dock in 1965. But Alaska's Katmai National Monument, then the largest park, also had a never-implemented Mission 66 plan calling for roads, trails, campgrounds, and other facilities (Norris 1996a). It remains a wilderness park today.

Mission 66 was frequently criticized for a perceived emphasis on developing parks to bring in more tourists; the reality was quite different (Rydell and Culpin 2006; Carr 2007). More tourists were coming whether the parks were developed or not; the goal of Mission 66 was to better prepare parks for the ever-increasing numbers. Environmental organizations such as the Sierra Club were strongly against Mission 66, as they perceived it largely as a road-building program (Carr 2007). Others were entirely opposed to any kind of park development and worked to ensure that new parks remained undeveloped. The Wilderness Act was one consequence; this allowed Congress to designate wildernesses of at least five thousand acres in which no motorized vehicles were allowed. However, as noted below, wilderness has been surprisingly slow to appear within national park units.

In many small national monuments, Mission 66 provided the very first visitor facilities. Almost every building, road, trail, or picnic table you see at Tonto, Montezuma Castle, Wupatki, Sunset Crater, and Natural Bridges, among many others, is a result of the program. Without Mission 66, the national park system would look vastly different, though whether that is good or bad is still being debated.

Since Mission 66

Many parks have been added to the system since the days of Mission 66, and plans for their development often owed much to the program as well as its backlash. When the NPS began planning for Arizona's Fort Bowie National Historic Site it initially planned a standard Mission 66 package, perhaps even with an aerial tramway to a mountaintop overlook, designed to serve the expected crowds (Gomez 1984). Others within the agency protested this in favor of a minimalist approach, with almost no development beyond a trail and a small ranger station. It was expected that relatively few visitors would make the journey to the fort, and the experience of getting there would be part of the appeal. These views prevailed, and today the fort ruins are reached by a desert hike during which you will likely encounter no one else.

The vast Alaskan parks added in 1978 received little or no development, and the park headquarters might be in a nearby or not so near town. Among the new units added, only Lake Clark had its headquarters within the park. Kotzebue and King Salmon, two towns unknown outside of Alaska, are among the communities from which distant parks are administered. The headquarters for Gates of the Arctic is in Fairbanks, 170 air miles away.

In the contiguous states, Mojave National Preserve typifies this minimalist approach. Created in 1994 in the California desert, the large park has a visitor center, but it is not a new building; rather, it is in a renovated 1924 train depot at Kelso, near the center of the park (Fig. 6.7). The Mission Revival–style building once housed a lunch counter for railroad employ-

FIG. 6.7 Kelso Visitor Center at Mojave National Preserve, formerly a train station for the Union Pacific Railroad on its line between Los Angeles and Las Vegas. It is a nice example of Mission Revival architecture.
Photo by Pretzelpaws, 2004. Wikipedia, Creative Commons Attribution–ShareAlike 3.0 Unported License.

FIG. 6.8 Map of Mojave National Preserve, 2021. Aside from the Kelso Visitor Center and campgrounds, there are few facilities in the park. The headquarters is far to the west in the town of Barstow.

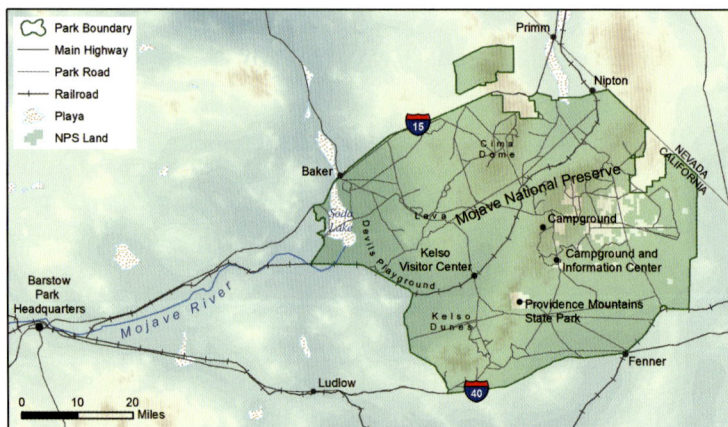

ees and had rooms upstairs for train crews on overnight runs. The lunch counter has been re-opened, and most of the building has been converted to a museum; outdoors, the grass and palm trees that made the depot an oasis remain a welcome sight after a long drive through the desert. The park headquarters is in Barstow, a drive of almost two hours to the west, in a non-descript office building well off the town's main street (Fig. 6.8). Park employees live there, in the smaller but closer town of Baker, or in even smaller desert communities ringing the preserve. The NPS did not build any roads in the park but simply used those that the railroad, miners, a few homesteaders, and the county highway department had constructed. No lodging was built in the preserve aside from two campgrounds; visitors are either self-sufficient or seek lodging in the small towns that ring the park. Considerable private land remains within the park, which is also crossed by a railroad and several power lines and pipelines. Visitors

FIG. 6.9   Craig Thomas Visitor Center in Grand Teton National Park, a striking example of postmodern architecture in the park system. Photo by Acroterion, 2007. Wikipedia, Creative Commons Attribution–ShareAlike 4.0 International License.

who enter the park expecting a central visitor complex and resort, as at nearby Death Valley and the Grand Canyon, will not find it, nor will the park roads lead them directly to the park's attractions.

Architectural styles have changed since the 1960s, with postmodernist architecture taking over as the leading style in the 1980s. This successor to the concrete-and-steel modernist style allows for a much wider range of materials and designs and draws inspiration from the past, which can include references to the rustic style, though with updated materials and construction methods. The 2008 Paradise Visitor Center at Mount Rainier is a good example, but perhaps most spectacular is the 2007 Craig Thomas Visitor Center in Grand Teton National Park (Fig. 6.9). This style will likely spread throughout the system in future years; the old rustic style, made out of logs and stone, will never return.

Changing Spatial Organization

Given how differently the CCC and Mission 66 developed parks, it is often easy to identify their imprint in different parks. It is not just architecture that differs but geography: the two programs often appeared in different locations.

Both programs had a large impact on Mesa Verde (D. Smith 2002). The CCC fought fires, built roads and trails, and constructed several rustic buildings on Chapin Mesa near the Spruce Tree ruin, one of the park's main attractions. Decades later, one of the primary goals of Mission 66 was to move development away from the ruins. The Far View Lodge and gas station opened in 1966, and the old Spruce Tree Lodge on Chapin Mesa shut down. The campground was likewise moved away from the ruins to a new location in Morefield Canyon. It was intended to move the park administration offices and employee residences off the mesa as well, but this did not happen. The Far View Visitor Center opened in 1973, replaced in 2012 by a new one at the park entrance. The park visitor heading south toward the ruins will be traveling back in time through earlier and earlier cultural landscapes, culminating in the Anasazi cliff ruins.

Petrified Forest National Park provides another example of changing internal spatial organization, with two distinct developed areas having been developed in each program. The original monument consisted of the south end of today's park, where most of the petrified wood can be found (Lubick 1996). In 1929 a ranger residence was constructed near Rainbow Forest, the first facilities in the park, and additional park buildings and employee houses were added later. They were built in a pueblo-inspired rustic style.

To the north of the monument, Route 66 carried a steady stream of highway traffic, with spectacular views of the Painted Desert, a vast expanse of colored lands stretching to the horizon. The privately owned Painted Desert Inn opened along the highway and provided meals and lodging, rivaling the Rainbow Forest Museum as a local tourist attraction. The NPS sought to expand northward and did so in 1932 with a boundary expansion and new road system (still in use). The NPS bought the Painted Desert Inn, but when Mission 66 arrived the park took the opportunity to build a substantial visitor complex nearby, alongside the newly built I-40, which had replaced Route 66. This visitor complex opened in 1962 and included a museum, park administration offices, a restaurant, and park housing. The restaurant and lodging in the Painted Desert Inn closed at the same time, ending overnight stays within the park.

The visitor complex has since become a Mission 66 landmark but unfortunately did not take into account soil conditions, and it has suffered severe deterioration. Portions or even the entire structure may have to be demolished and replaced. The employee housing within the park is not enough, and the park has had to obtain eleven additional housing units at an abandoned air force facility in a nearby town. The park headquarters have been moved out of the park as well.

Spatial Organization in Different Park Types

The internal structure of each national park unit has been carefully designed. When a new park unit is created there may be no facilities of any kind, perhaps even no roads. All facilities must be carefully designed in a long planning process. Resources must be protected, roads and visitor facilities built, and facilities for carrying out park operations developed. While parks are designed around resources that must be unique to warrant the creation of a park, visitor and operations facilities are heavily standardized. Most park units will have an elaborate entrance sign, road, visitor center with ample parking, trails and scenic overlooks, concessionaire facilities, and campgrounds. They may also have scenic roads, outlying ranger stations, bathrooms, and picnic areas. There will also be a largely unseen administration and operation infrastructure, with offices, housing, a maintenance yard with facilities for vehicle servicing, heavy equipment parking, plumbing, carpentry, and sign construction shops, a water and sewer system, electricity, communications facilities, schools, a library, garbage dumps, airports, and any other specialized services needed. Many of these may be located near each other in what is effectively a small town, though off-limits to tourists. The exact facilities needed vary depending on the size and specific type of park unit.

National Parks and Large Monuments

Large park units are spatially organized in several ways. The most basic division of the park's territory is between a developed frontcountry and undeveloped backcountry, some or all of which may be wilderness. The frontcountry contains the roads, buildings, and visitor services; this is the part of the park seen by park users either through the windshield or from scenic overlooks and short walks. It is patrolled by rangers, who may have to keep visitors and wildlife separated. In large units, there may be several developed areas with lodging, restaurants, water and sewer systems, maintenance areas, power lines or solar panels, and employee housing. Yellowstone has seven, Death Valley three, and the Grand Canyon two. Those at the Grand Canyon are separated by the chasm itself and function independently; few visitors are likely to see both during the same visit. There may be historic districts where the park's early buildings are preserved.

Backcountry is that part of the park that has not been developed and that few visitors see. It will likely have hiking trails and primitive campsites featuring only fire pits and outhouses. Much of the Grand Canyon and high sierra of Yosemite, as well as Sequoia and Kings Canyon National Parks, is backcountry. Rangers may infrequently patrol or be based at backcountry ranger stations. Visiting requires preparation by visitors, who cannot expect assistance if problems arise. Different rules may apply in the backcountry. The Saline Valley Warm Springs within Death Valley National Park have long been famous, in part because many bathers don't bother wearing clothes. Nudity is illegal here, but the valley's remoteness has meant the law has never been enforced.

At Arches most of the park away from roads and trails is backcountry, but the backcountry should also include the background behind Delicate Arch. The usual view faces southeast toward the La Sal Mountains, in Manti–La Sal National Forest. Though they are an integral part of the experience of Delicate Arch, they are not under NPS control. At Grand Teton the spectacular mountain range is the backcountry, and the valley is the frontcountry. This is not an accident; the valley floor was pursued by the NPS through a lengthy struggle to provide the foreground to views of the spectacular mountains.

The distinction between frontcountry and backcountry may be drawn into the boundaries of units. In 1968 several park units were created in the North Cascade Mountains of Washington. Because of long-running debates over whether a park in this area should be a traditional or "wilderness" park, several different units were created as a park complex under joint management (Louter 1998; Danner 2017). North Cascades National Park contains the scenery and was planned as a largely undeveloped area in two sections divided by Ross Lake National Recreation Area, which contains a reservoir, highway, campgrounds, and other facilities, while Lake Chelan National Recreation Area contains another reservoir with both scenery and limited development. While none of the three units has designated wilderness, the boundaries between frontcountry and

backcountry are drawn in the boundaries of the three units. When parks are expanded, as has happened at places like Death Valley and Great Sand Dunes, it is usually only the backcountry that is expanded (Geary 2016). Newly added sections will have few if any roads, and no visitor facilities are likely to be built in these areas.

Alaskan parks are in a special category for their size, remoteness, low visitation, and very limited facilities. Many Alaska parks today are like parks in the lower forty-eight states in the nineteenth century, with vast open spaces, few roads, and limited tourist facilities. Aniakchak, Bering Land Bridge, Cape Krusenstern, Gates of the Arctic, Kobuk Valley, Noatak, Yukon–Charley Rivers, Lake Clark, and Katmai have few or no facilities. They also receive few visitors, who must be self-sufficient. They do, however, still have cultural landscapes, consisting of place-names, favored landing spots, camping or fishing locations, and the traditional ways of Native Alaskans. Wrangell–St. Elias and Denali have some development allowing for independent exploration, but much less than visitors to other parks would expect. Denali is also unusual among park units in that many visitors arrive by train and see the park in tour buses. Visitor facilities are designed for huge crowds; enormous bathrooms are filled with crowds of people when buses and trains arrive but stand empty much of the rest of the time. Alaskan parks also tend to have far smaller budgets and fewer employees.

Island parks are similar. The Channel Islands are off the coast of Southern California near Los Angeles but have seen little development, and the visitor center is located on the mainland. Isle Royale in Lake Superior has two visitor centers with a range of services, but the island is otherwise undeveloped. It provides an Alaskan-style adventure within the lower forty-eight states. Isle Royale did receive a Mission 66 park headquarters but on tiny Mott Island, well removed from visitor facilities. Here park employees have their own secluded island on which to work and live.

Some of the backcountry may also be wilderness. Wilderness areas are designated by Congress and can be found within parklands, national forests, wildlife refuges, and BLM

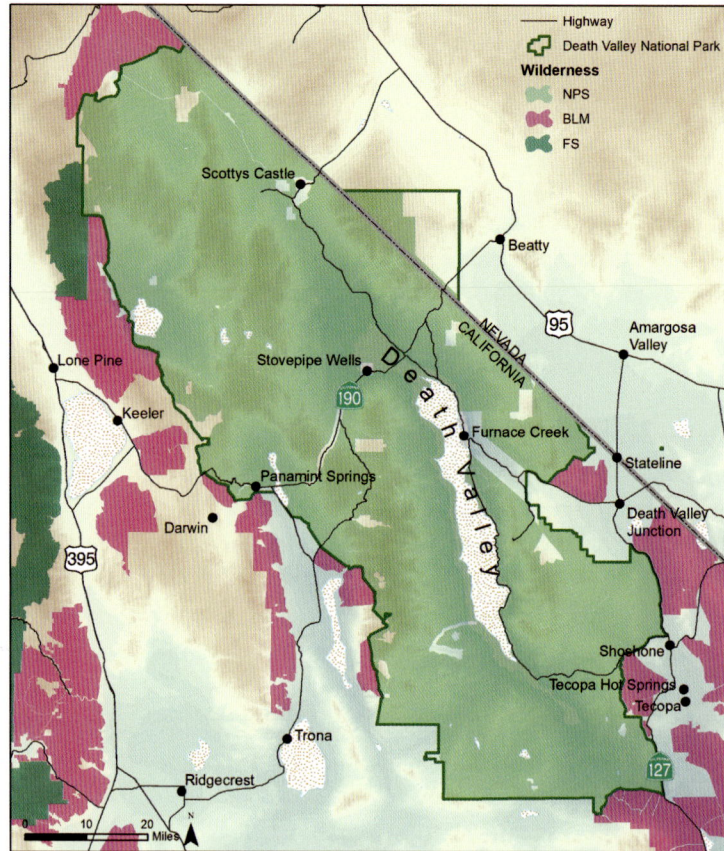

FIG. 6.10   Map of Death Valley wilderness, 2021. About 93 percent of the park is wilderness, excluding only roads and several developed areas or abandoned mines. Additional wilderness can be found just outside the park on other federal lands.

lands. They are administered by whichever agency controls the land. While wilderness areas seem like a natural fit with national parks, the NPS has not been completely supportive of them. Management may feel that a wilderness designation is unnecessary, as much of a park's territory may already be managed to a similar condition (Sellars 1997). They may also see wilderness designation as reducing the flexibility of park managers to deal with changing conditions. For that reason, wilderness areas are absent from many parks; Yellowstone is noteworthy for having none. Big Bend, Dinosaur, Glacier, Grand Teton, Great Smoky Mountains, and the Grand Canyon are other large wilderness-free parks.

Wilderness has been created in other parks. Death Valley represents one extreme; 91 percent of the park was designated as wilderness in 1994 (Fig. 6.10). Except for a few developed areas and former mining areas, everything beyond fifty feet of a road or power-line corridor is wilderness. Wilderness managed by the BLM and the USFS exists outside the park boundary, creating a larger wilderness landscape. Yosemite, Sequoia–Kings Canyon, Organ Pipe Cactus, Rocky Mountain, and Zion are similar. However, most parks that have wilderness have only one or a few sections. Lake Mead and Petrified Forest each has several separate wildernesses located in undeveloped corners of the unit.

A further overlay exists in most parks in the form of areas closed to the public. These may be in frontcountry or backcountry and serve to keep visitors away from sensitive resources such as archaeological sites, caves, springs needed by wildlife, and dangerous places. Many facilities needed for park maintenance, employee housing, and utilities such as water and sewer systems are also off-limits to tourists. These are usually well screened from the road and do not appear on park maps, and most visitors will never even know they exist. Restrictions may be posted in a superintendent's compendium, which also provides a set of rules for visiting the park. In Death Valley housing, water, sewer, and electric systems, caves, a Native American cemetery, and several recent mining areas are off-limits to tourists. The Lower Vine Ranch near Scotty's Castle is open only to guided tours. Copper Canyon, a large drainage in the Black Mountains, is also closed to protect a collection of fossilized footprints of ice age animals. This canyon is also protected by hidden cameras that photograph unauthorized visitors; several people have been arrested for chipping off rocks with footprints. At White Sands the western half of the park is within a joint NPS-military zone and off-limits to the public.

Large Recreation Areas

National recreation areas were designed to an even greater extent than many other units in the park system. Not only were roads and visitor services planned, but the primary feature of these parks, a reservoir, was also designed and constructed. The Bureau of Reclamation was responsible for the design and construction of most of these dams but left recreation to the NPS. The NPS provided a standard set of recreational facilities for reservoir units, including boat ramps, marinas, beaches, and campgrounds, and these can be distributed at appropriate locations around the lake, usually within sheltered coves or bays. A road system will be constructed to reach these scattered developed areas, with developed areas located on spur roads leading down to the lakeshore (Fig. 6.11). While many national recreation areas do have spectacular scenery, this is not the emphasis; if any wilderness is present, it will likely be in small isolated sections that do not include the reservoir. The reservoir itself may be subdivided into zones, including ones where no motorized boats or wakes are allowed.

Lake Mead National Recreation Area provided the template for this design, with several facilities built by the CCC and others added postwar. Six major developed areas were spaced at regular intervals along the lakeshore, with room for many others if needed. When the unit was developed it was thought to be remote, and a lodge was built for guests staying for a prolonged period. In the postwar era, the recreation area has received a tremendous number

of short visits from nearby Las Vegas, as well as those camping by the lakeshore or staying in houseboats, and the lodge was closed.

National recreation areas with large dams are designed to store water and produce tremendous amounts of electricity; the national park system generates more than 12 percent of the nation's hydroelectricity (though Lake Mead National Recreation Area buys its electricity from the local power company). This requires a massive infrastructure of substations and forests of power lines through the national recreation area to connect the dam with outside markets, creating an industrial landscape rarely seen in other national park units. These facilities are not readily compatible with tourism, and security checkpoints and mobility restrictions may exist.

A complication with reservoirs is that water levels will vary due to drought and the needs of electricity consumers, flood control, and agriculture. Changing water levels may leave beaches and boat ramps high and dry, as is the case at Lake Mead National Recreation Area, which has seen continual decreases in lake level since the 1990s due to prolonged drought. The shoreline has receded over a mile in some locations. The park has had to extend boat ramps or close marinas altogether, and park visitors must be wary of out-of-date maps. Lake Meredith in Texas has been nearly dry in recent years, with the shoreline having receded almost six miles and three boat ramps closed. The number of visitors has varied along with the water level, and the park has created special events to try and bring people back (Garcia 2018).

Seashores

Seashores are linear, focused on a stretch of coastline. Along the Atlantic and Gulf Coasts, seashores are frequently located on barrier islands, paralleling the coast. They may have a road running their entire length giving access to uninterrupted beaches, with bridges or ferries connecting them to the mainland at several locations. They are often quite close to major population centers and will make up the oceanfront for that city (as with Pensacola and Gulf Islands National Seashores).

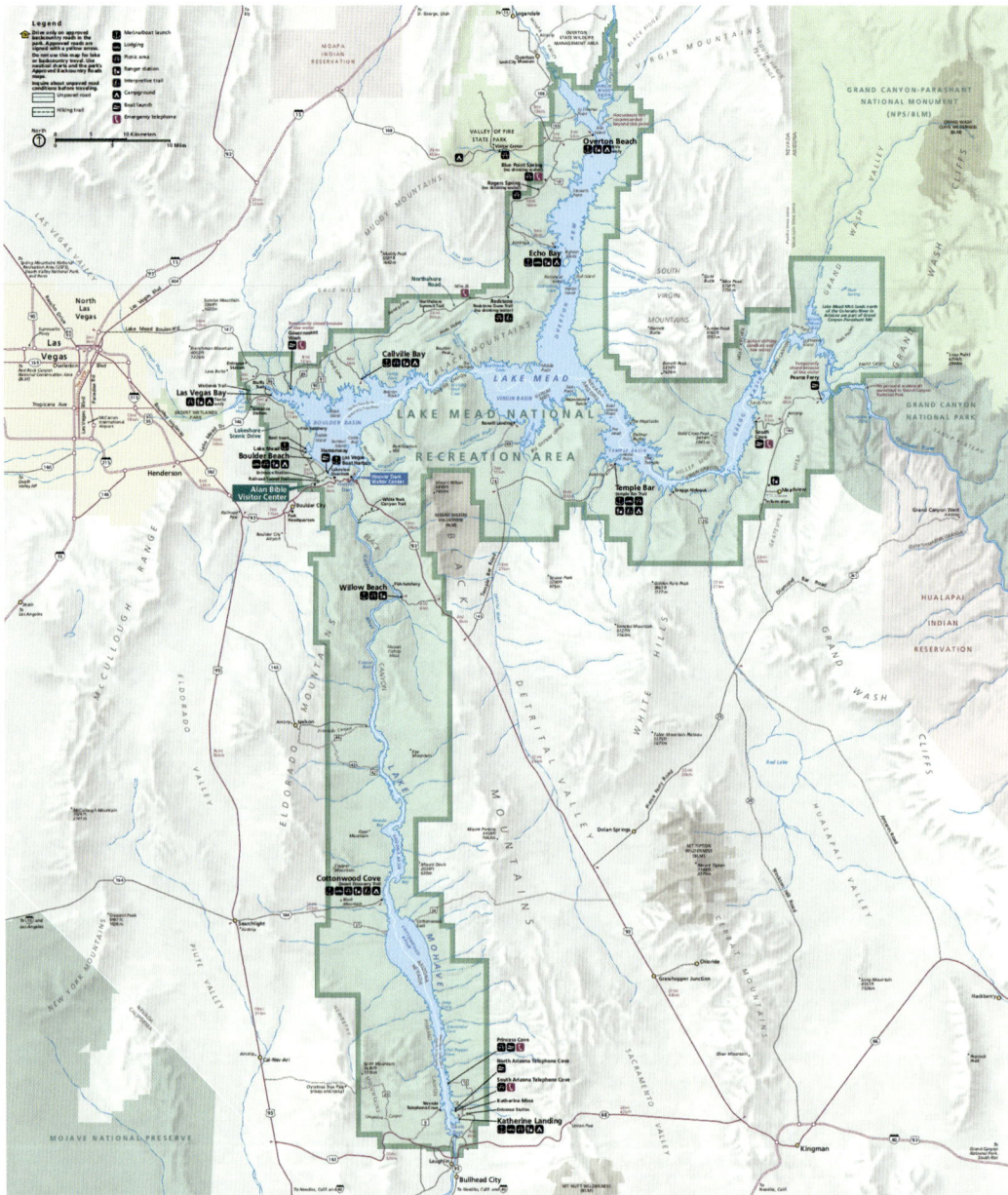

FIG. 6.11 Map of Lake Mead National Recreation Area, 2021. The lakes are the central features, with roads and facilities arrayed around them. The park has many similar developed areas, each serving visitors from different directions. NPS.

Seashores are recreational parks and will have campgrounds, boat ramps, bathrooms, and other developments, but these will not be as concentrated as they are within a national recreation area. Instead, most visitors will spread out wherever they please on the seashore's vast beaches. These are not favorable environments for building roads and buildings; storms and especially hurricanes regularly destroy them, and barrier islands are by nature impermanent. In the off-season they are little used and can have much the same emptiness and feeling of remoteness as large western parks.

Small Parks and National Monuments

Beginning in 1906, national monuments were created to preserve scientific features such as ruins, caves, and rock formations. They were thought of not as tourist attractions but as scientific preserves, and for many years the NPS

FIG. 6.12   Moses Carver house at George Washington Carver National Monument, Missouri. This small monument contains a visitor center and historic trail leading past several historic sites. Photo by RuggyBearLA, 2020. Wikipedia, Creative Commons Attribution–ShareAlike 2.0 Generic License.

largely ignored them as second-class units with limited visitor appeal (Rothman 1994a, 1994b). Facilities were gradually added, but it was not until the Mission 66 era that small monuments such as Montezuma Castle and Tonto received much in the way of visitor amenities (Protas 2002). Today these units often have an entrance road, a visitor center, one or two trails to the attraction, and perhaps a picnic area but no other services. Most are day-use only, with no camping or lodging. Arizona's Tonto National Monument is a good example; the 1.75-square-mile site has these standard facilities and is open from 8:00 a.m. to 5:00 p.m. during the summer months, though the longer and more strenuous of the monument's two trails closes at 1:00 p.m. There is little or no distinction possible between frontcountry and backcountry in such a unit, and wilderness is absent.

George Washington Carver National Monument in southwestern Missouri has a similar layout: all visitors enter on a single entrance road, pass through a post–Mission 66 visitor center with exhibits describing what the visitor is about to see, then walk down a trail to see several sites that interpret the life of Carver (Fig. 6.12). Visitors return by a different route to the visitor center and leave.

One set of small monuments is cave units, for which the primary attraction is underground. Opening the cave to tourism was implicit in the creation of these park units. (No park units were created to preserve an undeveloped cave.) Opening caves to visitors requires considerable effort to expand natural entrances or blast artificial ones, build walkways and stairs, and install lighting. This development can be thought of as the unit's frontcountry; undeveloped passages and the aboveground portion of the park are its backcountry.

In most cave units, there is relatively little interpretation of the aboveground environment, and hiking may be limited in order to keep the public away from cave entrances. Carlsbad Caverns was one of the most heavily developed caves in the system, with a long tour route leading to a great room containing an underground restaurant and an elevator to return people to the surface. Such extensive development has fallen out of favor both underground and aboveground: newer caves opened to the public at Carlsbad Caverns have been left undeveloped, and tours in these caves require visitors to bring flashlights and traverse rough, uneven, and slippery surfaces. Caves are a bit different from other units in that they are still being explored and mapped, though newly discovered areas will never be opened to the public except as a backcountry experience. These may be the only parks where the backcountry continues to grow.

Buck Island Reef and Virgin Islands Coral Reef are almost entirely water, but even they have different zones for boat anchorages (or where anchoring is prohibited) and designated diving sites. They also offer what might be the purest wilderness in their underwater expanse, which ensures that visitors will not long remain.

Parkways

The Blue Ridge and Natchez Trace Parkways are in a class by themselves, though the Blue Ridge shares many characteristics with Shenandoah. The parkways are carefully designed scenic roads within a narrow strip of parkland. The road itself is the focus of the park, and there is no central attraction or feature, as there is at a monument. Scenic views are provided, usually alternating to the left and right, with occasional pullouts for an attraction. The NPS had a goal of obtaining one hundred acres of land per mile of road, with a scenic easement on another fifty acres, producing a right-of-way that averaged 825 feet in width (T. Davis 2016). Every thirty or so miles this would expand in the form of "bulges," where restrooms, restaurants, lodging, and other attractions could be located. Anyone accustomed to long highway journeys understands and appreciates this regular spacing, which replicates that found on interstate highways (but without the trucks). Each parkway has several visitor centers at strategic locations, with the administrative offices halfway along the road. There is no backcountry or wilderness, and parkways also have far more entrances than other park units.

Although the Blue Ridge Parkway has many rustic design elements, most of the mileage of these roads was built in the postwar era (T. Davis 2016). Later construction became more modern in appearance, culminating in each road's signature bridge. The Natchez Trace includes a spectacular double arch bridge completed in 1994, while the Blue Ridge's signature feature is the Linn Cove Viaduct, which carries the road along the shoulder of North Carolina's Grandfather Mountain. Views of this bridge have become ubiquitous on websites, calendars, and other publications since it was completed in 1987. The graceful beauty of both bridges can only be appreciated from a viewpoint away from the road, thoughtfully supplied by the NPS.

Small Urban Units

National historic sites, many national monuments, and some national memorials are usually small urban sites often centered on a building associated with a historic site or person. Often no NPS parking lot exists. These parks are all frontcountry. Visitor facilities are likely to consist of little more than a museum and are unlikely to be major destinations in these cities; instead, they are one of many storefronts competing for your attention and time. Some may be well off the beaten path, left behind by urban development. Klondike Gold Rush National Historical Park includes commercial buildings in use during the 1890s mining boom, one in Seattle and the other in Skagway, Alaska. These parks include a museum interpreting the mining boom and each city's role in it. Each is open from 9:00 a.m. to 5:00 p.m. and might be confused with a local history museum were it not for the NPS logo, signage, and uniforms. Birmingham Civil Rights and Stonewall are different, encompassing city streets, sidewalks, and a park, and each has only one NPS-owned building. The Birmingham site does, however, include many privately owned buildings, and whether the agency will seek to acquire them remains to be decided.

You enjoy visiting national parks, but would you want one in your neighborhood? Parks in residential neighborhoods are hard to develop and operate; they need parking and attract large numbers of people that nearby residents are not always going to appreciate. Eugene O'Neill National Historic Site offers one development possibility, with visitors accommodated at an off-site visitor center and then brought by bus to the house. At Harry S Truman National Historic Site (the NPS spells it without a period after the S) in Independence, Missouri, the NPS was able to locate a visitor center in a vacant building five blocks down the street in the town's commercial district, allowing the home to remain within an intact residential neighborhood. Cities are constantly changing environments, and if you wait long enough, opportunities will arise. William Howard Taft lived in what was once a fashionable neighborhood in Cincinnati but has long since emptied out and become more commercial. The NPS preserved the house the president grew up in and was able to buy adjacent lots, demolish the homes located there, and build a small museum and parking lot next door to the house. Medgar and Myrlie Evers Home National Monument consists of a small suburban house on a 0.15-acre lot on a quiet residential street in Jackson, Mississippi. The new park will not be an easy one to develop, with no room for parking or a visitor center and long-term residents who do not want to see their neighborhood disrupted.

The National Mall and Its Memorials

Washington, D.C., is built within and around national park units; in no other city do park units provide the basic layout or structure. It is one of the most powerful cultural landscapes in the country, instantly recognizable to all Americans. The existence of the National Mall dates to Peter L'Enfant's 1791 plan for the city of Washington, which specified a four-hundred-foot-wide Grand Avenue extending westward from the Capitol and intersecting another running south from the President's Palace (or White House) (Benton-Short 2016). This formal plan was derived in part from that of Paris, with traffic circles on hills connected by main avenues. Since the 1930s it has been perceived as comprising several main elements: the Capitol, the Washington Monument and Lincoln Memorial along the primary east–west axis, with the White House and Jefferson Memorial at right angles.

The original vision for the National Mall has been altered and continues to change. The mall has never been complete, and the NPS is only one of many agencies that have their own plans for it. Much of it was covered by "temporary" office buildings left over from World War II as late as the 1970s. The Smithsonian Institution and the National Gallery of Art have large and well-known buildings rivaling NPS sites in visitor appeal. The completion of the Vietnam Memorial in 1982 unleashed a memorial-building boom that shows little signs of abating (Benton-Short 2016). Although these are units of the national park system, the NPS does not design, build, or pay for the construction of memorials. Rather, Congress authorizes a new monument and a private foundation that will fund it. A lengthy process of selecting a design and location ensues, with construction not allowed to start until the foundation has raised the necessary money (which will likely be well in excess of $100 million). Each new memorial has reduced the open space of the mall, and locations for some, especially the World War II Memorial, have been controversial. Battles over future memorials will likely last longer than the wars they commemorate.

Battlefields

National battlefields and military parks may be large but are quite different from other large park units in that they are commemorative rather than scenic in nature. Battlefield sites have been marked with a variety of statues, columns, cannons, plaques, or other markers commemorating a specific military unit that served at that location or where events in the battle took place (T. Smith 2006). Unlike a scenic park, the signage at a battlefield refers to events that may have unfolded over hours or days. The visitor must mentally reconstruct what happened at each location and how this place fits into the battle's timeline. These parks are usually experienced via a scenic loop road, though this may not directly correspond to the sequence of events, as at Shiloh.

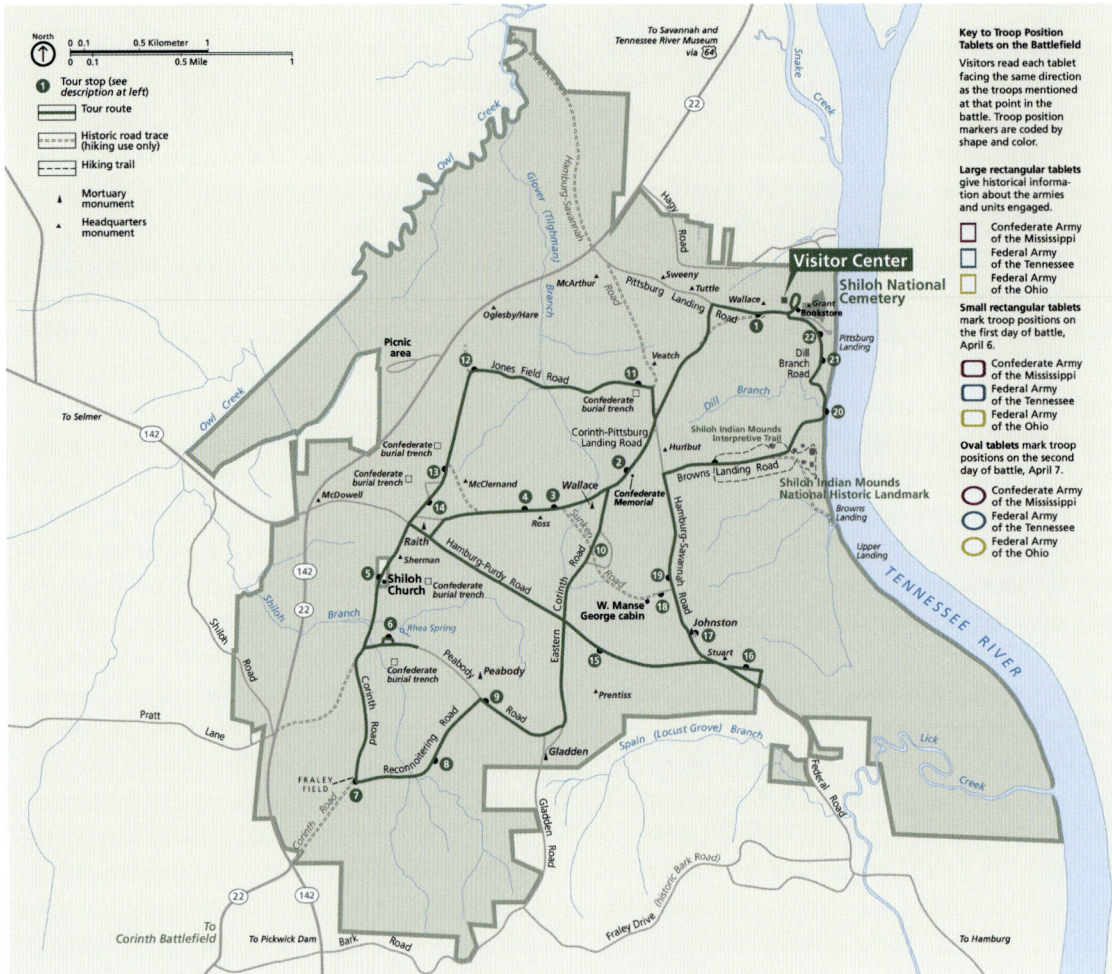

FIG. 6.13   Map of Shiloh National Military Park, 2021. The green line and numbered circles correspond to the battlefield driving tour. NPS.

Shiloh National Military Park has a convoluted road tour that doubles back on itself in a maze of roads and requires the viewer not just to read interpretive signs but also to mentally compute what area, direction, and moment of time in the battle they are looking at (Fig. 6.13). The commemorative columns, statues, and cannons may detract from the viewer's understanding, though they have now become part of the park's historic and cultural landscape. And within this, the visitor will come across a Native American mound complex that is completely unrelated to the battle but worthy of preservation in its own right. There is also a battlefield museum, but few visitors are aware of it.

Another difference from large scenic parks is that the landscape present at the time of the battle has been reconstructed to the extent possible. In eastern states, which account for al-most all battlefield sites, this means cutting trees from areas where they were not present and preserving buildings that were known to have been there. This differs from the goal in other parks of re-creating the landscape before white settlers arrived. These differences from scenic parks may make battlefield sites much less easy to navigate or understand for the viewer accustomed to scenic parks. You will have to spend considerable time studying the park before your visit to fully understand what you are looking at.

A subset of military parks is forts, of which there are two very different kinds. Eastern forts are massive brick coastal structures, as seen at Fort Jefferson (Dry Tortugas), Fort Pulaski, Fort Sumter, Fort Monroe, and old Fort Wood underneath the Statue of Liberty, among many others (E. Lewis 1970). They are in or near cities and can be quite crowded. Western frontier forts have no fortifications and look more like small towns, with officers' quarters facing barracks across an open parade ground. These can be seen at Forts Union, Bowie, Scott, and Larned, among others, but Fort Yellowstone is perhaps the most heavily visited of all. They are rural and peaceful places.

Other Units

The higher the elevation, the more severe the climate and the shorter the tourist season. Many mountain parks have visitor facilities defined by altitude. In Alaskan parks, which contain the tallest peaks in North America, development is at low elevations. Those park units with high-elevation roads or visitor facilities are at much lower latitudes, where those elevations have more pleasant conditions. The central Rockies, the scattered peaks of the Great Basin, the parks of the Colorado Plateau, and Hawaii have the highest roads and facilities. Hawaii is in the tropics and little affected by seasons, but those in the Colorado Plateau, the Great Basin, and the western mountains can be extreme. Parks have short seasons and may close entirely in the winter. Mission 66 had a limited impact at Crater Lake because the high altitude and very high snowfall made for a short summer season. It was not considered worthwhile to develop it for tourism (Harmon 2002). Devils Postpile and Cedar Breaks, which shut down in the winter, were also skipped over.

Just as the north face of a mountain is harder to climb due to colder and icier conditions, visitor facilities are likely to be on the south side of mountains, as at Mount Rainier, Crater Lake, and even Mount Rushmore. If you go to the end of the Denali road and catch a glimpse of the great one, you will be looking south, but from seventeen miles away and sixteen thousand feet lower, an equivalent distance at Mount Rainier would put you far outside the park.

Finally, several parks exist as multiple units. This is common for small archaeological units that contain several sites on small plots of land. In such units, one unit will be the primary unit, as at Hovenweep's Square Tower Unit with a visitor center and campground,

and outlying units may have few or no facilities or public access. Salinas Pueblo Missions National Monument in central New Mexico is another example, with a visitor center in the town of Mountainair and the monument's three seventeenth-century Spanish mission ruins all within twenty-five miles. The scattered units and visitor center function as a single park, with the state highway department furnishing the roads and businesses in Mountainair providing services. It is a contemporary form of the "Antietam model" for parks, with only those pieces of land necessary included in the park. It can succeed because Salinas is an archaeological park, and the NPS does not need to purchase vast scenic views.

In other cases, facilities may be duplicated, as with the East and West Units of Saguaro, each with its own visitor center, trails, roads, and scenic and historical attractions. For all intents and purposes, there are two Saguaro National Parks. Klondike Gold Rush National Historical Park is similar, split between Seattle, Washington, and Skagway, Alaska. The Grand Canyon is one unit, but the North and South Rim visitor services are 215 road miles apart. They also function as separate units, with few of the canyon's visitors making it to the North Rim.

Conclusions

All national park units have a similar design style achieved through the implementation of standards for signs, roads, architecture, interpretation, and perhaps especially brochures and maps. The Harpers Ferry Service Center is located within Harpers Ferry National Historical Park and handles cartography and graphic design. This center provides cartographic support for all national park units, using the standardized Unigrid brochure format and cartographic styles. The park maps and brochures familiar to every visitor are designed here. (Some of their maps appear in this book.)

Other government land agencies have standardized approaches as well. The national park system provided the model for their development. Many national forests were developed by the CCC in the same style as parks, and the Forest Service later launched a program similar to

Mission 66 called Operation Outdoors to build new roads, campgrounds, and visitor centers (Carr 2007). The agency with the most standardized facilities does, however, have a different look: military bases. Regardless of where they may be in the United States or foreign countries, military bases often have the look of suburban neighborhoods, almost invariably with a fast-food restaurant and a large Walmart-like post exchange.

That many park units also have very similar layouts is also no coincidence. While Americans love to express their loathing of homogeneous homes, shopping centers, restaurants, and motels, the standardized appearance of parks is viewed positively. The architecture and cultural landscapes of national park units are often appreciated (Kaiser 1997; Carr 1998, 2007; McClelland 1998; Allaback 2000; Davis and Davis 2011; Walklet 2004). Guidebooks have been written about these elements of western national parks (Kaiser 2002, 2003; Davis and Davis 2011). Old rustic lodges are a popular attraction for many visitors (Scott and Scott 2012). Given the vast range of conditions and amenities offered at these lodges, a guide is useful; many park accommodations lack television or even telephones, a blessing for many but perhaps a shock to others expecting far more for the high prices.

Despite the public's awareness of and appreciation for park structures, it should also be noted that there are no national park units dedicated to architects, a strange omission, given the existence of many important architects in the nation's history. Falling Water, Frank Lloyd Wright's best-known masterpiece, is a private museum in Pennsylvania. However, memorials to some architects remain within the park system. Richard Neutra designed many houses, churches, and schools, as well as the Mission 66 Petrified Forest Community Complex and Get-tysburg Cyclorama visitor center. This latter building was demolished in 2013.

The landscape architects who designed early parks have received some recognition within it. Frederick Law Olmsted Sr., who designed Central Park, is commemorated by a national historic site in Massachusetts. His son, Frederick Law Olmsted Jr., was involved with the early development of Yosemite, Acadia, Everglades, and many NPS sites within Washington, D.C. The Olmsted Point overlook in Yosemite is named after him, but perhaps his most spectacular legacy is the enormous roof built to shelter Casa Grande Ruins National Monument in Arizona.

With the passage of time, the modernist architecture of the Mission 66 years has begun to be appreciated and is increasingly regarded as historic and worthy of preservation (Hine 1990; Hess 2004a, 2004b). Many modernist commercial buildings along the roadside have long since been demolished or altered to keep up with today's styles; some of the best remaining examples of preserved modernist architecture are in national park units. Dinosaur's iconic Quarry Visitor Center was stabilized rather than being demolished and replaced. Even the once-hated Canyon Village at Yellowstone is being preserved and restored to its former glory. Perhaps Richard Neutra and other modernist architects will someday have their own units.

An important limitation of Mission 66 (and earlier efforts to carefully design the interior of national parks) is that it ignored everything outside the boundaries of a park. Decades of land use and population growth outside parks have made this perspective untenable; park planning must now be coordinated with local land use in order to protect resources. Issues arising from growth outside park boundaries are discussed in a later chapter.

SEVEN

Transportation and National Parks

How do people get to parks, and how do they move around within them? Transportation is a fundamental and critical part of the national park system. Getting people to parks was the first major challenge for the early park system (Kraft and Chappell 1999). For many parks, this is still a challenge, though largely one of cost and time rather than physical capability. The means of transport used to get to a large park like Yellowstone or Yosemite has changed tremendously over time, though many of the spatial patterns each means of transport created have endured. Trains brought the first great wave of tourists to parks, followed by cars, with airplanes and boats connecting us to the farthest parks and those without roads.

With improving transportation have come more visitors and pressure on parks to develop the infrastructure necessary to handle these volumes. The relationship between transport improvements and visitation growth is not accidental. The National Park Service worked hard to open parks up to the public, especially to the middle class. Railroads and then highways were central to these efforts. It was the combination of the twentieth century's automobile with the national park idea from the nineteenth century that truly created the national parks that we know and love today (Louter 2006).

Transport, especially roads, is part of our experience of the national park system. While parks and roads may seem like opposite concepts, they are in fact closely linked, and "for many people, what they see from the road *is* the national park experience" (T. Davis 2016, 3). Our park system depends on roads: they helped democratize parks, and without the car,

parks might have remained the preserve of the elite.

This chapter discusses the connections between transportation and national parks, the ways that people get to parks, the options for moving around within parks, and how these interact with visitation, economic activities, and the transport problems parks face.

Transportation and the Geography of National Parks

The Railroad Era

The success of the national park idea was not possible without the enthusiastic partnership with railroads (Valliere et al. 2002). Yellowstone was created only three years after the first transcontinental railroad was completed, connecting the Atlantic and Pacific coasts (an event later commemorated at Golden Spike National Historical Park in Utah). More western railroads quickly appeared, following closely the creation of other western parks. Many parks saw rails and passengers appear not long after they were created. This was not a coincidence; Stephen Mather, the first NPS director, worked tirelessly to persuade railroad executives to support the national parks (Waite 2006). Railroads saw tourists as a new source of profits and played critical roles in both establishing and developing the national park system (Runte 2011).

Yellowstone is the best-known example of this partnership. Travel to and within the parks was difficult, costly, and time-consuming in the nineteenth century. Travel was conducted via various modes of transportation such as horseback, wagon, and train (Heacox 2001). The 1871 Hayden party traveled by train to Corinne,

FIG. 7.1   A stagecoach carrying tourists in early Yellowstone. NPS.

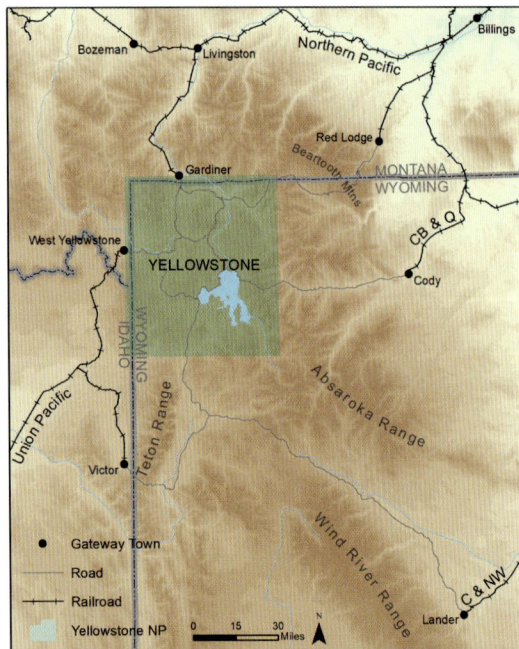

FIG. 7.2   Map of all railroad lines built to Yellowstone National Park. These once provided the principal means of reaching Yellowstone, giving travelers many options.

Utah, before switching to wagons for a roughly four-hundred-mile trip to Bozeman, Montana, and then rode horses the rest of the way to the Yellowstone plateau (Waite 2006). Few tourists would put up with such a journey, and this situation did not last long (Fig. 7.1). Within a decade the first of several railroads had been built near the park, providing a much faster and more comfortable journey (Shankland 1951; Waite 2006; Runte 2011).

There are five entrances to Yellowstone National Park, and each was served by a railroad (Fig. 7.2). The Northern Pacific Railway provided the first service to the Yellowstone area when it opened a line to Livingston, Montana, in 1882. This town was still fifty miles north of the park boundary, but by 1903 trains could be taken all the way to Gardiner, just outside the north entrance (Fig. 7.3). The railroad advertised itself as the Yellowstone Park Line and published its Wonderland series of colorful brochures and magazines from 1886 to 1906. A stone entrance arch was built in 1903 to give visitors arriving by train a memorable entrance to the park.

No. 183   NORTHERN PACIFIC DEPOT—GARDINER MONT.   HAYNES-PHOTO.

FIG. 7.3   Northern Pacific Railway depot at Gardiner, Montana. The entrance to Yellowstone National Park is adjacent, marked by the Roosevelt Arch. NPS.

FIG. 7.4   Union Pacific Railroad advertisement for Yellowstone National Park, 1921. Railroads were eager to attract customers by showing off the wonders that could be reached via their rails. NPS.

In 1898 the Chicago, Burlington and Quincy Railroad arrived in Cody, Wyoming. From here, stage service provided access to the east entrance of the park. The Union Pacific Railroad provided service to the town of West Yellowstone, Montana, in 1907, just outside the west entrance to the park. This quickly became the most popular route to enter the park and remains so today. Almost 48 percent (twenty-three thousand or more) of visitors arriving by train from 1920 to 1925 came via this entrance (Waite 2006). Gardiner was the second most popular entrance (35 percent of visitors arriving by train), and 16 percent of visitors arrived at the east entrance. A small number came via the south entrance after the Chicago and North Western Railway provided service to Lander, Wyoming, in 1922. The Union Pacific also served this entrance from 1929 via Victor, Idaho (Fig. 7.4). The park's northeast entrance also had service on the Northern Pacific Railway via Red Lodge, Montana, though this accounted for less than 1 percent of visitors.

Hot Springs National Park in Arkansas is another example. In 1875 the twenty-six-mile

FIG. 7.5 Malvern, Arkansas, train station in 1876. This station served nearby Hot Springs National Park, a major tourist destination of the era. Authors' collection.

Hot Springs Railroad, or Diamond Jo Line, was built to bring visitors comfortably to the baths at Hot Springs (Butler 2007). Partnering with the St. Louis, Iron Mountain and Southern Railway company, the Diamond Jo Line also attracted tourists to Hot Springs (Fig. 7.5). The company not only expanded its lines to St. Louis to connect with other railroads but also became the first to produce a booklet advertising for Hot Springs with descriptions and photographs of famous landmarks, hotels, and bathhouses to increase its popularity (Butler 2007).

Likewise, the Southern Pacific Railroad lobbied Congress for the creation of Sequoia National Park in 1890 (Kraft and Chappell 1999). In 1893 the Great Northern Railroad completed a main line from the Great Lakes to the Pacific coast, passing just south of what would become Glacier National Park. The Great Northern promoted visitation and constructed several lodging facilities, including the Many Glacier Hotel, still in operation.

Similarly, the Atchison, Topeka and Santa Fe Railway worked hard to make the Grand Canyon a popular attraction and eventually a national park. The railroad completed a rail line from Williams, Arizona (on the railroad's main line), north to the South Rim of the canyon in

1901, after which the railroad built El Tovar luxury lodge directly on the rim, adjacent to the train station, to accommodate the visitors the railroad brought to the canyon (Richmond 1995; Butler 2007).

These examples show that railroads not only brought tourists to parks but also helped publicize and develop them. At smaller parks and monuments the influence of railroads could be even greater. Zion, Cedar Breaks, and Bryce Canyon were small and isolated national monuments in southwestern Utah with little development. The first NPS director, Stephen Mather, tirelessly persuaded the president of the Union Pacific Railroad to develop the parks. In the 1920s the Union Pacific built a spur to Cedar City, Utah, and set up a company to build lodging at these parks and to bus tourists between them and on to the North Rim of the Grand Canyon in Arizona (Fig. 7.6). There was a long drive between Zion and the North Rim, and Mather went looking for an intermediate stop. A pioneer Mormon settlement at Pipe Spring, Arizona, fit his needs, and he used his own money to purchase the stone fort located there and donate it to the government (Shankland 1951; McKoy 2000). It was proclaimed as a national monument by President Harding in 1923, but

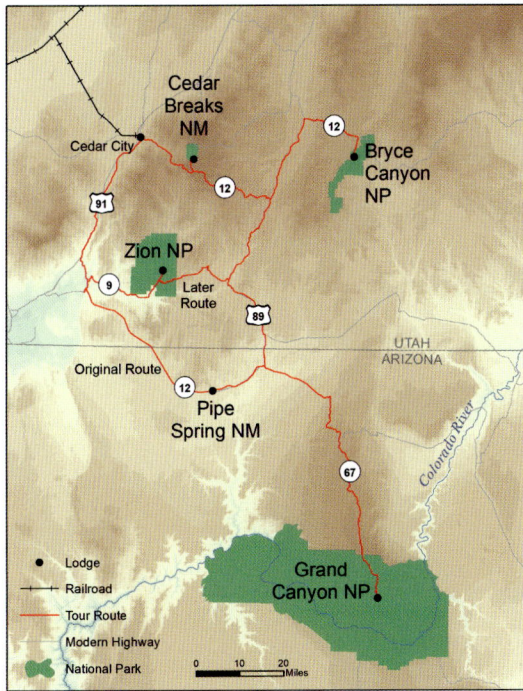

FIG. 7.6 Map of Union Pacific tours in Utah and Arizona in the 1920s and 1930s. The Union Pacific Railroad once operated scenic tours of several park units, providing lodging and transportation from the Cedar City railroad station.

the tiny monument was bypassed by the Union Pacific tours in 1930 when a new highway was built east from Zion.

After 1925 the number of visitors arriving at the parks by rail began to decline, followed by a more substantial decrease in following decades. Less than 1 percent of visitors arrived at Yellowstone by train in 1930 (Waite 2006). Railroads began dropping passenger services to parks and getting out of other park tourism businesses. The Union Pacific Railroad stopped its passenger service to Cedar City in 1959 and donated all its lodges to the park. The Zion lodge burned down in 1966, and the Cedar Breaks lodge was torn down in the 1970s, but those at Bryce and the North Rim of the Grand Canyon survive, as does the railroad station in Cedar City.

The park unit that saw the greatest railroad activity was not one of these. Most of the Great Smoky Mountains had been clear-cut before logging companies gladly sold the cutover lands to the government. To haul timber out of the mountains, logging companies built crude railroads up almost every canyon along the north

side of the mountains, nearly reaching the crest of the Smokies (Schmidt and Hooks 1994). The last of these was removed in 1939, and virtually no trace of them remains, though many roads and hiking trails between the Newfound Gap road and Cades Cove follow the paths of former rail lines. Visitors to the park could also reach Bryson City by train until service ended in 1948; the modern Great Smoky Mountains Railroad is a tourist train that carries passengers from Bryson City on round-trip excursions away from the national park.

The Early Automobile Era

At the end of the nineteenth century, the automobile quickly became part of experiencing the national parks for those Americans who could afford one. But parks were far behind the rest of the country in their road conditions (Shankland 1951). Most of the park roads were made not of asphalt or concrete but of packed dirt (on good days) or mud. However, the push for automobile access to the national parks continued during this time, and Stephen Mather, a fan of the new technology, was convinced to build more roads within national parks.

The first automobile in a park unit arrived on June 23, 1900, in Yosemite (T. Davis 2016). The primitive car vastly outperformed stagecoaches and was driven to several locations, including Glacier Point, where it was photographed parked on a hanging rock. The event apparently did not make much of a positive impression, as it was not repeated, and cars were banned from the park in 1907. The first park to specifically allow automobiles was Mount Rainier in 1908 (Catton 2006), at which time there were only 3,130 autos in the state of Washington and 194,400 in the United States (Public Roads Administration 1947). Automobile access was soon allowed at other national parks as well, such as Kings Canyon in 1910, Crater Lake in 1911, and Glacier in 1912. Yosemite reopened to cars in 1913, accompanied by Sequoia, Mesa Verde in 1914, and Yellowstone in 1915 (Shankland 1951).

In the early years of automobile travel, private organizations promoted long-distance highways, and national parks were a popular destination for adventurous motorists. The Na-

FIG. 7.7 Map of National Park to Park Highway around 1920. This road was promoted by a private organization seeking to encourage long-distance automobile travel at a time when few good highways existed. The NPS encouraged these efforts to attract more visitors to parks.

tional Park to Park Highway Association promoted a circular route, a drive of about 6,350 miles, connecting twelve western parks with the best roads in order to promote tourism (Fig. 7.7) (Whiteley and Whiteley 2003). Similarly, the Eastern National Park to Park Highway emerged to promote roads connecting the new Shenandoah, Great Smokies, Mammoth Cave, and smaller historical parks (T. Davis 2016). These were to be upgrades of regular state or county roads, but the idea later emerged as the Blue Ridge Parkway between Shenandoah and the Smokies.

With the passage of the Federal Aid Road Act of 1916, the highway era in America began in earnest, with states taking over highway construction (T. Davis 2016). In 1926 the state highway departments and federal Bureau of Public Roads agreed on a nationwide set of numbered

highways, and the National Park to Park Highway faded into history. When the first map of the new system was created, Yellowstone was of such importance that it was the only park shown and was reached by several main highways. However, befitting Yellowstone's status as federal land, the new numbered highways did not pass through the park; they ended on one side and began on the other, connected by the park's internal road network. This situation remains the case today.

The Postwar Era

In the postwar era, the national highway network provided the routes that ever-increasing numbers of Americans used to reach their favorite parks (B. O'Brien 1966; Youngs, White, and Wodrich 2008). Marian Clawson and Burnell Held (1957) reported that 92.4 percent of visitors to the Grand Canyon in 1954 came by car, along with 96.4 percent of visitors to Glacier and 97 percent of those arriving at Yosemite.

Roads helped to create a love affair between people and their automobiles, which did not end at the gates of America's national parks. The automobile afforded visitors the opportunity to experience and explore the parks more directly than was previously allowed. Driving for pleasure through parks ranked as one of the most popular recreation activities in the United States (Youngs, White, and Wodrich 2008). Many park officials also believed that the automobile could be used as a tool for educating visitors. Even early preservationists supported automobile touring in the California redwoods region with the hope that it would help educate the public about the natural landscape and generate interest in preserving a unique area.

Some national parks were explicitly established and designed to provide a recreational driving experience with scenic natural landscapes. For example, the Blue Ridge Parkway was designed for a slow-paced, relaxing drive that would reveal the natural and cultural history of the southern Appalachian Mountains. The roads in essence provide the park experience. The quality of the "drive-through" park experience often determines the overall visitor experience.

Today state highway systems furnish the routes by which the vast majority of park visitors arrive at their destinations. At some parks state roads are important for travel within parks as well; California Highway 190 crosses through Death Valley, with a Caltrans maintenance station within the park. Grand Teton has several highways running through it, created to carry visitors to Yellowstone before Grand Teton was created. One of the major attractions at New River Gorge National Recreation Area is a West Virginia state highway (U.S. 19) that crosses the river on a spectacular tall steel arch bridge, once the longest in the world. On the third Saturday in October, the bridge closes so visitors can parachute off it. For small units scattered across multiple locations, state highways function as the park road network. The units of John Day Fossil Beds National Monument are con-

nected by U.S. 26 and Oregon Highways 19, 207, and 218, while Salinas Pueblo Mission locations are linked by New Mexico Highway 55 and U.S. 60.

Air Transport

Air travel is essential to national park visitation, nowhere more so than to the Pacific, Alaskan, and Caribbean parks. The national parks of American Samoa, Hawaii, Puerto Rico, and the Virgin Islands can only be accessed by most visitors by air transport. Guam's Antonio Won Pat International Airport is the gateway to the island and War in the Pacific National Historical Park. The airport is a hub for United Airlines services to Asia, with direct flights to six cities in Japan, three in China, two in South Korea, and others in Taiwan, the Philippines, and neighboring island countries. United offers only one daily flight to a U.S. destination outside Micronesia, to Honolulu. The park can be more easily reached from Japan and Korea than from elsewhere within the United States and receives many of its visitors from these countries. With the exception of a local interisland carrier, the airport's other airlines are Korean, Filipino, Taiwan, and Japanese, along with occasional charter flights by an airline based in Uzbekistan. Manila can be reached in three hours and thirty-four minutes, Tokyo in three hours and forty minutes, and Beijing in five hours, but distant Honolulu takes seven hours (Guam International Airport Authority 2019).

Want to visit the National Park of American Samoa? It won't be easy. American Samoa's Pago Pago International Airport is connected to the United States by only two flights a week from Honolulu (with five hours and thirty-five minutes flying time). This is a recent phenomenon; as late as the 1980s the island was a stopover on many daily flights between North America and the South Pacific. But with newer, longer-ranged airplanes this stopover is no longer necessary, and the island dropped off most airline networks. It is a rare example of space-time divergence, when places become more remote from each other.

Hawaii has an enormous number of air connections to many countries on three continents, yet it also contains a park unit reachable only by small plane. Kalaupapa National Historical Park is on a coastal plain on the north side of Moloka'i; it is cut off from the rest of the island by cliffs towering two thousand feet above it. Access is possible only by a treacherous and difficult trail (washed out in 2009) or airplane. A small airport within the park has a scheduled airline service (one of two in the park system), with most flights connecting to a nearby airport and lasting only a few minutes. This service is not profitable and is subsidized by the Essential Air Service program, a federal program that pays airlines to operate to smaller cities and isolated airports that would otherwise be difficult to reach.

Puerto Rico and the Virgin Islands are much easier to get to, with several commercial airports having frequent flights to South Florida and major airline hubs in the northeastern United States, as well as to other Caribbean islands. A few flights also serve Central and South America and Europe.

Alaska is once again in a class by itself. Anchorage is the principal air hub of Alaska, with Fairbanks a distant second. Both are strongly connected to Seattle, which most travelers from the contiguous forty-eight states will have to pass through. Those bound to Sitka or Skagway might also pass through the Juneau airport; Klondike Gold Rush National Historical Park may compete with the National Park of American Samoa as one of the hardest park units to reach.

Few Alaskan units have airports, but most visitors to several units arrive by plane. At Lake Clark many arrive by floatplane to lakes or rivers within the park, landing on open ground, beaches, or gravel bars along rivers or on frozen lakes in winter. Planes must be hired at nearby towns such as Homer and Anchorage; there are no scheduled services. Lake Clark National Park has issued permits to twenty-eight operators flying from ten towns within Alaska that serve the park. At least one airplane trip is involved in getting to Aniakchak, with the few visitors braving the reliably poor weather landing either on a lake within the park or at a small village outside and trekking in. The small village of King Salmon (population 374) is a ma-

jor hub and gateway for these parks and other nearby federal lands (Fig. 7.8).

At Denali, many visitors take a flightseeing trip, and some land on glaciers in the spring and summer. (Those who touch down must pay the park's entrance fees.) Noatak National Preserve, located far to the north in the Brooks Range, is served by ten air taxi companies operating from Kotzebue and Bettles. These small towns must first be reached by air; each has scheduled service to Fairbanks and Anchorage.

Twenty-five national park units have airports or landing strips within them. Wright Brothers National Memorial preserves the site of the first airplane flight and has a small airstrip within its boundaries. Death Valley and Glen Canyon each has three airports. Gateway in New York City includes Floyd Bennett Field, no longer open to the public but once one of the country's leading airports. Within the contiguous forty-eight states only at Grand Teton National Park is there an airport within a park with scheduled air service (the Jackson Hole Airport). The airport opened before the park was created and remained in use. Today it has jet service by four airlines to twelve cities.

Those looking for an Alaskan airport experience without having to travel there could try the Chicken Strip in Death Valley's Saline Valley (Fig. 7.9). This remote, unlighted, and unmaintained rough dirt strip does not appear on aeronautical charts and is not for novice pilots, who must cross a ten-thousand-foot-high mountain range before quickly descending into a deep enclosed basin while keeping an eye out for low-flying military jets. The runway is frequently washed out, and high winds are a common occurrence; there are no services.

Other Forms of Transportation

Two-thirds of national park units cannot be reached without a car, but the remainder have at least some alternate transport possibilities (Table 7.1). These possibilities may involve a bus, train, or ferryboat, depending on the unique circumstances of each park. Only 1.7 percent of park units can be accessed by all three of these possibilities. About 14 percent of parks can be accessed by both bus service and

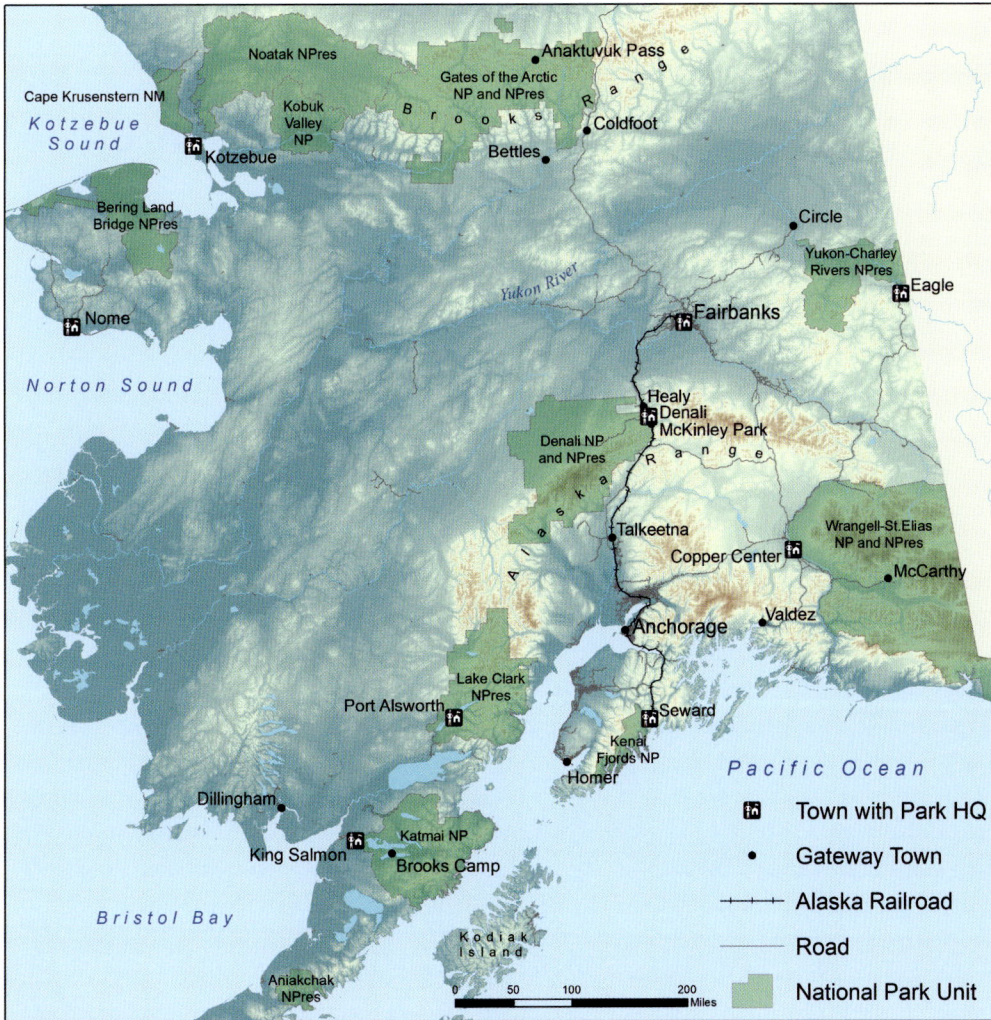

FIG. 7.8   Map of Alaska parks and gateway towns, 2021. Because of the absence of roads and a reliance on air transportation, many towns are quite distant from the parks they serve.

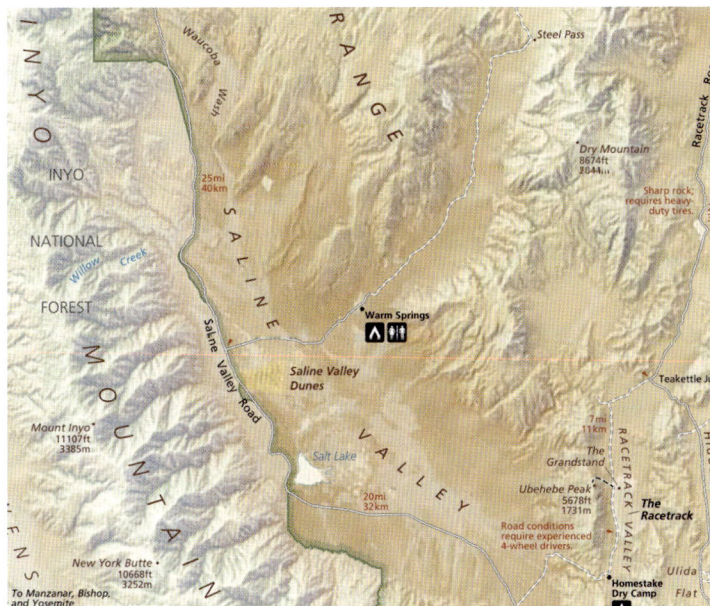

FIG. 7.9   Portion of a map of Death Valley National Park showing Saline Valley. The remote Chicken Strip airfield is adjacent to the clothing-optional warm springs. This is perhaps the most primitive airport in the national park system outside of Alaska. NPS.

TABLE 7.1   Transportation to and within the parks, 2021

Mode of transport	Transportation to parks		Mobility within parks	
	Number of parks	Percent	Number of parks	Percent
Car only	282	66.7	371	87.7
Bus/shuttle only	39	9.2	33	7.8
Train only	10	2.4	3	.7
Ferry only	23	5.4		NA
Both bus/shuttle and train	59	13.9		NA
Both bus/shuttle and ferry	3	.7		NA
All alternate modes	7	1.7		NA
Greenway			16	3.78
TOTAL	**423**	**100**	**100**	**100**

train, 9.2 percent can be accessed only by bus and shuttle service, 5.4 percent can be accessed only by ferry, followed by 2.4 percent that are accessible only by train. Shuttle systems are increasingly common, and they often strive to connect a park with surrounding communities. For example, there is a public-private partnership shuttle system in Wyoming that serves Yellowstone and Grand Teton National Parks, as well as nearby gateway communities (Kack and Chaudhari 2009).

A few parks can still be reached by train. The Denali Star on the Alaska Railroad has service to Denali (one train each way daily during the summer), the Grand Canyon Railway tourist train runs from Williams to the South Rim (two round trips daily), the East Glacier Station is on Amtrak's Empire Builder line between Chicago and Seattle (one train each way daily) (Amtrak 2019), and Yosemite can be reached by the Amtrak San Joaquin line to Merced, followed by a special Amtrak bus into Yosemite Valley. The White Pass and Yukon Railroad also serves Klondike Gold Rush National Historical Park in Skagway, but it is more likely to be used by day-trippers from the town. Travelers on Amtrak's Southwest Chief will get a brief glimpse of Petrified Forest as their line passes through the middle of the park; those on the Sunset Limited will likewise glimpse Amistad Reservoir as the train passes through Texas.

Some parks can only be reached by boat. This includes many island units: Channel Islands, Dry Tortugas, the Statue of Liberty, Alcatraz, Boston Harbor Islands, Isle Royale, Buck Islands Reef, and even Rainbow Bridge

National Monument in the Utah desert, where most of the 108,000 visitors arrive by tour boat on Lake Powell. Most of these boats are operated by concessionaires, but Isle Royale operates the Ranger III, at 165 feet in length the NPS's largest boat. It has been in service since 1958 (Fig. 7.10).

In Alaska cruise ships provide access to the glaciered coast, as well as offloading trainloads of passengers for the Alaska Railroad's Denali-bound trains at the port of Seward. Cruise ships belonging to eighteen different lines enter Glacier Bay and travel many miles inland. Two ships enter the bay each day in the summer, usually staying from 6:00 a.m. to 3:00 p.m. Park rangers board the ships when they enter the park, and the rangers point out the sites to passengers. The enormous ships are the largest vessels to operate within park units but will still be dwarfed by the tidewater glaciers at the head of the bay. These ships often stop at Skagway, allowing passengers to visit Klondike Gold Rush National Historical Park.

The park system has become more urban over time, and in the largest cities, especially Washington, D.C., San Francisco, and New York City, taking public transportation to the parks is not usually a problem. In these cities, getting to a park unit involves a subway ride, a walk along city streets, a drive to the local beach, or a ferry trip to an island. NPS directions for Pullman National Historical Park on the south side of Chicago include driving, bus, and commuter train routes. The National Mall in the nation's capital has a subway station and is crisscrossed by bus routes and busy

FIG. 7.10   The Ranger III, the largest boat operated by the NPS. It carries many visitors to Isle Royale National Park.
Photo by Joe Rossi, 2011. Wikipedia, Creative Commons Attribution–ShareAlike 2.0 Generic License.

streets. However, transit services (and parking) are extremely limited in the western half of the mall, between the Washington Monument and Lincoln Memorial (Benton-Short 2016). Visitors here may walk much farther than they had expected.

There are even parks that can only be reached by trail. Although Arizona's Fort Bowie National Historic Site has a road through it following a nineteenth-century stagecoach route, visitors wanting to see the park's visitor center and the fort ruins must hike about 1.5 miles one way to do so. Summer temperatures are high, but the visitor who makes it will likely have the park to themselves and an experience to remember.

Mobility within Parks

How do people move around within parks? Many small units require only a short walk to see the park's attractions, as at Montezuma Castle and the Lincoln Memorial, or even an elevator ride, as at the Washington Monument. But in larger parks, some form of transport is required to see the attractions. In the early years of the

park system, railroads were proposed as the means to do so but were rejected (Waite 2006). Instead, roads were built for the purpose, and it is hard to imagine parks without them.

Death Valley National Park has more miles of roads than any other park unit, about 1,287 miles, or about 13 percent of the total NPS mileage. Several parks with large mileages are also large western units, including Lake Mead, Mojave, Yellowstone, Big Bend, Canyonlands, and the Grand Tetons (Fig. 7.11). The Blue Ridge and Natchez Trace Parkways are also among the top ten, each with the main park road of 469 and 444 miles, respectively, along with a much shorter mileage of service roads. Great Smoky Mountains rounds out the top ten units by road mileage, which altogether make up about half of the NPS's road mileage. Another park with substantial mileage for its size is Ebey's Landing National Historical Reserve, located on Whidbey Island in Puget Sound. It preserves a rural area, with a mix of federal, state, and private lands, and contains the town of Coupeville, as well as agricultural land, giving it a substantial road mileage more typical of small towns than small parks.

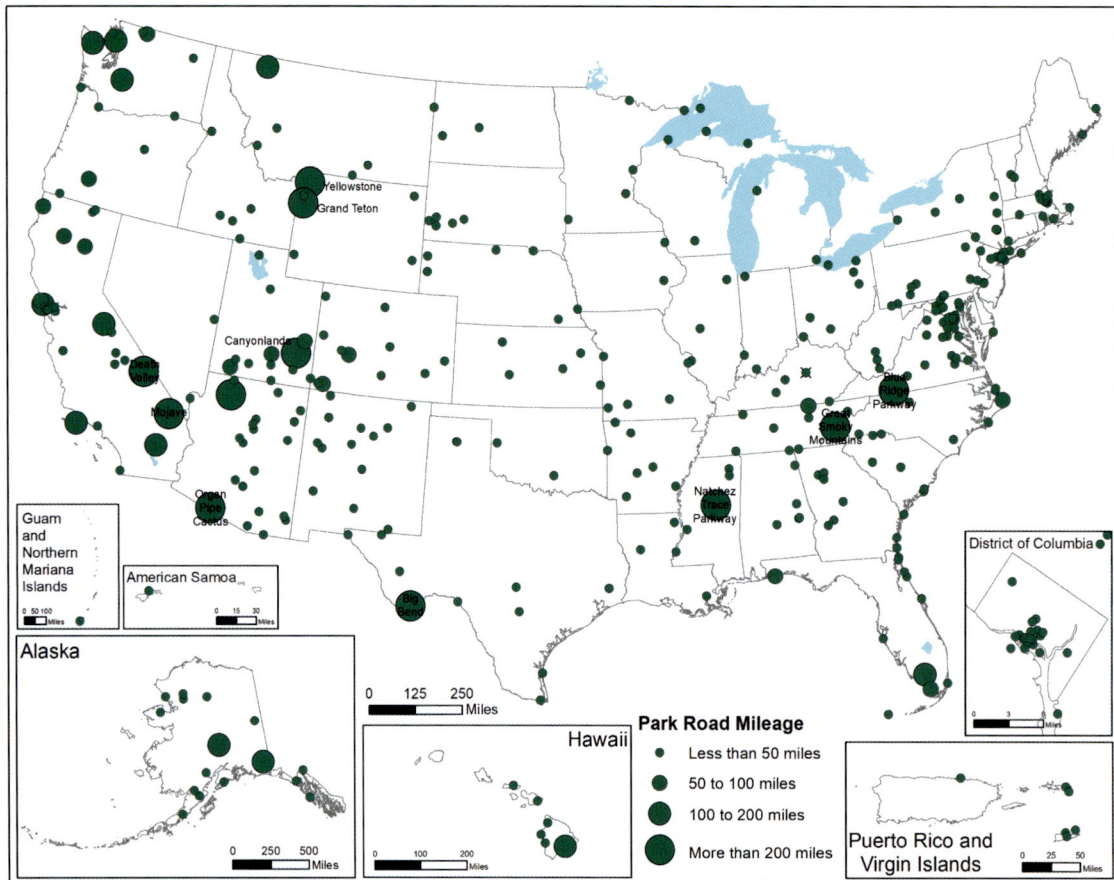

FIG. 7.11   Map of road mileage by park unit, 2020.

At the other extreme, some parks have no roads: the seven islands of the Dry Tortugas are one, and Rainbow Bridge, reached only by boat, is another. Alaska parks usually have few or no roads. Others have few roads for their size. Olympic and North Cascades were both viewed as roadless wilderness parks in contrast to the more developed Mount Rainier (Louter 2006), though Olympic later added the popular Hurricane Ridge road. Although it has over 450 miles of roads, Yellowstone has a unique claim: the southeast corner of the park is the farthest point from a road anywhere in the forty-eight contiguous states (Watts 2005).

Developing Park Road Systems

At several parks, the road networks predate the invention of automobiles. The basic road network of Yellowstone, with the Grand Loop and branching entrance roads, was planned in the 1870s (B. O'Brien 1966; Culpin 1994; Haines 1996; T. Davis 2016). Some of these roads were built between 1878 and 1882 through the efforts of the first park superintendent, and the army finished the basic network between 1883 and 1918, constructing or rebuilding three hundred miles of roads.

The NPS began building roads in 1918 and has mainly rebuilt them over time to increase capacity, which included straightening, widening, and paving roads. The agency had very limited funds available to do so, and Stephen Mather bought the privately owned Tioga Pass Road with his own money to add to the park's road system (Shankland 1951). In 1924 more money was provided (T. Davis 2016). The agency's initial goal was to build one good road in every park rather than run roads everywhere.

Park roads are designed to provide spectacular experiences yet be as unobtrusive as possible. This is not an easy requirement to balance, but the NPS has carefully refined the art of road building in a wide range of physical environments and for different types of vehicles.

A key element in this success was hiring landscape architects to help design roads in harmony with the landscape. These were skillful designers of the land, working with natural or modified terrain and using vegetation to create attractive views and hide unappealing features. Later attempts at collaboration between the NPS and engineers with the Bureau of Public Roads (now the U.S. Department of Transportation) were frequently contentious because the two agencies had vastly differing understandings and standards for road design. By the 1920s the state of the art for building efficient modern highways exceeded what the NPS preferred for park roads.

The NPS designs its roads using a different approach from state or county highway departments, including the use of slower speed limits, an attempt to integrate the road with the terrain, and the use of rustic guardrails and signs (Davis, Croteau, and Marston 2004). Forty-five miles per hour is the standard maximum speed limit in parks. Long straight sections of roads (tangents) are avoided in favor of smooth curves. Park roads are not fenced to keep animals off the road, as are many other rural highways; this allows the bear jams of Yellowstone and other wildlife encounters. Highway departments are increasingly building wildlife overpasses to keep animals off busy roads, but these would not be popular in parks. Since the 1960s bicycle, tour bus, and recreational vehicle (RV) usage has increased, requiring adaptation by the NPS to ensure that these vehicles can be safely accommodated (National Park Service 1984).

The traffic on many park roads is highly seasonal. Many roads in summer parks are kept open year-round with expensive snow removal efforts on main roads, while roads in winter parks may be closed in summer to keep tourists off lightly used backroads during the summer heat. Park road traffic can also show extreme peaks during weekends and popular times of the year, creating what is known as the peak-period problem. Park planners must build facilities appropriate for these peak conditions even though they will be empty and unused much of the year. Enormous bathrooms serving bus tours on the Denali road and the vast Sunset campground (a giant parking lot for RVs ar-

riving for the winter) in Death Valley are examples. A visitor who sees them when they are empty will wonder why the parks seem to have wastefully overbuilt these facilities.

Small units developed during the CCC and Mission 66 eras usually have carefully designed road systems, with roads combining scenic views with modern engineering standards, characterized by curvilinear patterns. A visitor to Arches can see the entire park, with frequent stops and short hikes, without making a single left turn, thanks to the Mission 66 road network. In a heavily traveled park with drivers paying more attention to spectacular scenery than the road, this is a notable safety feature. Unfortunately, there are no detour routes in the event of congestion or an accident, causing problems for park management, especially during emergencies.

In contrast, other parks did not have planned road systems and simply used whatever roads already existed when the park was created. Shiloh is one, with a maze of local roads passing through the park. The park has a driving tour of the Civil War battlefield, but it uses these local roads, with the route requiring many stops and left turns, and it even doubles back on itself several times. The roads do, however, have historical continuity with the time of the battle.

In the stagecoach era, views to the side were often emphasized by road designers, with vistas alternately opening up to the left and right (T. Davis 2016). With cars, a focus on the view ahead became important, with the road providing a succession of dramatic vistas. Perhaps the most spectacular example was in Yosemite, where the NPS built a new Wawona road to replace the old one but wanted to keep the stunning first view of the valley seen from the old road's Inspiration Point. The solution was to use the rock removed from the Wawona tunnel to create a new spectacular overlook where visitors from the south will first see Yosemite Valley.

Park planners prefer the use of one-way loops and three-way intersections. One-way loops are popular with campground designers (Young 2017) and are often used for short scenic loops, as in the East Unit of Saguaro or Natural Bridges. The Natural Bridges road, like most park loops, runs counterclockwise, so vehicles do not need to cross in front of oncoming traffic at the beginning and end of the loop. The road runs along the top of a plateau whose edges are deep canyons. The monument's three natural bridges are visible in the canyons below to the right, with a short walk required for the best view. Only right turns are possible on this road. After the last bridge is passed the road returns to the visitor center, away from the canyon edge. The Saguaro loop road is unusual in that it runs in a clockwise direction, requiring entering and exiting traffic to cross each other's path, but this was chosen because it allowed park staff to develop a linear narrative that fit the terrain and views of cacti. Another advantage of one-way roads such as the Saguaro loop, Artists Drive in Death Valley, and the Roaring Fork Motor Nature Trail in the Smoky Mountains is that roads can be built narrower, allowing them to fit into more restricted terrain and creating less of an ecological impact.

Today the NPS has 9,550 miles of roads, 57 percent of them paved, along with 1,414 bridges (Federal Highway Administration 2008). These totals are small compared to those maintained by the military (26,564 miles), the Bureau of Land Management (68,880), and the Forest Service (99,100 miles) and are dwarfed by the road mileage of most county and state highway systems. Parks contain only 0.24 percent of the four million miles of road in the country. Other transportation features within parks include 1,451 bridges, 63 tunnels, 17,872 miles of trails, bike paths, waterways, and extensive parking facilities.

The emphasis on integrating roads with landscapes has some interesting consequences. Although it accounts for about 0.24 percent of the nation's road mileage, the national park system contains sixty-three highway tunnels, about 10 percent of all the highway tunnels in the country. This is partly a reflection of the challenging terrain many park roads are built in, but it is also due to NPS road-building techniques, which discourage large cuts through hills and ridges. This is evident with the twenty-six tunnels along the Blue Ridge Parkway, which has more tunnels than any other highway in America. In other parks, such as Cumberland Gap and Golden Gate, the tunnel is a bypass underneath the park and keeps heavy traffic away from the park landscape enjoyed by visitors. At Cumberland Gap the tunnel has allowed the old pioneer trail through the gap to be restored to its original condition, which was not possible while it was in use by a heavy stream of cars and trucks.

The most spectacular tunnel in the park system is surely the Zion–Mount Carmel Tunnel in Zion National Park. This curving, 1.1-mile-long tunnel was completed in 1930 and has several large openings looking out into a colorful canyon, with parking for early visitors to enjoy the view. Today's travelers aren't allowed to stop, and since the tunnel was designed in the 1920s, when vehicles were smaller and less numerous, traffic is now one-way, with larger vehicles requiring an escort. This tunnel highlights a dilemma for park planners: How should roadways built before modern vehicles and traffic conditions and now considered historic be improved without substantially altering them?

The inventory of NPS tunnels does not include drive-through trees; there were once seven of these in sequoia or redwood trees in California. Two were in Yosemite, but one fell in 1969, and the other can now only be walked or biked through (United States Forest Service 2006). Sequoia National Park has a tunnel cut through a fallen sequoia, and elsewhere in the state three drive-through standing redwoods can be found on private land. A relic of a time when cars were much smaller than today, they live on in the popular imagination.

Mission 66 is often perceived as having been a road-building program, and 2,767 miles of road were built during its run (T. Davis 2016). However, many of these miles replaced older, substandard roads, which were then abandoned, and only 1,197 miles were new. The Mission 66 program built a new road bypassing the Old Faithful area in Yellowstone, with the bypass a four-lane divided road connecting to a local road with a trumpet interchange. It is often said that this road is the only four-lane road in the park system, but this is far from true.

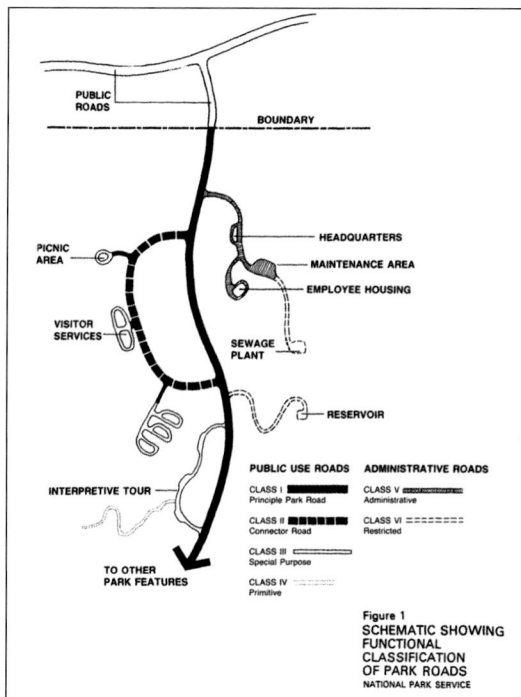

FIG. 7.12 Diagram from the 1984 National Park Service road manual, showing different types of park roads. National Park Service 1984.

Several units in the national capital parks are freeways, and there are many four-lane streets within the District of Columbia's parklands. Other units have freeways or multilane highways through them, including I-40 through Petrified Forest, I-10 in tiny and unvisited Hohokam Pima, and U.S. 101 through Golden Gate National Recreation Area. It is also said that the interchange at Old Faithful is the only one in the system, but this is also not the case. There are at least 173 interchanges in the national park system, including 137 on parkways (the only points of access between these roads and others), 11 in Golden Gate National Recreation Area (and 4 in other recreation areas), and 15 within national parks. Most of these are where a major highway passes through a park unit and connects to a park road, as in the Petrified Forest.

Today the NPS groups roads into different categories depending on their function, much as do state highway departments (Fig. 7.12). There are public use roads, administrative park roads, and urban parkways and city streets (National Park Service 1984). Public use roads are those used by visitors. These in turn are grouped into

four classes. Class 1 includes principal park roads. Class 2 is connector roads, which provide access to overlooks, campgrounds, and other areas. Class 3 is special-purpose roads, which allow circulation within campgrounds, visitor center parking lots, concessions, and similar areas. These are often one-way loops. Class 4 is primitive roads, which are usually dirt and may require high-clearance or four-wheel-drive vehicles. Administrative park roads allow park employees and residents access to administrative areas and housing, as well as access to the park's utilities and communications infrastructure. These may be administrative access roads that are open to the public or restricted roads that are gated and used to reach utilities or communications facilities. Urban parkways and city streets are most common in the national capital region, where they are part of the District of Columbia's transportation system.

The standard park maps produced by the Harpers Ferry Service Center reflect these categories; class 1 and 2 roads are often shown with red lines (sometimes with class 2 as a thinner or black line), class 3 roads appear only on insets as thinner red lines, class 4 roads are shown by symbols such as a white line with solid or dashed black borders, depending on road conditions, and administrative roads are usually not shown at all (Fig. 7.13).

Not all park roads were carefully designed by park staff; many units already had established road networks before the NPS came along. This is especially true at Death Valley, which has one of the longest associations with the automobile among national park units. Automobiles first arrived in this vast desert region in 1905 during a mining boom, and a few years later one of the first transcontinental auto trips passed through the future national park unit (McConnell 2000). In 1908 several drivers on a New York to Paris auto race even passed through the valley (Fenster 2005). Many roads were built by prospectors, miners, and promoters of early tourism. The California state highway department and Inyo County also built several roads. Although the park was created in 1933, mining was allowed, and more mining roads were added through the 1970s, of-

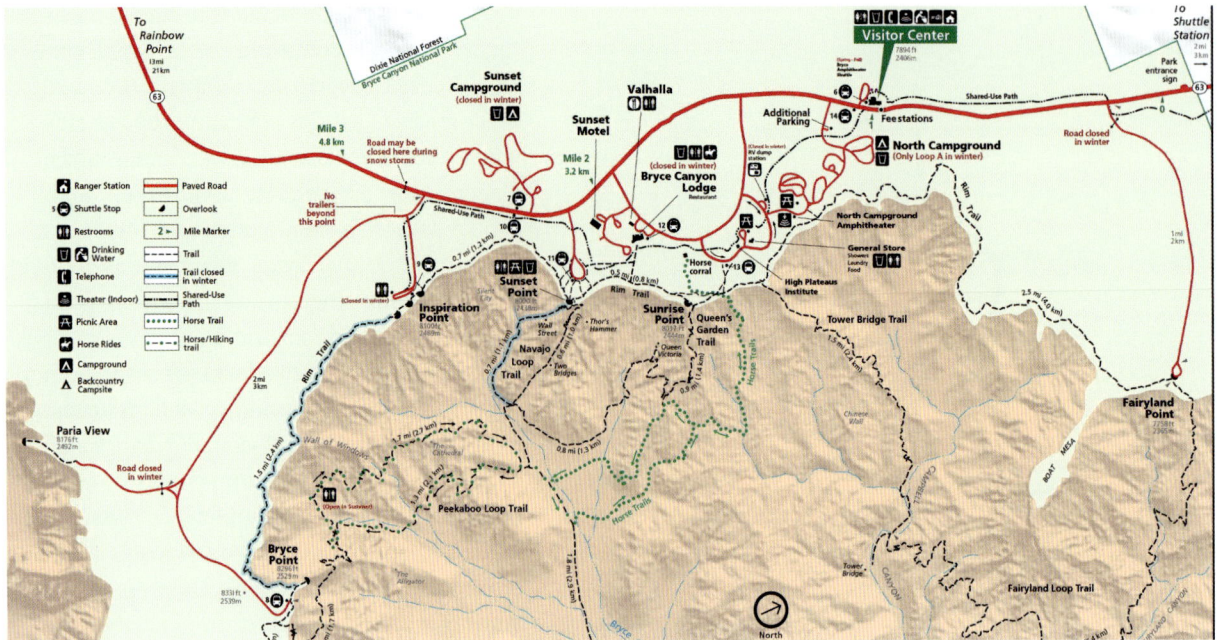

FIG. 7.13 Portion of the Bryce Canyon National Park map, 2021, showing the different types of roads that appear on park maps. NPS.

ten without the knowledge of the NPS. The NPS built only a few miles of roads to new tourist attractions or to service facilities. Since the 1970s about seven hundred miles of lightly used roads were abandoned, but the park's road network remains a collection built by miners, tourist promoters, and the state highway department. These were not built to modern standards or in good locations, and a considerable mileage of road is washed out by flooding every year, in some cases several times. Over 90 percent of the park is now wilderness, yet because wilderness boundaries typically follow roads, most of the park's wilderness is based on the crude roads graded by prospectors or tourist promotors as far back as 1907.

Saguaro National Park provides a contrast between early and NPS-built road systems, with the East Unit having an NPS-designed loop road built by the CCC and the West Unit inheriting a road network left over from mining days (Burtner 2011). Visitors to the East Unit pay an entrance fee at a single entrance and move slowly along a single roadway, with a linear interpretive narrative. Those driving into the West Unit may do so at one of six entrances and select their own route and destination; there is no central focus. Several roads run

between points well outside the park boundary, and one is frequently busy with locals commuting to and from Tucson, creating conflicts with slower sightseeing traffic.

Scenic Highways

The park system is well known for its stunning scenic highways. Among these roads, the Going-to-the-Sun Road in Glacier stands out. This was the premier park road project at the time (T. Davis 2016). It was built between 1921 and 1932 and runs fifty-one miles across the park. The high point is 6,646-foot Logan Pass, where snow can accumulate up to 80 feet deep; it is only open about four months of the year. Speed limits vary from forty-five down to twenty-five miles per hour (Davis, Croteau, and Marston 2004). The road was not fully paved until 1952, and its annual snow clearance is a mammoth affair.

The Trail Ridge Road in Rocky Mountain National Park is another example (T. Davis 2016). This road was built by the Park Service to replace the Fall River Road, completed by the state highway department in 1920. By the time it had been built, it was obsolete and considered dangerous. The NPS wanted a showpiece road, and it was built between 1930 and 1932 through the tundra, with great pains taken to minimize environmental harm. The road reached a peak elevation of 12,183 feet above sea level, with four miles entirely above 12,000 feet and eleven miles above 11,000 feet. This is the highest paved road in the entire park system; Cedar Breaks, Great Basin, and Haleakalā are the only other units with paved roads over 10,000 feet. (The highest paved road in the country is the Forest Service's road up Mount Evans in the Colorado Rockies, reaching 14,130 feet.)

Zion received a new road that included a spectacular 1.1-mile-long tunnel with gallery windows overlooking the canyon. The road's asphalt was mixed with local red sandstone, giving the park's road a distinctive red color. Mount Rainier, Sequoia and Kings Canyon, Crater Lake, Shenandoah, and Yellowstone also received scenic roads (T. Davis 2016). Shenandoah's Skyline Drive was the first to run along a ridgeline, setting a precedent followed by several later roads. It was not just big parks that received scenic roads; tiny Scotts Bluff National Monument has a road built by the CCC, using hand labor as much as possible, that climbs to the top of the bluff. The road was carefully hidden from the flat lands below, requiring three tunnels (Davis, Croteau, and Marston 2004).

The Blue Ridge and Natchez Trace Parkways are park units in and of themselves. Neither was originally intended as a park unit; instead, each parkway was built by the NPS due to its involvement with the CCC and other Depression-era work programs (T. Davis 2016). These roads were completed in 1987 and 2005, respectively, long outlasting the Great Depression. They were designed using the NPS approach to harmony with the landscape, with a succession of alternating views to the left and right, rustic design features, and a mix of nature and culture. The NPS had a goal of obtaining one hundred acres of land per mile of road, with a scenic easement on another fifty acres, producing a right-of-way that averaged 825 feet in width. Every thirty or so miles this would expand in the form of bulges, where restrooms, restaurants, lodging, and other attractions could be located. The Natchez Trace includes the spectacular double arch bridge over Tennessee Highway 24 near Nashville, completed in

1994, while the Blue Ridge's signature feature is the Linn Cove Viaduct, carrying the road along the shoulder of Grandfather Mountain. Views of this bridge have become ubiquitous since it was completed in 1987, making it one of the most recognizable places in the park system.

Other scenic roads can be found throughout the system. The twenty-seven-mile-long Park Loop Road in Acadia, Maine, constructed between 1921 and 1958, is one of the most magnificent recreational driving experiences for park visitors. It allows motor vehicles access to the park from the east side of Mount Desert Island, connecting Acadia's lakes, mountains, and the shoreline. It provides access to popular areas such as Sieur de Monts, Sand Beach, Otter Point, Jordan Pond, and Cadillac Mountain. The thirty-mile-long Ross Maxwell Scenic Drive in Big Bend, Texas, was built in 1968 through spectacular desert landscapes (Fig. 7.14). It runs through the western slopes of the Chisos Mountains, climbs up to one of the park's most outstanding views at Sotol Vista, then descends to parallel the Rio Grande at the Castolon Historic District before winding up to the Santa Elena Canyon trailhead, where the pavement ends. Lake Mead's Northshore Road is another scenic desert road. It is a very well-built postwar highway, carries hardly any traffic, and offers distant views of the blue lake, as well as a dizzying array of mountain colors, shapes, and geologies, from dark volcanic rocks to flaming red sandstone. The serenity of the road is even more impressive, given the teeming activity on the other side of the mountains in Las Vegas. It should rank near the top of any list of spectacular western park roads

Off the Road

For many visitors today, the preferred way to experience a park is on foot, though the vast majority will arrive at their trailhead by car or perhaps on a bus. Walking is appropriate at Klondike Gold Rush National Historical Park, which contains the American end of the Chilkoot Trail. This runs thirty-three miles from Alaska into Canada along the route followed by those heading to the Klondike gold rush in the 1890s. (Hikers today do not have to carry the two thousand pounds of supplies required

FIG. 7.14 **Ross Maxwell Scenic Drive in Big Bend National Park.** Photo by Steewen1, 2019. Wikipedia, Creative Commons Attribution–ShareAlike 4.0 International License.

by each person to survive a year in the Yukon during the gold rush.)

Most large parks have a substantial mileage of trails forming an interconnected network. At the Grand Canyon and Sequoia–Kings Canyon, these networks allow more of the parks to be visited than from the limited road networks. Others, such as Death Valley, have few formal trails but almost unlimited opportunities for cross-country hikes wherever visitors may want to go.

Many park trails were built by tourist promoters, miners, and other visitors long before these places were parks, as at the Grand Canyon. The NPS has long designed and built trails within parks to allow for easy and safe travel, as well as for environmental protection. Switchbacks with retaining walls, stone steps, and even bridges and tunnels are provided to keep hikers safe, or at least safer. The Angel's Landing trail at Zion is one of the more spectacular, with a series of twenty-one switchbacks called

Walter's Wiggles and a final ascent along a narrow spine, with only a chain handhold to protect against a fourteen-hundred-foot fall. (According to a helpful sign at the trailhead, at least six people have fallen to their deaths since 2004.)

Some park units do not have any trails or do not allow hiking. This is by necessity with small parks such as George Rogers Clark National Historical Park. Though the tiny memorial park in downtown Vincennes, Indiana, was created to commemorate a 180-mile trek, there is no room for hiking today. This is also the case at most historic sites, small monuments, and memorials, especially in urban locations. Limits to hiking and camping can also be found in bigger parks. Mesa Verde National Park has only a few trails for those who wish to wander; the few ruins must be experienced on a ranger-led tour or on a paved path. Hiking is not allowed in order to protect the park's fragile archaeological sites. Saguaro allows no off-trail hiking below 4,500 feet elevation to protect the sensitive cryptobiotic desert soil crust; this affects almost all of the West Unit and the desert portions of the East Unit.

Early Public Transport

The first park visitors relied on some form of what we now call public transportation, involving stagecoaches or carriages that are operated by a private company and that transport large numbers of travelers. Yellowstone was developed with this form of transport, with several luxury hotels near major attractions connected by roads, with stagecoaches carrying tourists between them (B. O'Brien 1966; T. Davis 2016). In 1909 there were 1,372 horses stationed in the park to pull the 393 stagecoaches and carriages maintained for this circuit (Haines 1996). Additional freight wagons were required to deliver hay throughout the park. After the Grand Loop road system was completed in 1915, most visitors were handled on a counterclockwise route, with a night at each of the luxury hotels.

All of this changed when the automobile arrived on the scene. In 1915, when the first cars were allowed into Yellowstone, 85.6 percent of visitors used the stagecoaches; this percentage shrank very quickly. The public was strongly in support of automobiles at the time, and they quickly became the preferred means of entering parks. By 1911 90 percent of visitors entering Mount Rainier did so in a car, while 60 percent of those coming to Yosemite did so in 1921 (T. Davis 2016). Besides the need to improve or replace roads better suited to automobiles, their presence was a problem because they didn't co-exist well with horses. For that reason, Yellowstone's stagecoaches were retired at the end of the 1916 season and replaced by 116 buses (Haines 1996; T. Davis 2016). Motor vehicles were so popular that visitors were banned from entering the park with horses in 1917. With the arrival of buses, there was no longer any reason to have more than a thousand horses living in the park, and they were sold to the British army, a small portion of the more than one million shipped from the United States to Europe during World War I (Shankland 1951). It is doubtful that any of Yellowstone's horses survived the horrors of European battlefields.

Automobile tourists had different needs from those traveling by carriage. Camping had been negligible in a time when well-heeled visitors arrived by train and were taken on tours beginning and ending in luxury hotels. Camping out along the road was common for those traveling in their own vehicles, whether by design or by necessity in the days of unreliable cars and bad roads. The NPS reacted by building campgrounds for this new kind of visitation. New campgrounds sprang up, and the first gas station in Yellowstone opened at Mammoth in 1919 (Haines 1996). Stores were also opened to supply motorists and campers with their essentials. With these changes, the era of the great lodges slowly faded.

The Emergence of Alternative Transportation Systems

The personal automobile has long been the most prevalent (and usually the only) method of travel in the parks (Table 7.1). Cars allow visitors to travel to and within the park on their own schedule, maximizing their freedom and enjoyment. The Park Service worked hard to add new roads and parking areas to accommodate the needs of increasing visitors (Heacox 2001). The reliance on roads and cars led to outcomes similar to those experienced in the cities many visitors came from. Many national parks began to suffer from traffic problems such as congestion, lack of parking, and air and noise pollution (Eck and Wilson 2001). At Muir Woods National Monument the park's only road was closed in 1919. It is perhaps the only park unit that has permanently banished cars, but at most other units park planning has often revolved around how to accommodate more cars.

In many parks, wildlife sightings may exacerbate this problem, leading to the well-known "bear jams" and "moose jams." Here, traffic comes to a halt as visitors near an animal stop to take photographs, blocking the road and preventing others from experiencing the encounter. This often leads to visitors in the rear leaving their cars to approach on foot, creating the potential for dangerous interactions with wildlife and automobiles. Visitors further delay traffic flow until they walk back to their vehicles after the event. Pollution in this setting not only is a health hazard for humans but also may damage the park's environment. Stress is perhaps harder to quantify but certainly does not add to the enjoyable experience of visiting the park (Daigle 2008).

As this problem continued to plague many parks, park planners began to struggle with the conflicting missions they were charged with when the agency was created. This conflict between public access and conservation has long been a burden for the NPS, which must provide ways for large numbers of people to visit while still protecting wildlife and the environment (Winks 1997). There is no simple solution for either building or removing a park transportation system.

Public transport has returned to many parks, though for reasons quite different from those of the early years. Given the large and ever-increasing numbers of visitors in many parks, some form of mass transit has become necessary to move the crowds. This occurred first at some of the most famous parks: Yosemite and Denali (Daigle 2008).

Yosemite began its first shuttle service in 1970 when roads in the east end of the valley were closed to private cars (T. Davis 2016). Denali National Park implemented a bus system in 1972 to address concerns about increased visitation when the Parks Highway from Anchorage to Fairbanks was paved (Stroud 1985). The park's single road was unpaved and narrow, with numerous high drop-offs and no guardrails due to permafrost problems, creating unsafe conditions for the large numbers of visitors expected to use the park. Rather than improve the road to accommodate cars, the NPS decided to prohibit all privately owned vehicles beyond mile 16.

The Grand Canyon began a transit system at the South Rim in 1973 using shuttle buses (Anderson 2000). Although passenger service over the railroad line to the South Rim shut down in 1968, it was restarted in 1989. It was once again possible to not only reach the canyon by train but also travel throughout the South Rim without a car and the seemingly endless search for a parking space. In the late 1990s a management plan for the South Rim of the Grand Canyon

called for a light rail system along the rim to reduce traffic congestion and associated problems (T. Davis 2016). The rail system, a very expensive project, was considered necessary to preserve the Grand Canyon environment. This proposal ultimately went nowhere, but visitor circulation in the park was altered, and a revised bus system was instituted.

Similarly, Zion National Park in Utah was suffering from congestion along the six-mile Zion Canyon Scenic Drive due to increasing visitation, with nearly five thousand vehicles driving through the canyon and competing for four hundred parking spots each day (Upchurch 2015). The quality of both the visitor experience and the environment was deteriorating, and the park started an optional free shuttle service in 1999. With ridership increasing, the park administration closed the road to private automobile use and instituted a free shuttle service in the canyon in 2000. The new system, with twenty-nine propane-fueled buses, eliminates about twenty-five hundred vehicles per day from the main canyon, equal to 4.5 million vehicle miles not driven and contributing to a safer place for pedestrians, bikers, and wildlife (Retzlaff 2000).

Glacier, Rocky Mountain, Bryce Canyon, and Acadia are among other parks with bus systems, usually during the peak season and only when parking is in short supply (Fig. 7.15). Bus systems are commonly found in canyon parks (Yosemite and Zion) where the role of geography is particularly obvious. These parks have a linear pattern that works well with a transit system; larger parks like Yellowstone would be much more challenging due to the range of features, differing travel patterns, and vast size.

But transit use brings problems. A fleet of buses must be purchased, paid for out of the park's budget or entrance fees, and then operated and maintained. Visitors do not always enjoy riding in buses. There is less opportunity for an individual experience when you are exposed to conversations all around you, and it seems that dirty windows are a frequent occurrence. Transit also produces problems for hiking trails: on busy days there will be a surge or pulse of

FIG. 7.15   Acadia National Park shuttle bus, 2020. This is one of a growing number of transit systems that operate within national park units.

dozens of visitors disembarking from a bus and beginning a hike all at once, leading to a suddenly crowded trail. Visitors wanting a quiet trail must either be the first off the bus and keep ahead of the herd behind them or seek out a trail without its own bus stop. Those trailheads next to bus stops may receive much more foot traffic than before, requiring more maintenance and trail improvements. Transit requires much the same kind of hardening (development) as seen in the Mission 66 program, even if transit is designed to counter the road building associated with that program.

In 2005 Congress authorized funding for the Alternative Transportation in Parks and Public Lands (ATPPL) Program (Hodges and Faulk 2007). The program was designed to protect natural and cultural resources by reducing air, noise, and visual pollution while promising to improve the visitor experience. The program is administered by the Federal Transit Administration, in partnership with the Department of the Interior and the Forest Service. Since then, many parks have invested in alternative transportation systems to accommodate escalating visitation and to handle growing problems such as air and noise pollution related to car use in parks (Ament et al. 2008). Alternative transportation services are available in 12 per-

cent of the national park units (Table 7.2). As of 2019, the NPS had ninety-five transit systems with 45.9 million passengers boarding in sixty park units (National Park Service 2020d). Shuttles, buses, vans, and trams make up 59 percent of these transit systems, followed by boats and ferries (33 percent), trains and trolleys (4 percent), and aircraft (1 percent of transit systems). Transit system operations in park units generally vary according to seasonal visitation trends. While approximately 34 percent of the transit systems operate year-round, 12 percent of systems operate seven to ten months, and about 53 percent of the transit systems operate only three to six months of the year.

Among the top transit systems, the Statue of Liberty has the highest ridership: more than 10.3 million riders, followed by the Grand Canyon South Rim and Zion shuttle services (Table 7.3). Ridership on park transit systems has been increasing over the years. For example, there were 45.9 million riders in 2019 compared to 33.6 million riders in 2012, and in 2019 there were 3.8 million more riders compared to 2018 (National Park Service 2020d). While the top ten transit systems accounted for 84 percent of the 45.9 million riders in 2019, the parks experiencing the largest increases in ridership are the National Mall and Memorial Parks, Gulf Islands

TABLE 7.2   Parks with transit service, 2021

No.	Park Name	Type	State	Bus/Shuttle	Train
1	Acadia	NP	MN	■	
2	Adams	NHP	MA	■	
3	Bandelier	NM	NM	■	
4	Biscayne	NP	FL	■	
5	Bryce Canyon	NP	UT	■	
6	Cape Lookout	NS	NC	■	
7	Cuyahoga Valley	NP	OH	■	
8	Denali	NP	AK	■	
9	Devils Postpile	NM	CA	■	
10	Devils Tower	NM	WY	■	
11	Eisenhower	NHS	PA	■	
12	Eleanor Roosevelt	NHS	NY	■	
13	Eugene O'Neill	NHS	CA	■	
14	Fire Island	NS	NY	■	
15	Gettysburg	NMP	PA	■	
16	Glacier	NP	MT	■	
17	Golden Gate	NRA	CA	■	
18	Grand Canyon	NP	AZ	■	
19	Harpers Ferry	NHP	MD, VA, WV	■	
20	Independence	NHP	PA	■	
21	Indiana Dunes	NP	IN	■	
22	Jean Lafitte	NHPP	LA	■	
23	Lowell	NHP	MA		■
24	Manhattan Project	NHP	NM, TN, WA	■	
25	Mount Rainier	NP	WA	■	■
26	Muir Woods	NM	CA	■	
27	Pictured Rocks	NL	MI	■	
28	Point Reyes	NS	CA	■	
29	Rocky Mountain	NP	CO	■	
30	Sequoia–Kings Canyon	NP	CA	■	
31	Tumacácori	NHP	AZ	■	
32	Vanderbilt Mansion	NHS	NY	■	
33	Virgin Islands	NP	VI	■	
34	Yosemite	NP	CA	■	
35	Zion	NP	UT	■	

National Seashore, Yosemite Winter Ski Shuttle, and Blue Ridge Parkway Sharp Top Mountain Shuttle.

Through the implementation of alternative transportation systems, parks are accommodating visitors without adding more capacity to existing infrastructure and allowing more automobiles. Perhaps someday high-speed "green" trains will cut down even more on auto pollution and traffic jams, and we can develop "safari systems" to make the visits cleaner, more expansive, and more relaxed. Besides, the train is just plain more fun.

Flightseeing

Jules Verne's 1886 novel *Robur the Conqueror* (also known as *The Clipper of the Clouds*) revolves around the mysterious Robur and his flying machine; it includes a flight over Yellowstone National Park, which must surely be the first instance of flightseeing (aerial sightseeing) in the national park system. In 1917 a biplane landed and took off near the Firehole Hotel in the Lower Geyser Basin, perhaps the first nonfictional flight to Yellowstone (Haines 1996). Regular flightseeing began there in 1936 (Waite 2006).

By the 1950s overflights of national parks had become commonplace, though in the absence of any air traffic control to monitor the skies and keep aircraft away from each other, flights could also be dangerous. Airplanes are like stagecoaches in that passengers have views out the side of the vehicle. This means

TABLE 7.3   Passenger boarding for the ten highest use transit systems, 2019

Rank	Park	System name	Boarding	Business model	System purpose
1	Statue of Liberty NM	Statue of Liberty Ferries	10,370,679	Concession contract	Critical access
2	Grand Canyon NP	South Rim Shuttle Service	7,644,231	Service contract	Mobility to or within park
3	Zion NP	Zion Canyon Shuttle	6,777,100	Service contract	Critical access
4	National Mall and Memorial Parks	DC Circulator	5,565,092	Cooperative agreement	Transportation feature
5	Yosemite NP	Yosemite Valley Shuttle	3,161,758	Concession contract	Mobility to or within park
6	Golden Gate NRA	Alcatraz Cruises Ferry	1,680,553	Concession contract	Critical access
7	Pearl Harbor NMem	USS *Arizona* Memorial Tour	1,133,784	Cooperative agreement	Interpretive tour
8	Sequoia and Kings Canyon NP	Giant Forest Shuttle	940,164	Cooperative agreement	Critical access
9	Bryce Canyon NP	Bryce Canyon Shuttle and Rainbow Point Shuttle	774,010	Service contract	Mobility to or within park
10	Rocky Mountain NP	Bear Lake and Moraine Park Shuttle and Hiker Shuttle to Estes Park	764,423	Service contract	Mobility to or within park

Source: **National Park Service 2020d.**

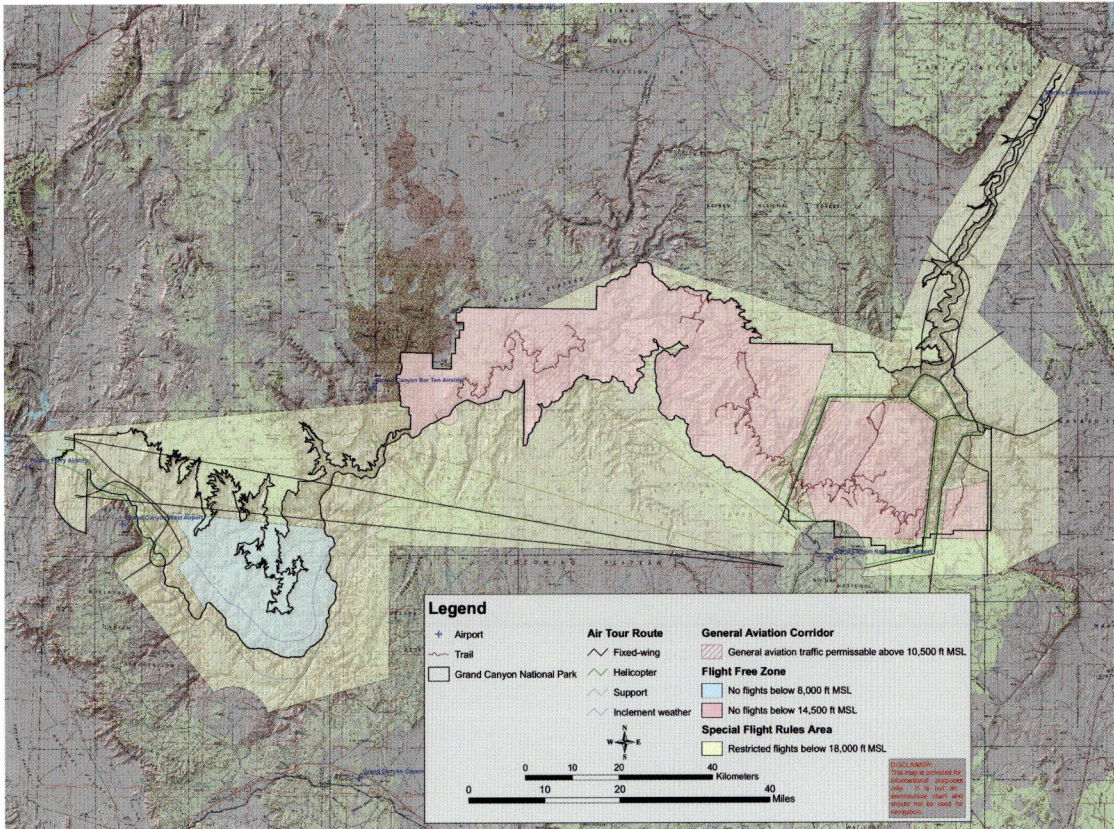

FIG. 7.16 Map of the Grand Canyon Special Flight Rules Area, 2021, a set of restrictions over where and how low aircraft can fly over the canyon. NPS.

that flightseeing airplanes must swing around so that both sides can see. In 1956 two airliners collided in the skies over the Grand Canyon while doing exactly this. All 128 on board the two planes were killed when the wreckage crashed near Temple and Chuar Buttes within the canyon. It was the deadliest air crash in the nation up to that time.

The skies above the Grand Canyon are now perhaps the most tightly regulated civilian airspace in the country, part of the Grand Canyon National Park Special Flight Rules Area (Fig. 7.16). This covers the park from the upper end of Lake Mead to Lees Ferry and restricts unauthorized flights below 18,000 feet above sea level or below 14,500 feet over the central part of the canyon. Flights are restricted below eight thousand feet where the Colorado River forms a big bend near Grand Canyon West. Despite these rules, crashes are common in and around the canyon; seventy-seven people have died in twelve crashes since 1986. Even when they are not crashing, these airplanes and he-

licopters are quite noisy. Airplane or helicopter noise is audible in much of the park, sometimes frequently (Ambrose and Florian 2008). In 2011 the park completed an air tour management plan to limit flightseeing aircraft to designated routes, altitudes, and hours, but this plan was overturned by politicians receiving large financial contributions from air tour operators.

The Federal Aviation Administration has received flightseeing proposals for 106 park units (Federal Aviation Administration 2019). At least fourteen other parks have developed or are pursuing air management plans. Most of them are large western, Hawaiian, or South Florida parks. One geographic grouping of parks not on the list are those on or near the National Mall in Washington, D.C.; overflights of these units are prevented by the air defense identification zone, which protects these memorials and nearby buildings from aerial attack. The airspace within thirty miles of Washington National Airport is included within this zone; the only way to get a spectacular aerial view of the mall is to get an airline ticket to Washington National Airport.

Conclusions

Although we like to think of parks as unchanging places, how we get to them and then move around within them has changed enormously over the years. Parks have mirrored changes in American society during this time, with railroads playing an enormous role before being replaced by private automobiles and, for long-distance travel, airlines. Some of these changes have been directly represented within the national park system, as at Golden Spike National Historical Park and Wright Brothers National Memorial. Transportation has become part of the landscape of many parks (Youngs, White, and Wodrich 2008). The stone arch at the north entrance of Yellowstone is only one of many features within national parks that have

become part of what we see and treasure. The grand lodges that survive in many parks, rustic bridges, overlooks with a rough stone wall, and even the aerial view of parks provided by airplanes are part of this landscape. Many roads are listed on the National Register of Historic Places, part of the country's historical heritage. The Federal Highway Administration, the successor to the Bureau of Public Roads, has a scenic byways program to identify the country's scenic and historic roads. There are now more than 120 of these roads, including several in the national park system. The Blue Ridge, Colonial, George Washington, and Natchez Trace Parkways are included, as are California Highway 190 through Death Valley National Park and Arizona Highway 67 at the North Rim of the Grand Canyon.

The means of transportation we use to explore a park affect how we perceive it. The boundaries of national park units are shown on

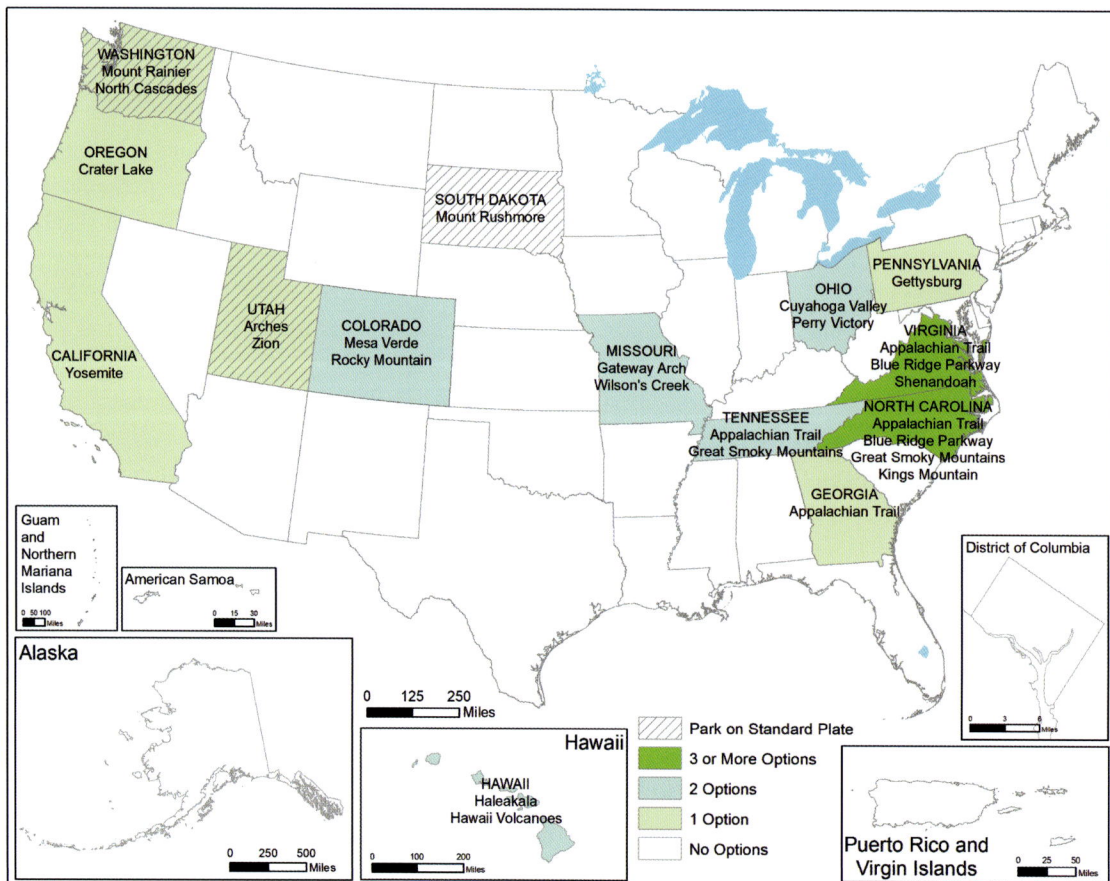

FIG. 7.17   Map of states with national parks on their license plates, 2021. Only a few states make national park images standard, but many offer them on optional versions.

the colorful maps distributed by the NPS. But these do not always correspond to what visitors see or experience when they visit. Although Petrified Forest National Park encompasses a large acreage, park regulations state that a person can enter the park only at either of the two road entrances, where toll booths await to extract a twenty-dollar fee from each car. The park is open only during daylight hours, and the entrance gates are shut in the evenings. There is no camping, and most of the park is closed to hiking or exploration on foot. Many travelers are making a quick loop through the park on their way west or east on I-40 and have no time to linger anyway. Although the park appears large when seen on the map, from the perspective of the average visitor it is a single winding road with a few side roads to overlooks.

Farther south in Arizona, Chiricahua Na-tional Monument is much smaller but has many hiking trails, allowing visitors to explore the rugged backcountry of hoodoos, or rock spires. A campground allows visits to stretch over several days, and those days would be necessary to hike the park's trails. Others may want to enjoy the solitude and peacefulness of Bonita Canyon, a quiet paradise far from the worries of the outside world. From the perspective of the visitor, Chiricahua will seem like a much larger park than Petrified Forest, despite the latter being eighteen times larger.

Finally, just as park visitation has been supported by mobility based on the family car, we can use our cars to support parks even if we can't get to one. Seventeen states and territories have license plate options that help support one or more parks (Fig. 7.17). Paying extra for one helps protect these places.

EIGHT

National Park Visitation

Parks would be nothing without people, and people have been coming to parks in large numbers. How many people visit, and how are they counted? Who are they, where do they come from, what do they do, and when do they come? The National Park Service has been concerned with these endless questions since it was created. This chapter addresses a range of topics related to park visitation, with a primary emphasis on examining national park visitors and their origins; the geographic, seasonal, and sociodemographic issues that influence their visitation patterns; and racial and ethnic diversity among park visitors. The chapter concludes with a discussion about shortcomings in park visitation data collection and the implications that these shortcomings have for documenting visitation patterns.

Like many other tourist activities, going to a national park is highly seasonal. Most parks have a summer peak when the weather is nice and families have the time to travel. A few parks have winter peaks, while some have no pronounced seasonal pattern. Who visits parks? Where do visitors come from? There is little information available about the origins of park visitors and their race and ethnicity, but a few geographic principles and the United States' historical relationship with its minority population can help fill in much of the story. Although parks may be "America's best idea," they are not visited equally by all Americans. A substantial number of visitors come from foreign countries, especially from Europe. A trip to a park has long been more common among the country's white population than among its minorities. While there are many explanations for lower at-

tendance rates among African Americans, Hispanics, and others, an important one is geography. There are many more opportunities to visit nearby parks in some areas of the country than others, and most minorities are concentrated in certain areas of the country. With rising diversity among American populations, attendance may decline if parks fail to attract diverse populations.

Visitation Trends

In 1904 the first accounting of visitors recorded a total of 120,690 visitors to national parks. At the time, the system contained only a handful of remote park units, located in the American West. Visitation did not change much over the next two decades but showed a substantial increase in the 1930s (Fig. 8.1). Though the country was in the midst of the Great Depression, it was also a time of rapid expansion in the park system, with the transfer of the National Mall and its memorials to the NPS, along with battlefields and national monuments under the control of other agencies. The NPS also worked to create new eastern parks closer to most of the population and put the Civilian Conservation Corps (CCC) to work developing parks for tourism.

Unfortunately, this growth abruptly ended with the coming of war in 1941. Potential park visitors instead went into uniform or factories to support the war effort. Even when people had time for vacations, the war made travel very difficult due to crowded trains and gas and tire rationing for those traveling by road. The park system also largely shut down for the duration, with budgets shrinking, the CCC pro-

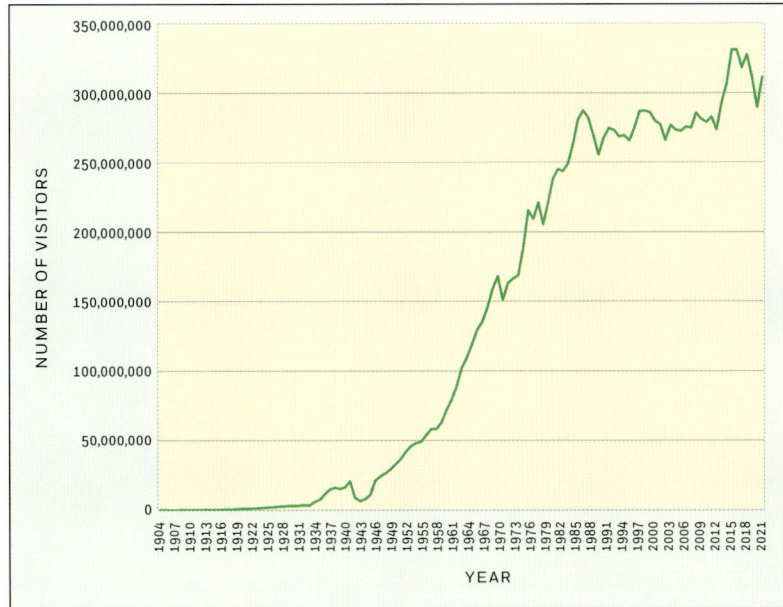

FIG. 8.1   Graph of total visitors to the national park system, 1904–2022. NPS.

gram ending, and many park employees trading their ranger uniforms for those of the army or navy. After peace returned in 1945, the parks reopened, and visitation rapidly returned to prewar levels. Then visitation continued to rise in a decades-long pattern of ever-increasing levels that the NPS has struggled to keep up with. The Mission 66 program was developed in part to handle this surge of visitor numbers and prevent parks from being overrun by the public. This ever-increasing growth came to an end around 1985, followed by a period of relative stability that was punctuated in 2015, when park visitation jumped to over three hundred million for the first time. At current levels, an average of 1.46 million people visit national park sites every day.

What are the most visited units in the national park system? The Golden Gate National Recreation Area is the most visited national park unit (more than fifteen million visitors), followed by the Blue Ridge Parkway (Fig. 8.2). Eight of the ten most-visited parks in the United States in 2019 are designated as national recreation area, parkway, or a historic unit (Table 8.1). Only two units that are designated as national parks are included in this list: the Great Smoky Mountains and the Grand Canyon. Looking beyond the top ten, the sixty-two national parks received the largest percentage (27 percent) of visitors annually, followed by national

recreation areas (16 percent), national memorials (12 percent), and national historical parks (11 percent) (Fig. 8.3).

Visitation trends reflect many factors; they show the negative effects of 1970s gas shortages but were almost unaffected by the 2007–9 recession. The multiple government shutdowns between 2015 and 2019 received tremendous attention but were of minor significance to visitor numbers. Visitation numbers for 2018 declined from 2017, reflecting several factors, including lingering economic impacts of hurricanes, large wildland fires in California, a volcanic eruption in Hawaii, and the longest government shutdown in U.S. history, lasting thirty-five days, from December 22, 2018, to January 25, 2019. Visitation rebounded in 2019, with almost 3 percent growth from the previous year.

The 2020–21 COVID-19 pandemic produced short-term increases in visitation but then severe decreases as parks shut down for much of the year (Fig. 8.4). There were decreases of 50 percent or more in 189 units, meaning they received less than half their usual numbers. Alaska's Glacier Bay, where most visitors arrive during the short summers on crowded cruise ships, saw a 99 percent decline in 2020. Blue Ridge Parkway, the second most popular unit, experienced a loss of eight hundred thousand visitors that year.

Fifty-eight units had increases; many of

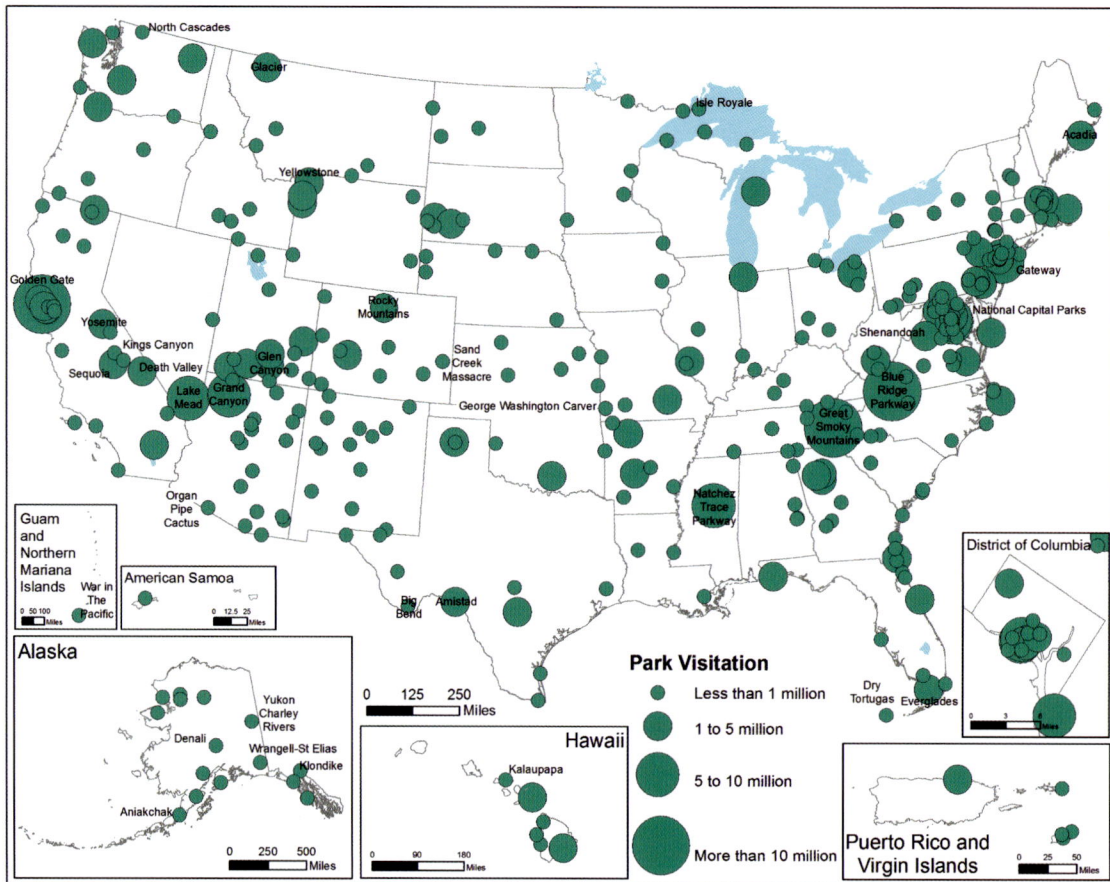

FIG. 8.2   Map of national park unit visitation, 2019. NPS.

TABLE 8.1   The twenty most-visited national park units, 2019

Other national park units	Number of visitors	National parks	Number of visitors
Golden Gate National Recreation Area	15,002,227	Great Smoky Mountains National Park	12,547,743
Blue Ridge Parkway	14,976,085	Grand Canyon National Park	5,974,411
Gateway National Recreation Area	9,405,622	Rocky Mountain National Park	4,670,053
Lincoln Memorial	7,808,182	Zion National Park	4,488,268
Lake Mead National Recreation Area	7,499,049	Yosemite National Park	4,422,861
George Washington Memorial Parkway	7,487,265	Yellowstone National Park	4,020,288
Natchez Trace Parkway	6,296,041	Acadia National Park	3,437,286
Gulf Islands National Seashore	5,600,240	Grand Teton National Park	3,405,614
Chesapeake & Ohio Canal National Historic Park	5,116,787	Olympic National Park	3,245,806
World War II Memorial	4,831,327	Glacier National Park	3,049,839

Source: National Park Service 2020e.

these were smaller units with modest visitation, but among them were Lake Mead, one of the most-visited park units, as well as several other large recreation units (Cape Hatteras, Cape Lookout, Assateague Island, Indiana Dunes, Lake Roosevelt, Upper Delaware, Padre Island, Sleeping Bear Dunes, and Cuyahoga Valley). Lakes appeared to have been popular getaways for urban populations eager to get outside, though other lakes and seashores experienced large declines. There are no clear patterns of relationship between visitation numbers and increases or decreases in COVID cases in the country, but parks that attract visitors from farther distances and require airplane travel to get to generally saw a decline in visitors. Once

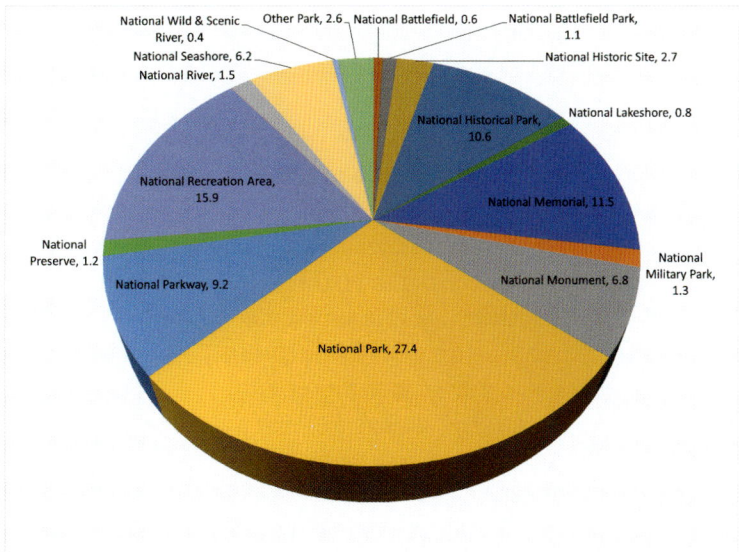

FIG. 8.3   Chart of visitors by type of park unit, 2019.
NPS.

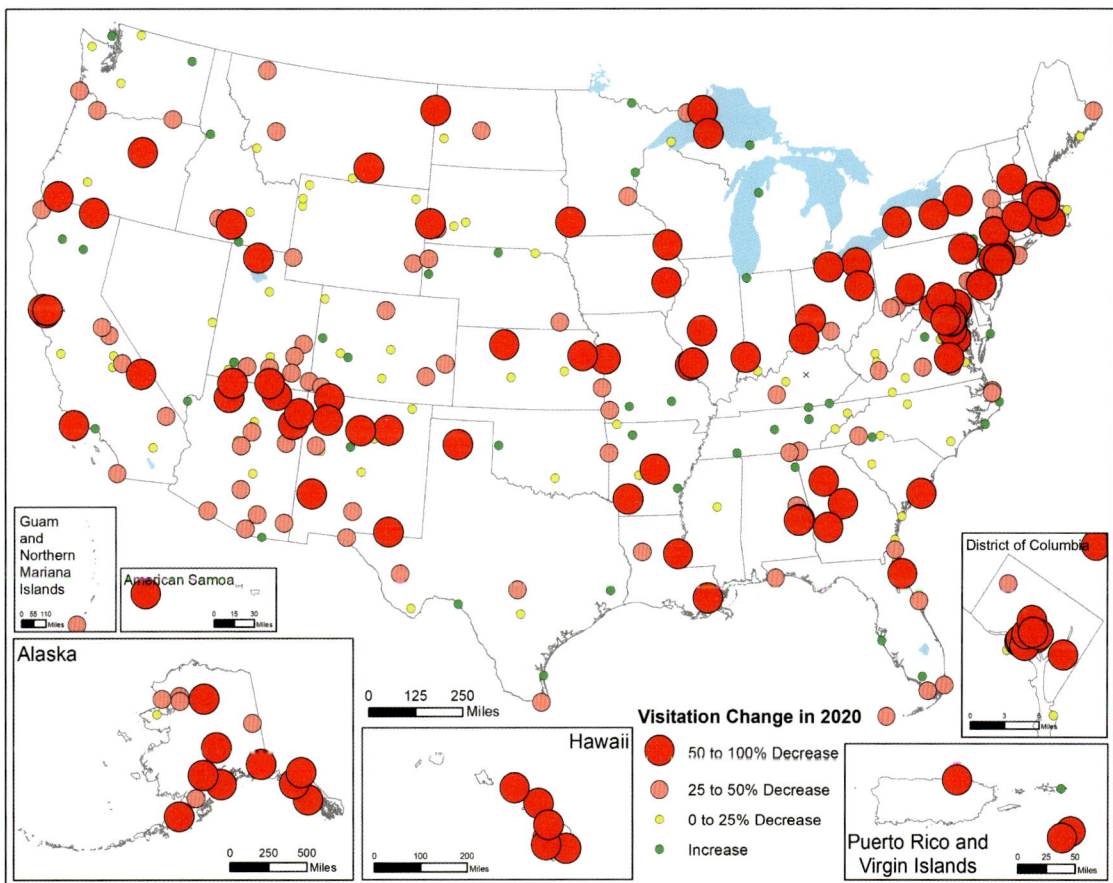

FIG. 8.4   Map of change in park visitation in 2020.

travel restrictions were lessened, parks saw an increased number of visitors (Kupfer et al. 2021). In fact, during the summer of 2021, many national parks experienced a surge of visitors that showed a substantial increase over previous years. For example, Yellowstone counted a record number of visitors (2,668,765) in July 2021 alone, which is 374,074 visitors more than compared to those counted in July 2019.

Those numbers are just recreation visits; park staff also collect statistics on nonrecreation visits, which encompass those passing through the park on a state highway, employees and their families, and deliveries to the park. These can be large; George Washington Memorial Parkway, an urban freeway in Washington, D.C., had 37.2 million nonrecreation visits in 2017 compared to 7.5 million recreation visits. In fact, only 4 percent of the National Capital Parks East group is counted as recreation visits. These urban parks and roads are crisscrossed by all manner of trips every day. The same is true in other cities. Atlanta's Kennesaw Mountain likewise has 25.2 million nonrecreation visits but only 2.5 million recreation visits. Similarly, 90 percent of visitors to Guilford Courthouse National Military Park in suburban Greensboro, North Carolina, are nonrecreational. These numbers highlight the challenge of running a small battlefield park within a large metropolitan area.

How long do visitors stay in national parks? The NPS estimates visitor recreation hours based on the number of visitors and the average time each spends in the park. By this measure, the Blue Ridge Parkway is again the most visited park unit, followed by the Grand Canyon, Yellowstone, Great Smoky Mountains, Yosemite, Lake Mead, Glen Canyon, Sequoia, Glacier, and Rocky Mountain. This list more closely matches the expectation that the big western parks will absorb the greatest share of park visitors and explains why they can be so crowded.

When recreation hours are divided by visits, Rio Grande Wild and Scenic River is the unit with the longest average visit, followed by Aniakchak, Yukon–Charley Rivers, and Wrangell–St. Elias, three vast Alaskan parks with a small visitation of dedicated wilderness explorers. Isle Royale in Lake Superior is

another remote and rarely visited unit, while Kings Canyon, Sequoia, Big Bend, Yellowstone, and North Cascades are more conventional and easily reached parks.

What is the least visited park unit? This is a trickier question. Many of the newest park units have not yet started counting visitors, and there are others (such as Boston Islands National Recreation Area) that never did. Hohokam Pima National Monument and Saint Croix Island International Historic Site are closed to the public and have no visitation. At any given time, some units are temporarily closed for renovations; neither Clara Barton nor the Washington Monument was open to the public in 2017 and so officially had zero visitors that year. Hamilton Grange National Memorial, the home of Alexander Hamilton, was closed in 2006 while the park was moved; the entire house was moved by the NPS several blocks to a nearby location and reopened in 2011.

Among those units where visitation is counted and that were open to the public, Aniakchak National Monument and Preserve is the least visited of all national park units, counting just one hundred people (though the average visitor spends eighty-four hours there, far more than almost every other park). By comparison, about five hundred people make it to the top of Mount Everest each year (though they usually stay at the top for less than an hour). Of the eleven units receiving fewer than five thousand visitors, three are in Alaska, while Rio Grande Wild and Scenic River is in a remote southern corner of Texas. (The NPS counts visitors by the number of river-rafting permits issued.) The remainder are mainly small historic sites, many of them little known (Table 8.2). Four rarely visited units are devoted to African Americans, and two are dedicated to women. The Mary McLeod Bethune Council House National Historic Site and the Thaddeus Kosciuszko National Memorial are also the two units with the fewest recreation hours and not located near the main museum hotspots in their cities.

Several units have limited hours or limits on visitation. Isle Royale, Devils Postpile, and several others are open in the summer months only; Thomas Stone National Historic Site closes in

TABLE 8.2 Parks with fewer than five thousand visitors annually, 2019

Rank	Park	Location	Number of visitors
1	Aniakchak National Monument and Preserve	Alaska	100
2	Rio Grande Wild and Scenic River	Texas	324
3	Port Chicago Naval Magazine National Memorial	California	830
4	Yukon–Charley Rivers National Preserve	Alaska	1,114
5	Thaddeus Kosciuszko National Memorial	Pennsylvania	1,921
6	Carter G. Woodson Home National Historic Site	Washington, D.C.	2,381
7	Bering Land Bridge National Preserve	Alaska	2,642
8	Eugene O'Neill National Historic Site	California	2,944
9	Nicodemus National Historic Site	Kansas	3,540
10	Mary McLeod Bethune Council House National Historic Site	Washington, D.C.	3,788
11	Clara Barton National Historic Site	Maryland	4,100

Source: **National Park Service 2020e.**

FIG. 8.5 Port Chicago Naval Magazine National Memorial, Concord, California. Located within a military base, it requires advance planning to visit. Photo by Luther Bailey, NPS.

the winter and the rest of the year is open only four days a week, while Carter Woodson National Historic Site is only open three days a week. Only three hundred people are allowed on Cumberland Island at a time. Kalaupapa National Historical Park allows in only one hundred people each day, and every visitor must have a permit from the state health department and be over the age of sixteen.

Several parks are not directly accessible to visitors. At these interested visitors are required to make prior reservations to be transported by a shuttle service from a nearby location. For example, Eugene O'Neill National Historic Site is open to the public but needs to be accessed through a private gated road owned by a community of neighboring residents. The Park Service won't tell you where

it is; visitors must obtain an advance reservation and then go to a local history museum, where they board a bus that takes them to the park. (For those interested, the park is at 1000 Kuss Road in Danville, California.) Reservations are also required if you want to tour the White House; you need to contact your congressperson at least twenty-one days in advance for a ticket. If you want to climb the interior of the Statue of Liberty, you will need to buy your tickets six months in advance.

One of the most restricted parks is Port Chicago Naval Magazine National Memorial, which commemorates an explosion that killed hundreds of African American workers at a segregated loading facility during World War II (Fig. 8.5). It is located within an active navy base, the Military Ocean Terminal Concord, a port for

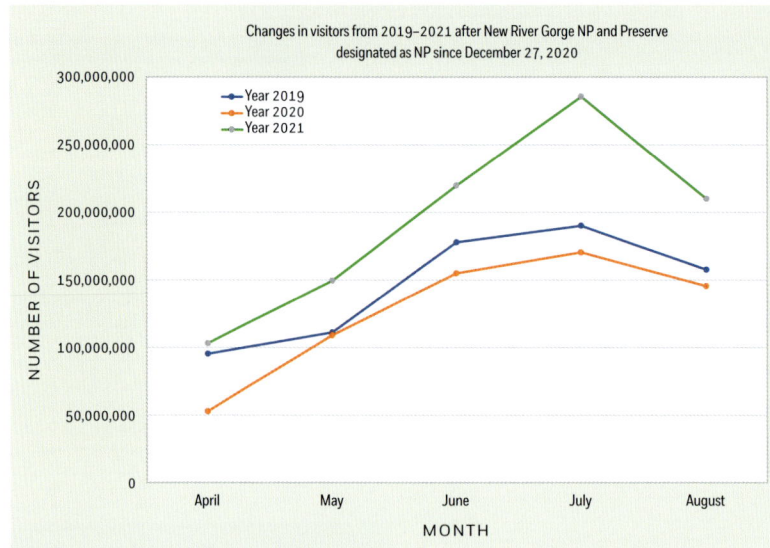

Changes in visitors from 2019–2021 after New River Gorge NP and Preserve designated as NP since December 27, 2020

FIG. 8.6   Chart of summer visitation at New River Gorge National Park and Preserve before and after the 2020 redesignation.

shipping ammunition and explosives to overseas bases. As a result, it may be visited only on guided tours, and prospective visitors must arrange the guided tour with at least two weeks' notice (and must provide a government-issued photo identification).

Redesignation and Visitation

Many have suggested that changing a national monument to a national park, even with no other changes to boundaries or facilities, will give the park greater visibility and increased visitation. Does this work? The answer to this question is yes, but only for a short while. Studies of Black Canyon of the Gunnison and Great Sand Dunes found that redesignation from a monument to a park produced an increase in visitation for several years before attendance fell to a new but higher base level (Seidl and Weiler 2001; Weiler and Seidl 2004).

Death Valley National Monument saw a surge in visitation in the early 1990s as news of it becoming a national park became common, and after this happened in 1994 visitation soared to a new level more than double that of the 1980s (Wines 2019). But after 1999 visitation declined until 2007, when it once again began growing to a new record high in 2018. Reasons for declines or growth can obviously be complicated, but in Death Valley the recent growth may stem from media attention to spring wildflower booms, as well as the 2013 anni-

versary of the 1913 record high temperature (134° F). The recession that began at the end of 2007 did not appear to create a downtown, and neither did the closure since 2015 of Scotty's Castle, one of the park's main attractions. A massive jump of almost four hundred thousand visitors between 2017 and 2018 (the largest single-year change recorded) may simply be due to more accurate counts, as one of the park's main entrance highways finally received a traffic counter.

The New River Gorge National Park and Preserve in West Virginia found itself in the national spotlight after it was redesignated a national park in December 2020. The park's website saw a 90 percent increase in traffic, and while the park experienced a slight decline in visitors during 2020 due to COVID-19, the park has been experiencing a surge in visitors since its redesignation. There were about 1.2 million visitors at the park in 2019, while as of August 2021, more than 1 million people had already visited the park that year (Fig. 8.6). The park experienced the highest surge of visitors (a 50 percent increase) in the month July and was 32 percent higher than 2019 from April to August of 2021.

What frequently does not change with redesignation is the visitor experience. Aside from increased traffic, the expansion and/or redesignation of a park may bring little immediate change except for new entrance signs. Death

Valley was expanded and redesignated in 1994, but aside from several relocated entrance signs, the only development within added lands has been the paving of a parking area near Rainbow Canyon. A similar story can be told at Great Sand Dunes, greatly expanded and redesignated in 2004 (Geary 2016). Aside from a new sign and remodeled visitor center there are no visible changes, and most tourists still go to the same established locations. Many new sections of the park are not even open to the public, a situation that also existed at Petrified Forest following its 2004 expansion. Similarly, while traffic in New River Gorge increased in 2021, it did not receive additional funding after being redesignated as a national park. The area has few accommodations and other amenities such as restaurants, and there is a shortage of parking at the historic Thurmond Depot visitor center. The future of this park is now in the hands of the Park Service if it would like to see a continued increase in visitors.

The expansion that often comes with redesignation may, however, greatly increase the size of the backcountry in a park. Only serious backcountry hikers would ever be able to see much of the new lands, and any increased visitation that occurs is concentrated in the same small developed core of the park. An increase in visitation due to a redesignation or expansion will therefore likely increase congestion and crowding in that park.

How Are Visitors Counted?

Parks employ a range of methods for counting visitors, and almost every park does it differently. (The specific methods used can be seen on the NPS visitor use website.) Most park visitors arrive by car, so most parks simply use traffic counters laid across the entrance road. Alabama's Russell Cave National Monument has an induction traffic counter to count vehicles entering the park, with the number divided by two to account for those leaving on the same road and then multiplied by 3.1 to account for the average number of people per car. (Bus passengers are counted individually.) Guilford Courthouse has several traffic counters with a multiplier of 2.3. Yellowstone follows a similar practice but with more traffic counters at the various entrances and multipliers that vary by entrance and month.

Bigger parks naturally require more methods. The Grand Canyon records the number of people entering the park with four traffic counters at several entrances, along with counts of passengers on buses and trains and people rafting the Colorado River. (Those seeing the park on a flightseeing trip are not counted.) The Great Smoky Mountains has sixteen traffic counters on different entrance roads, with visitors entering at another road simply estimated and bus passengers counted separately. The traffic counts must all be multiplied by persons-per-vehicle factors, which vary by season, and the final count of visitors is reduced by 12 percent to account for those being counted more than once. Golden Gate National Recreation Area requires the summation of twenty-four different numbers representing traffic counters on various streets and parking lots; the number of Alcatraz ferry passengers; counts of pedestrians, bicyclists, and joggers at various beaches and open areas; and diners served at two restaurants within the unit, all with appropriate multipliers for people per vehicle and season.

What if you drove through a park and didn't stop? Many parks are crossed by main highways, and in some parks, these vehicles are counted as visitors (as in Death Valley and the Great Smoky Mountains), but in others they are not (as at Golden Gate, Redwoods, and Petrified Forest). Lake Mead counts vehicles at five traffic counters, bus passengers at four scattered locations, Colorado River rafters, people arriving by airplane at the park's two airports, visitors taking a boat tour, and those walking into the visitor center. But the park does not count those who drive through the park on busy U.S. 93 nor those who visit Hoover Dam; the dam is a Bureau of Reclamation facility within the national recreation area. Counting dam visitors would add about 700,000 more to the unit's count, while multiplying annual average daily highway traffic for U.S. 93 by the 3.3 multiplier the park uses would yield an additional 24,692,250 visitors. Hohokam Pima National Monument is in central Arizona on the Gila Indian Reservation. It was created in 1972 to protect an archaeological site known as Snaketown, which had been excavated between the 1930s and 1960s. The monument is one of several owned by an Indian tribe, which in this case has decided against opening the site to the public; there is nothing to see, as the excavation was carefully backfilled. But the monument does have visitors: the I-10 freeway runs through a corner, and this busy highway carries twenty-eight thousand vehicles every day, meaning that more than ten million people pass through the tiny monument every year.

Each of the park units located on the mall has its own methods. The Washington Monument counts the number of elevator riders to the top viewing room. When the elevator was closed for maintenance from 2016 to 2019, no visitors were counted. Counts for 2016 and 2019 were lower due to the closure of the memorial during part of the year, while no visitors were recorded for 2017 and 2018. This may come as a surprise to those who have walked up to it or even touched it but either chose not to stand in line for the elevator or were there when it was not open. Officially, those visitors didn't visit the Washington Monument. At the nearby Lincoln Memorial, a park employee must count the number of people inside the memorial six times a day between 7:00 a.m. and midnight, dividing these counts by six to get an average and then multiplying this by sixty-four (used because there are sixty-four fifteen-minute periods daily when the memorial is open) for a final count. The Secret Service contributes some of the visitor data used to calculate White House visitation numbers.

The National Mall is in a category by itself. Visitors to this unit are tallied separately from each of the memorials located within it. The mall counts visitors in the gift shops of the Lincoln and Jefferson Memorials and Washington Monument, as well as the number of people using the snack bars at these memorials and even in three of the Smithsonian museums. Because of the Park Service's unique responsibilities in the District of Columbia, it operates several city parks, tennis courts, golf courses, marinas, and athletic fields. Visitors using these facilities and spectators at sporting events are counted. The number of people participating in special events on the mall, such as the annual Cherry Blos-som Festival, is also included (though the NPS stopped estimating the sizes of crowds attending political rallies on the mall after 1995 due to controversies over these numbers).

Denali is Alaska's most famous park unit and receives far more visitors than Glacier Bay, Kenai Fjords, and other easily accessible sites. But it is only the second most visited park unit in the state. What is the most visited park unit in Alaska? Tiny Klondike Gold Rush National Historical Park in the town of Skagway in the state's panhandle, which had more than a million visitors in 2019. How is this possible? The NPS visitor counting guide reveals the answer: during the summer months the unit counts every cruise ship and marine ferry passenger as a visitor, along with everyone who arrives in Skagway by air, train, or highway. In other words, every person who enters the town between May and September for any purpose is counted as a park visitor!

Finally, how about visitation by pets? The NPS does not keep statistics on this, and some units ban pets altogether. Those that do allow pets have many rules about where they can go and under what conditions. Service animals are allowed wherever visitors can go, but these are limited to professionally trained dogs, and some parks also impose additional requirements on them to protect park resources. Service dogs headed for the Channel Islands require a permit and must be vaccinated and quarantined before they can travel to the islands; this is to protect wildlife from domestic diseases. Similar requirements exist at Isle Royale. Pets and even service dogs not only are threats to wildlife but also can be threatened. Many parks contain carnivores eager to make a meal out of family pets, and bears and wolves are more likely to attack humans if they are accompanied by dogs. The number of park pet visitors is best kept low.

The Geography of National Park Visitors

While all parks count visitors, parks do not commonly track where visitors are coming from. This requires additional data collection efforts, which have been carried out at few parks and only on rare occasions. Visitor sur-

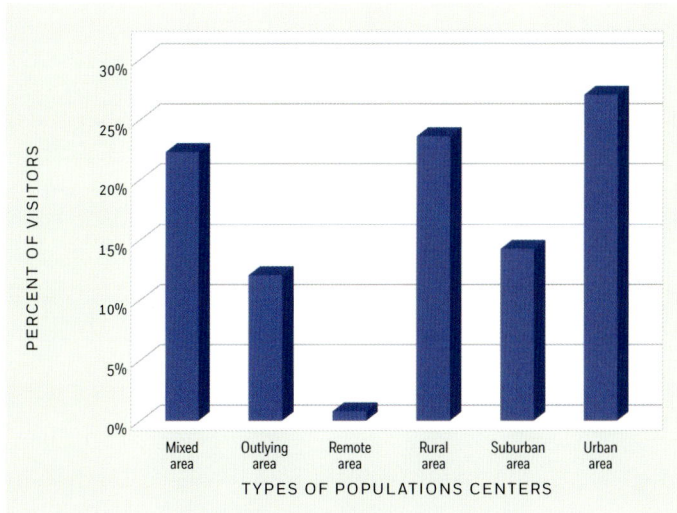

FIG. 8.7   Chart of percent of visitors to parks by location of park unit, 2019.

veys have been carried out at several dozen units over the last few decades, and they reveal that most park visitors are local, usually from the state where the park is located. This is the case with Acadia, Congaree, Gettysburg, New River Gorge, Rocky Mountain, Shenandoah, and the Great Smoky Mountains. Even the most famous parks also have an overwhelming number of visitors from their own state. Most Yosemite visitors are from California, Mount Rainier visitors are largely from Washington, and drivers of the George Washington Memorial Parkway are usually from the Washington, D.C., area. California has the largest population of any state, so Californians often make up the largest groups of visitors to parks in states with smaller populations, as at Haleakalā, Denali, and even Yellowstone.

The geographic proximity of national parks relative to population matters greatly to park visitation (Weber and Sultana 2013). Given that most of the population in the United States lives in a big city, it is no surprise that urban parks make up 27 percent of all park visitors, and the majority of the top ten most visited parks are in major urban centers (Fig. 8.7). Golden Gate National Recreation Area, the largest urban park in the world, was the most-visited park unit in 2019. It is in San Francisco, a metropolitan area with more than 6.7 million people. Similarly, Gateway National Recreation Area is the largest urban park in the New York City area, sur-

rounded by over 20 million urban residents. The Lincoln Memorial and Chesapeake and Ohio Canal are in the Washington, D.C., metropolitan area (more than 6.3 million people), while Lake Mead is adjacent to Las Vegas (2.2 million residents and more than 40 million visitors).

Parks located in rural areas but easily accessible by paved highways, air, or water make up the second-largest group of parks by visitor numbers (24 percent). The largest nature-oriented parks designated as national parks are generally located in such locations, including the Great Smoky Mountains, Blue Ridge Parkway, Grand Canyon, and Rocky Mountain National Parks. Parks located in mixed areas, which are rural but adjacent to metropolitan areas with a population of less than one million, have the third-largest share of visitors (22 percent).

Remotely located parks and those that require special travel arrangements have the lowest share of visitors (less than 1 percent). Isle Royale National Park in Michigan and Katmai National Park, Yukon–Charley Rivers National Preserve, and Bering Land Bridge National Preserve in Alaska are infrequently visited. Those who take the time and make the extra effort to reach them will appreciate their serenity and beauty, both due to the lack of crowding.

The NPS groups all parks into one of six geographic regions. If park visitation is examined within each region, the pattern remains

TABLE 8.3   Domestic national park visitors by NPS region, 2019

Region name	Visitors from the Intermountain Region (%)	Visitors from Alaska (%)	Visitors from the Midwest Region (%)	Visitors from the Northeast Region (%)	Visitors from the Pacific West Region (%)	Visitors from the Southeast Region (%)
Intermountain	45	0	26	9	15	5
Alaska	16	16	27	14	20	8
Midwest	7	0	84	3	3	3
Northeast / District of Columbia Region	7	0	6	83	0	4
Pacific West	5	0	4	6	82	3
Southeast	2	0	17	12	2	67

Source: National Park Service 2020e.

similar: Americans are more likely to visit parks nearby than those farther away (Table 8.3). The largest share of visitors in every park service region, except for Alaska, is from within the same region. Of all the visitors to large midwestern parks such as Wind Cave and Indiana Dunes, 84 percent come from the Midwest; 83 percent of visitors who visit parks in the Northeast live in that region; and 82 percent of visitors to Pacific West parks are residents there. This trend is not as strong in the Intermountain West (which includes the Four Corners states of Arizona, Utah, Colorado, and New Mexico, as well as Wyoming, Montana, Texas, and Oklahoma) and southeastern parks, where larger and more famous parks attract people from wider geographic areas. Yet even in these regions, most visitors are local (45 percent and 67 percent, respectively).

If most national park visitors are local, then an equal distribution of parks within each region of the country would allow all Americans, regardless of where they live, an equal opportunity to visit a park within driving distance of their homes. This, however, is not the case. The largest nature-oriented parks designated as national parks are concentrated in the Pacific West (which includes Hawaii) and the Intermountain West, far removed from the densely populated eastern United States. The Northeast (which includes Virginia, West Virginia, and every state to the northeast), for instance, only has two national parks, Acadia in Maine and New River Gorge in West Virginia, for 21 percent of the American population. The Southeast and Midwest, where almost 60 percent of the U.S. population live, have multiple states without a single national park (though with many battlefields, seashores, lakeshores, and historic units). Florida is the only state east of the Mississippi River to have more than one national park, while there are ten states west of the Mississippi that do.

Visitors from Abroad

Almost every park unit receives visitors from a wide geographic area, and many parks are international destinations. According to a U.S. Travel Association analysis of Commerce Department data, national parks are an increasingly attractive destination for international visitors to the United States (Fig. 8.8). In 2015 35.4 percent (13.6 million) of all international travelers to the United States visited national parks and monuments. This percentage continued to increase to 36.5 percent (about 14.6 million visitors) until 2017, which was up 7.3 percent from 2015. International visitors to national park units declined 1.5 percent despite their increasing presence in the country for other purposes. Overall, there was a slight decrease of international travelers, about 0.7 percent, to the United States in 2019 due to international travel restrictions, which in turn might have led to a further decline of international visitors to national park units.

Overall, park visitation varies widely depending on the world region (Table 8.4). Oceanic, European, and Nordic visitors to the United States have park visitation rates of greater than 40 percent, while visitors from the Caribbean, Central America, and Asia have national park visitation rates at or below 25 percent. These differences in park visitation by

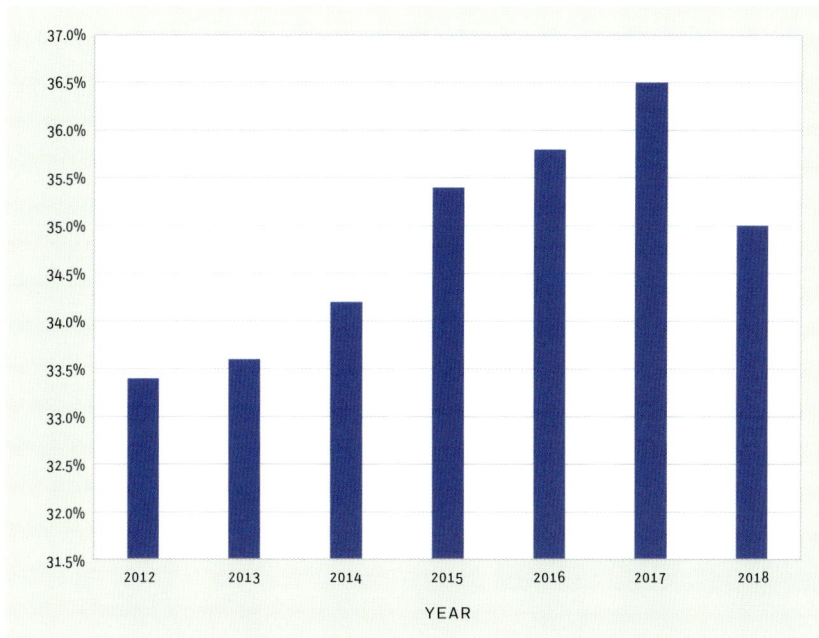

FIG. 8.8   Chart of percent of international visitors to the United States who visit national parks, 2012–18.

TABLE 8.4   International visitors by world region, 2018

Origin region	Number of arrivals	Respondent sample size	Percent visited national parks
Africa	598,000	797	32.1
Asia	11,874,000	13,911	25.3
Caribbean	1,793,000	2,228	19.4
Central America	1,283,000	1,276	23.2
Europe	15,424,000	14,284	42.4
Middle East	1,197,000	909	37.9
Oceania	1,687,000	1,758	52.3
South America	6,027,000	6,436	29.8

Source: Visa 2023.

world region may be explained by cultural differences, fiscal constraints, or some other factor that is not yet clearly understood.

The National Travel and Tourism Office within the U.S. Department of Commerce also tracks visitation to cultural and ethnic heritage sites managed by the National Park System (Table 8.5). Unlike visitation to national parks and monuments, there is little variation in ethnic and cultural heritage site visitation between international tourists from different regions. While Oceanic visitors once again report the highest park visitation rates and Caribbean visitors report the lowest rates to cultural sites, the spread between them is much smaller than it was for national parks and monuments. Visitors

from all the other regions report visiting cultural and ethnic heritage sites at rates between 10 and 20 percent.

The national park label is a very important part of attracting international visitors. The largest and the most famous national parks, such as Grand Canyon, Grand Teton, Yellowstone, Yosemite, and Zion, are far more likely to attract international visitors (Table 8.6). These famous parks receive approximately 45 percent of their visitors from abroad. These parks have another attribute in common with each other: location. All are within driving distance of multiple parks and can easily be seen on one visit to the United States. According to the Visa (2023) database, 35 percent of international vis-

TABLE 8.5   International cultural and heritage site visitors, 2018

Origin region	Number of arrivals	Respondent sample size	Percent visited cultural/ethnic heritage sites
Africa	598,000	797	13.9
Asia	11,874,000	13,911	11.6
Caribbean	1,793,000	2,228	7.7
Central America	1,283,000	1,276	10.6
Europe	15,424,000	14,284	18.3
Middle East	1,197,000	909	15.6
Nordic countries	1,333,000	1,077	11.6
Oceania	1,687,000	1,758	22.9
South America	6,027,000	6,436	12.1

Source: National Travel and Tourism Office 2023.

TABLE 8.6   Top ten national parks that received the most international visitors, 2019

Rank	Name of the park	Region	Percent international visitors of total park visits
1	Grand Canyon	Intermountain	14
2	Yosemite	Pacific West	11
3	Glacier	Intermountain	9
4	Yellowstone	Intermountain	7
5	Bryce Canyon	Intermountain	7
6	Zion	Intermountain	6
7	Death Valley	Pacific West	6
8	Acadia	Northeast Region	5
9	Teton	Intermountain	4
10	Olympic	Pacific West	4

Source: Visa 2023.

itors to the Grand Canyon also visited multiple nearby parks. This trend is even higher among the European travelers: 40 percent included parks located in California, Utah, or Nevada in their trip.

Some parks are simply easier to get to by foreign visitors. War in the Pacific National Historical Park on Guam has received large numbers of Japanese and Korean visitors since the 1990s (Evans-Hatch and Associates 2004); about 90 percent of visitors to the island are from Japan. The island has very strong airline connections to those countries and few to the rest of the United States. Hawaii is also well connected to Asia, and about 20 percent of the visitors to the USS *Arizona* Memorial were foreign. On the other hand, a 1997 survey showed that only 5 percent of visitors to Virgin Islands National Park were foreign; New York and Massachusetts, strongly connected by frequent

air services, were the most common origins (Visitor Services Project 1997).

A few other park units impose some conditions on international visitors. All park units except Port Chicago Naval Magazine National Memorial and the Oak Ridge unit of Manhattan Project National Historical Park are open to visitors regardless of their origin; those two units are open only to U.S. citizens. Organ Pipe Cactus National Monument and Amistad National Recreation Area both have road entrances from Mexico; these entry points are controlled by U.S. Customs and Border Protection. Visitors leaving these parks to the south will likewise require approval by the Mexico Customs Authority, meaning that park visitors who are not North Americans may need a visa to leave these parks by these exits. Those foreign visitors wishing to take the White House tour are asked to contact their embassies to get tickets;

nine countries have no embassies in the United States to arrange this, so citizens of these countries are unable to visit this NPS site (though it seems unlikely there will be many park tourists from North Korea anytime soon).

Seasonality in Park Visitation

Many parks are in places with climatic extremes and are highly seasonal in their visitation patterns, with a peak season that attracts the most visitors. For most parks the peak season is summer (Fig. 8.9). July has consistently been the busiest month at Yellowstone, reaching nearly a million visitors in some years. This means that more people visit Yellowstone in July than visit most park units in an entire year. August in Yellowstone is lower, followed by June, September, and May. December, the least visited month, sometimes only sees one-twentieth the visitation of July (which is still

more than three dozen parks get each year). The Grand Canyon and Yosemite also show this pattern, as does the Great Smoky Mountains and New River Gorge, though with a second peak when the leaves change color in October. The busiest visiting season for Shenandoah, Virginia, begins in June and continues until a peak in October.

Acadia is a northern park with a late summer peak (reaching a high in August) that exhibits a slow tapering effect until October, also due to fall color. Denali receives most of its visitors from June to August, as do Glacier Bay and Katmai, while Kenai Fjords has a longer summer season. Aniakchak receives its entire visitation between June and September in most years. While Alaskan parks are open year-round, some parks close most or all of their facilities during winter months due to harsh winters at high elevations. Crater Lake National Park, Cedar Breaks National Monument, and

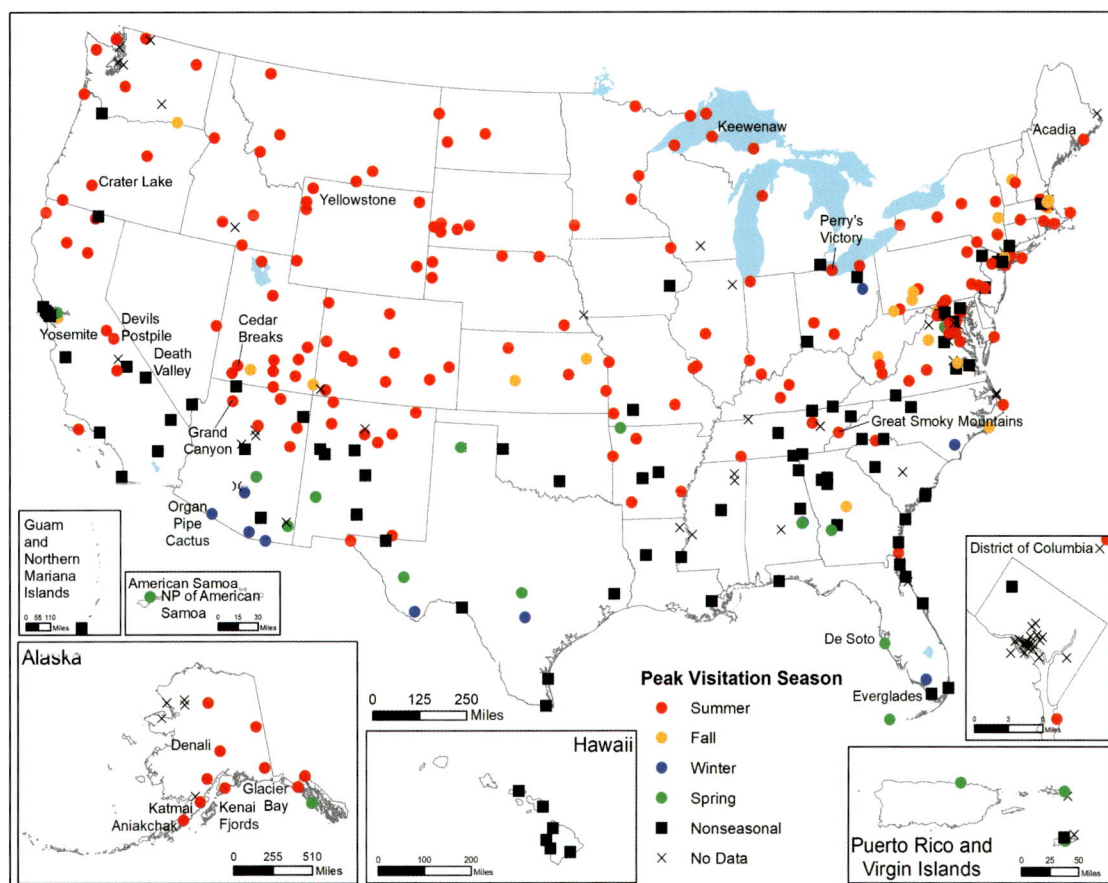

FIG. 8.9   Map of peak visitation season by park unit, 2019. Most parks have their highest levels of visitors during the summer.

Devils Postpile National Monument are other examples of parks that are restricted by the extremes of elevation.

For some units in the desert Southwest or subtropical Southeast, the main season is winter, a time of cooler temperatures coinciding with the annual arrival of vacationing snowbirds from the frigid North. Everglades and Organ Pipe Cactus show this pattern well, with peak visitation from January to March. The National Park of American Samoa is also a spring park, but in the Southern Hemisphere, this falls from September to December. Death Valley, the hottest park, is surprisingly nonseasonal. It has seen uniform visitation throughout the year, though often with a peak in the spring, a pleasant time to visit and when the valley often has abundant wildflowers on display. But in the past ten years, the peak month has often been in the summer, with people coming to experience the world's hottest temperature. In fact, in 2017 March was only slightly ahead of August for recreation hours, and when the park received a record number of visitors in 2018, it was August that was the busiest month (Wines 2019).

Smaller parks focused on history and culture also tend to exhibit seasonal visitation trends tied to weather and climate. The Keweenaw National Historical Park in the Upper Peninsula of Michigan, for instance, sees its peak visitation in the summer and early fall. The park experiences a precipitous drop-off in visitation during the harsh northern Michigan winter. On the other hand, the De Soto National Memorial in Florida sees its lowest visitation during the hot Florida summers, with higher visitation rates during slightly cooler times of the year when there are more vacationers in the area.

While almost every historical or cultural park has an off-peak visitation season, the parks that are closer to larger cities tend to see smaller drops in visitation compared to parks that are in difficult-to-reach areas. Visitor numbers at Guilford Courthouse, located in Greensboro, North Carolina, do not vary much between June and January. The situation at Perry's Victory and International Peace Memorial, a small park located on an island in Lake Erie, is quite different. The memorial sees between twenty thousand and forty thousand visitors a month during the peak summer tourism season, but there have been months in the winter where the park did not even record two hundred visitors. While this is an extreme example, it illustrates just how significant the changing of the seasons is to park visitation.

Visitor Activities

What do people do in parks? Park activities obviously vary by park type, and the resources within each park are often very specific. The most common activities reported in surveys are usually sightseeing and viewing scenery, with hiking often a close second (Vaske and Lyon 2014). Wildflower viewing is also popular, along with picnicking. However, these activities vary by park region. For example, while sightseeing is the major activity in parks located in the Alaska region (reported by 93 percent of visitors), hiking is the most popular thing to do in the parks located in the national capital (Fig. 8.10). Water activities are popular in parks located in the Midwest, Northeast, and Intermountain West, but not in parks located in Alaska and the national capital. Parks in Alaska are major destinations for those interested in the creative arts, but visitors with such interests did not visit parks in the national capital. Interest in history is popular in the Midwest and Northeast, where historic sites are found in abundance. Fishing and hunting are major activities in Alaska; the state has excellent fishing, and the many national preserves are among the few places in the park system where hunting is allowed. Bicycling is popular in the parks located in the Midwest, national capital, and Northeast but less common in the parks located in Alaska and the Intermountain West. Climbing is a popular activity in the parks located in the Pacific West and Northeast, but it is less common in other regions' parks, especially in flat midwestern parks.

Visitors often must match their chosen activity with a park, as some parks are better for certain activities than others. At Katmai National Park, 79 percent of visitors were there to watch bears, while at Haleakalā, 43 percent of visitors came to see the sunrise, with another 13 percent visiting to watch the sunset. At small historic sites or monuments, viewing exhib-

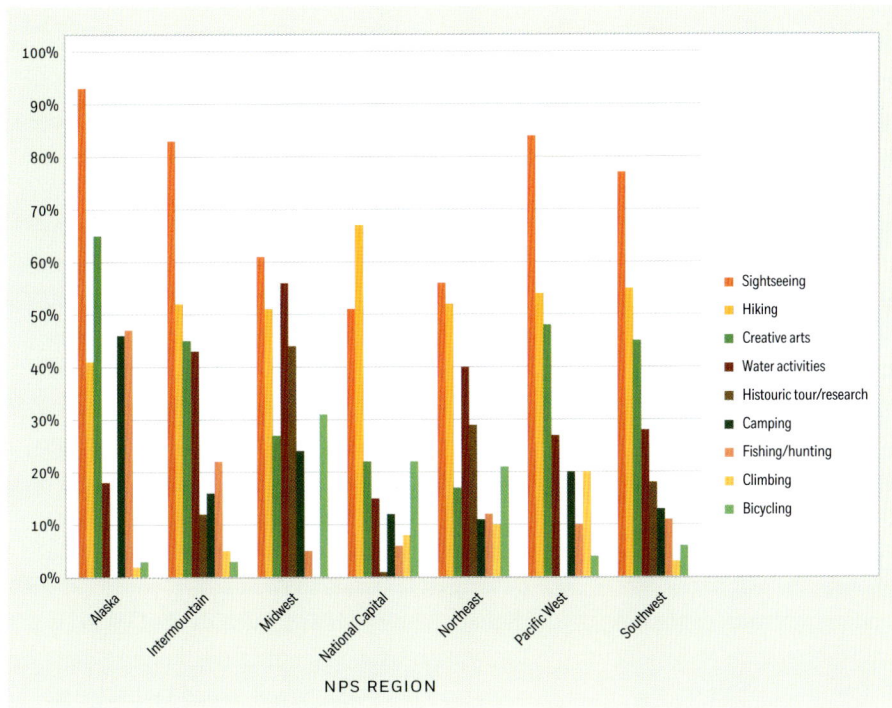

FIG. 8.10   Visitors' Activities in Park Units by NPS Region, 2019

its in the visitor center on the park's resources and learning about the history of the site may be the only possible activity. Boating, fishing, swimming, camping, rock climbing, sightseeing, picnicking, socializing, and enjoying the peace and quiet were common activities in recreation areas. Ten percent of those who went to Stones River National Battlefield reported genealogy or historical research as their primary motivation.

Given that 40 percent of America's wilderness is managed by the NPS, many of the national parks are home to iconic climbing destinations for hikers and mountaineers. These visitors seek out challenging activities such as mountain climbing to experience the most incredible panoramic views of America's wilderness. Climbing Denali is one of the toughest challenges in the park system and is only recommended for the most experienced climbers. Climbing is done as part of a team and typically takes several weeks. In 2018 1,114 climbers attempted the feat, with 45 percent making it to the summit. June is the most popular (and successful) month. The mountain's lethal combination of altitude, ice, below-zero temperatures, and triple-digit wind speeds make it one of

the most dangerous in North America. In 2018 thirty medical emergencies required NPS action, but it was the first season in fifteen years without a fatality. Up to that year, ninety-six people perished while battling the mountain. Mount Rainier may be a warm-up for those contemplating Denali, but it is a difficult and deadly mountain as well—over four hundred people are known to have died on it. Longs Peak in Rocky Mountain National Park is much easier but has still taken fifty-eight lives (Apt and Turnbaugh 2015).

Certain activities are prohibited in all parks or nearly all: no flying drones, no feeding animals, no collecting rocks (though Native Americans can quarry stone at Pipestone National Monument with which to make ceremonial pipes), no driving off-road, no landing airplanes away from an airport (except in Alaska), and no hunting (except in national preserves, with the necessary permits). Urban battlefield parks prohibit Frisbees, kites, and sunbathing in what is intended as a memorial park.

While these restrictions are generally sensible, they can also ostracize certain communities of visitors, especially those who may have historical ties to the land or an affinity for certain activities. A study of Congaree National Park in South Carolina, for instance, found that National Park Service restrictions on the traditional fishing activities of local African American residents have negatively altered their recreational visitation patterns and opinions of the park (Janae Davis 2015).

The NPS cannot always impose limits on visitor activity. A ban on boating on rivers within National Park Service units in Alaska was overturned by the U.S. Supreme Court in *Sturgeon v. Frost II* after a hovercraft pilot and the state government took the Park Service to court over watercraft rights within parklands. It is important to note that not every park has contentious land restrictions, and in many cases, the restrictions put in place at parks are heralded by the local community and visitors. A study of the iconic Cadillac Mountain at Acadia National Park in Maine found that visitors are willing to accept restricted access to the site in order to protect the soils and vegetation of the sensitive landmark (Bullock and Lawson 2008). In a positive example of Park Service cooperation with local communities, the National Park of American Samoa is operated in partnership with local Samoan communities and landowners, from whom the Park Service leases land (Raynal, Levine, and Comeros-Raynal 2016). While the National Park of American Samoa has restrictions on hunting, fishing, and other activities, the Park Service relies on the local community to self-enforce these regulations. Involving local communities and visitors in the process of managing park activities and land uses may help increase visitation to parks, although this possibility has not yet been studied.

Handicapped Visitors

A growing concern for parks and their visitors is accessibility for the disabled. How many of a park's facilities are accessible to people in wheelchairs, the blind, or the otherwise disabled? As much as 10 to 15 percent of the population has a disability, and the NPS has been working to ensure that the entire population can enjoy a park visit. Not much is known about park visitors who have disabilities, but some park visitor surveys (see below) have asked about them. As many as 35 percent of park vis-

itor groups report that at least one member has a physical condition that makes it difficult or impossible for them to engage in a park activity. Most park buildings were built long before modern standards and laws were written into building codes. Fort Sumter in South Carolina's Charleston Harbor was built in the early nineteenth century and survived two battles and decades of neglect. Not all the fort is accessible today, and there is no wheelchair-accessible restroom on the island. (Visitors needing one must return to the tour boat.) It may take $120 million to bring all park buildings up to these standards.

National park units vary tremendously in terms of their accessibility, which is sometimes described in detail on the park websites. Those that are heavily visited, such as the Grand Canyon and Yellowstone, have the most services and facilities, which include wheelchair-accessible buildings, lodging, campgrounds, and interpretive trails. (Scorching hot Death Valley also has accessible swimming pools.) Smaller parks tend to have fewer facilities; at Chiricahua, the visitor center is wheelchair-accessible, as is one campground site and a picnic area, but not the trails that lead into the monument's spectacular wonderland of rocks. At nearby Fort Bowie, normally reached by a 1.5-mile hike through the desert, people with a disability are allowed to use a service road that ends a few hundred feet from the visitor center, and they may even be allowed through a normally locked gate to drive through the ruins of the fort to the visitor center itself. They may also be able to use roads at the Grand Canyon that are otherwise accessed by shuttle bus or on foot.

The Rock Cut Overlook trail on Trail Ridge Road in Rocky Mountain National Park ends at 12,570 feet above sea level, the highest wheelchair-accessible point in the national park system. The lowest is at Badwater in Death Valley, 282 feet below sea level (though the restrooms there are not wheelchair-accessible). While the NPS has put efforts into making these parks more accessible, the vast majority of parkland and all of the backcountry are unreachable by those in standard wheelchairs. The same is true of most large parks, especially in Alaska.

Parks after Dark

The most spectacular sight to be found anywhere in a national park isn't found peering over a cliff or across a meadow but looking straight up at night. The night sky has been a constant companion and inspiration until the last hundred or so years, when electric illumination lit the night sky and banished the stars to the darkest reaches of the planet. Many of these darkest locations are now in national parks. Visitors awed by the sight of the Grand Canyon or Death Valley at sunset will be even more awed if they stick around a little longer and witness the night sky (though they may also be surprised at how dark the resort areas are without the overhead streetlights and advertising signs they are accustomed to back home). The nighttime sky is a growing part of the visitor experience. Visitors can see stellar sights invisible to urban residents: How many Americans have seen the Milky Way with their own eyes, let alone the faint glow of the gegenschein? The long night of Alaskan parks in winter is lit up by

the northern lights, and outside of Alaska Voyageurs National Park is one of the best places in the country to see them. The National Park of American Samoa offers the best opportunity to spot the Magellanic Clouds, among our closest neighboring galaxies, without leaving the country.

Many people like to spend the night during their national park visits, whether camping out or staying in lodging in the park. Many parks are too big to explore in one day and too far from a gateway community to leave the park and return. Parks offer many possibilities in the form of hotels, houseboats, and camping in either campgrounds or the backcountry. The hotels run the spectrum from modern facilities to grand lodges built during the railroad era; national parks are among the few remaining places where you can stay in one of these old hotels.

Overnight stays are counted by the NPS and are greatest in the large rural western parks and recreation areas and in eastern national parks, lakeshores, and seashores (Fig. 8.11). Among

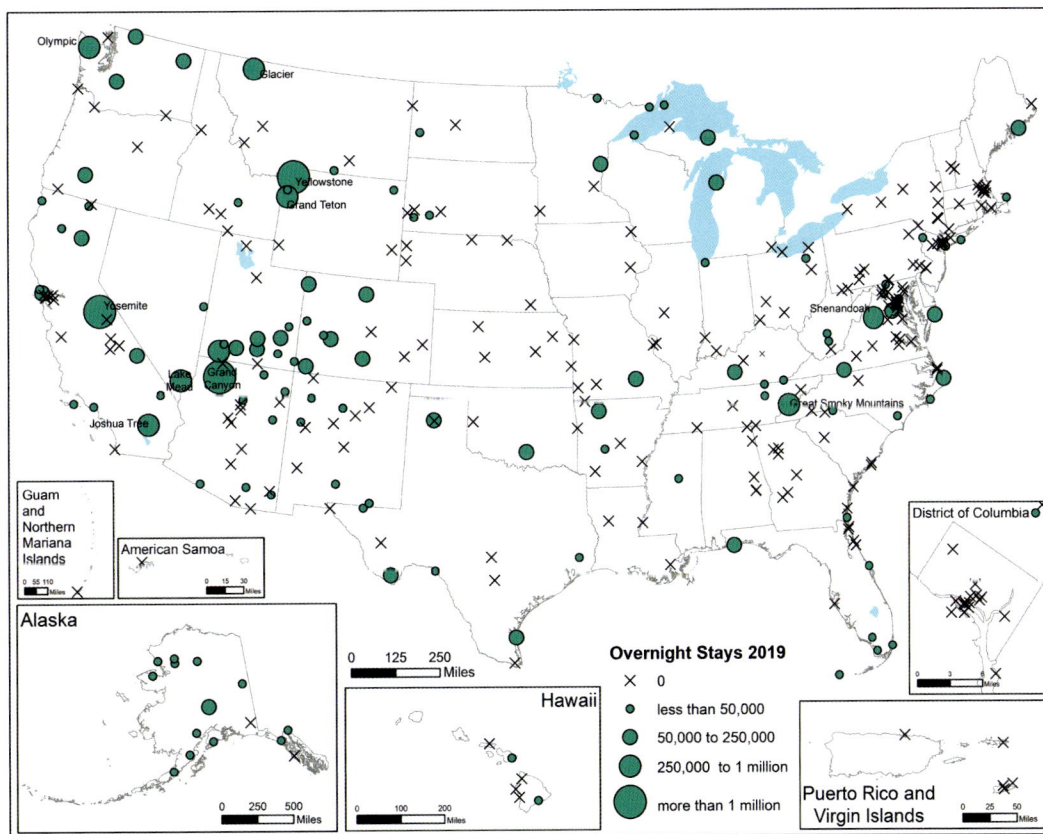

FIG. 8.11   Map of overnight stays by park unit, 2019.

overnight visitors, a minority engage in back-country camping. Relatively few park units have any backcountry, and most of these are large western parks. Foremost among these are reservoir units such as Lake Mead, with countless dirt roads giving access to the lakeshore and where boaters can tie up anywhere for the night. Among metropolitan areas, Seattle and Las Vegas are the closest to the most popular backcountry camping destinations. In the former, these are mountain parks, while near Las Vegas are vast lakes and canyons. The second cluster of backcountry camping can be seen around Lakes Michigan and Superior. Several Appalachian parks allow backcountry camping. Ozark National Scenic Riverways and Buffalo National River are additional backcountry parks. Outside the forty-eight contiguous states, conditions vary. Alaska parks are almost entirely backcountry, and Hawaii has some backcountry camping in Haleakalā and Hawai'i Volcanoes, but the park units in Pacific and Caribbean territories are too small.

Spending the night is not possible at most park units, which are open during daytime hours only. Battlefields, historic sites, and many monuments lack any facilities for overnight stays and shut down promptly at 5:00 p.m. And while the White House is a national park unit, you can't sleep there, at least not without making a major political donation.

Demographic Characteristics of Park Visitors

While parks belong to us all, not all Americans regularly visit them. Park visitors have clear demographic characteristics. Visitors tend to be more highly educated, with the majority (66 percent) holding a bachelor's or graduate degree, while only 13 percent of the population with a high school education or less visit national parks (Vaske and Lyon 2014). Americans with incomes between $50,000 and $150,000 are the primary park visitors (62 percent). Only 6 percent of visitors make $24,000 or less, and the same is true of visitors making $200,000 or more.

Most park visitors are middle-aged and above. Over 65 percent are forty-five years old or older, and 15 percent of that group represent the sixty-five or older age group. National historic sites and national preserves are especially popular with this older age group, where they make up more than 20 percent of the visitors. For example, more than 25 percent of the visitors at Eisenhower National Historic Site in 2017 were from this group. Similarly, more than 25 percent of visitors at Big Cypress National Preserve were sixty-five years old or older. People from the eighteen to twenty-four years old age group make up less than 10 percent of park visitors. However, this proportion varies considerably depending on location. A recent survey based on five Pacific West units found that 22 percent of park visitors are Generation Z (born after the late 1990s) (Crowell 2020). While there are equal numbers of male and female visitors, national historical parks, national preserves, and national military parks have more female visitors. In contrast, more male visitors are found in national recreation areas.

Visitors to parks are overwhelmingly white (Jacobs and Hotakainen 2020). Roughly 93 percent of visitors are non-Hispanic whites, only 2 percent are Asians or Native Americans, and 1 percent or less are Black / African American or Native Hawaiian / other Pacific Islander. The remaining 7 percent of visitors identify themselves as Hispanics of any race (Vaske and Lyon 2014). Surveys conducted at specific parks show similar findings. A 2014 Chickamauga and Chattanooga National Military Park survey showed that 95 percent of park visitors are white, while only 1 percent of visitors are African American. Similarly, non-Hispanic whites make up 87 percent of visitors at the George Washington Memorial Parkway, followed by 9 percent Hispanic of any race, 6 percent Blacks, 5 percent Asians, and 1 percent or less in other categories. Blacks or African Americans made up less than 2 percent of visitors in Great Smoky Mountains National Park at the beginning of that park's foundation in 1938, and that number remained unchanged in 2021. The last visitor survey conducted for New River Gorge National Park and Preserve was in 2004, and while that didn't include race variables, anyone visiting the park in 2021 could tell that nonwhite, especially Black / African American, visitors are rare in this park. The 2011 Shenandoah visitor study reported similar patterns: 92 percent of visitors were white, while 3 percent identified as Hispanic, 6 percent as Asian, 1 percent as Black / African American, and less than 1 percent American Indian or Alaska Native.

The Diversity Challenge in Park Visitation

Park visitors do not represent the demographic mix of the American population. Seven percent of park visitors are Hispanic, while 18 percent of the U.S. population identifies as Hispanic. Likewise, 1 percent of visitors are African American, while they make up 14 percent of the national population. This is considered a problem for the national park system. Given the sentiment that visiting national parks is a fundamental American experience, the expectation is that each racial or ethnic group will, on average, visit parks relative to their proportion of the population. The fact that they don't is considered to indicate access and equity problems and could impact the long-term future of the parks (Weber and Sultana 2013; Scott and Lee 2018). A multitude of factors contribute to the underrepresentation of nonwhite visitors at parks. The five most common explanations are (1) geographic location and accessibility, (2) systematic and structural racism, (3) socioeconomic resources, (4) ethnocultural preferences, and (5) cultural assimilation.

Geographic location and accessibility have a significant impact on nonwhite park visitation (Weber and Sultana 2013). Distance was identified as the most common reason for not visiting a park by Hispanics and African Americans (Solop, Hagen, and Ostergren 2003). Many large nature-oriented parks are highly concentrated in the American West and rural areas, which makes diverse urban communities far removed from those parks. Further, city dwellers' geographic proximity to wilderness varies substantially (Ewert 1998). Most park visitors are locals, and naturally, park units closest to the concentrations of minority populations have a higher level of visitation by those groups (Weber and Sultana 2013). For example, a California study (Ewert 1998) reported that even though whites made up most visitors

to parks, closer areas were more diverse than those that were farther away: 76 percent of visitors at closer areas were white, compared to 89 percent at parks that were more distant. The number of small national historic sites that are closer to urban settings has greatly expanded in the last half century. These park units tend to be close to Black / African Americans in particular, given their historically higher concentration in the eastern United States. However, geographic proximity to nature-oriented parks located in rural settings in the Southeast does not entirely explain the lack of Black / African American visitors in the parks (Weber and Sultana 2013 ; Sultana, Merced, Weber, Allen, and Carlton 2023). Black Americans' relationships with wilderness are much more complicated (Finney 2014).

Systematic and structural racism posits that institutional and overt discrimination are responsible for the lack of nonwhite visitation (Stanfield et al. 2006; Jacobs and Hotakainen 2020). From its inception, the National Park Service itself may have created a troubled historical relationship with minority communities by largely focusing on white American narratives of "America's best idea." The creation of parks such as Yellowstone and Yosemite required the displacement of resident Indigenous populations (Keller and Turek 1998; Spence 1999; Burnham 2000). Native Americans remain involved in legal battles at sites such as Devils Tower and Rainbow Bridge National Monuments (known by local Native Americans as Mato Tipila and Nonnezoshi, respectively) to restrict visitors from locations or activities that interfere with traditional religious practices (Sproul 2001; Taylor and Geffen 2004).

The Park Service also has a legacy of taking racially discriminatory actions against African Americans. Several parks created in the southeastern United States in the 1930s imposed constraints on African American visitation (Young 2009). Even though Black laborers were utilized to build the parks, there was a strong push among interest groups for segregated facilities in those park grounds due to Jim Crow laws. For example, the early development and planning of Virginia's Prince William Forest Park had two cabin camps designated for

African American visitors (Fig. 8.12). The largest segregated facility was the Lewis Mountain Negro Area in Shenandoah National Park (Fig. 8.13). This facility opened in 1940 with a campground and coffee shop. Park management at Great Smoky Mountains also discussed segregated facilities, but due to limited visitation by African Americans, these were seen as unnecessary. Talk of building new segregated facilities ended once the Lewis Mountain area was officially desegregated in 1942 following pressure from the headquarters office of the National Park Service. This facility remains in use today for all visitors.

Regardless of the facilities available at that time, long-distance travel was not easy for African Americans (Foster 1999). In the North, where segregation was less overt, racially tinged legacies still exist in many parks (Algeo 2013). One such park is Indiana Dunes National Park, which was at least partially created as a means of limiting white flight in the nearby community of Miller Beach (Hurley 1995). As part of the park creation process, during the 1950s and 1960s nearby white environmentalists claimed that widened roads and new recreational facilities would destroy the community's stability, and they successfully lobbied for these facilities to be scaled back to prevent their community from being infringed upon by outsiders.

The NPS is made up predominantly of white workers: 83 percent or more of agency employees are white (Jacobs and Hotakainen 2020), 20 percent higher than in other federal government agencies. Only 6 percent of NPS employees are African American, and 5 percent are Hispanic. This lack of diversity may send the message that parks are not safe or welcome places to visit, especially in remote rural settings (Carter 2008). In one study, Asians, Hispanics, and especially African Americans reported uncomfortable encounters in parks (N. Roberts 2007) and surrounding areas of the parks (Sultana, Merced, Weber, Allen, and Carlton 2023), while there is also some evidence that white park visitors experience some discomfort around nonwhite visitors (Stanfield et al. 2006). Ultimately, the national parks are no different from what our national character represents historically.

Among socioeconomic resources, money,

time, means of transportation, and, most important, limited sources of information about national parks may prevent certain groups from visiting (Washburne 1978; Floyd 2001; Johnson et al. 2007). For example, while a third of park visitors report that lodging and meals in parks are too expensive, a higher proportion of the minority population feels strongly about this: 47 percent Hispanic, 54 percent African American, 40 percent Asian, and 59 percent Native American (Taylor, Grandjean, and Anatchkova 2011). While median household incomes for Black (roughly $41,000) and Hispanic ($51,000) populations are lower than that of white households ($71,000), Blacks (about 20 percent) and Hispanics (17 percent) also live farther below the poverty level compared to the white population (8.1 percent). As noted earlier, Americans with an income between $50,000 and $150,000 are ten times more likely to visit national parks compared with poor Americans. Children living in poverty are less likely to be familiar with national parks (Erickson, Johnson, and Kivel 2009; Scott 2013), and this may be a reason for the underrepresentation of minority visitors.

Ethnocultural preferences hold that different groups have different values and interests, including views toward parks and nature (Floyd and Stodolska 2014; Finney 2014). This perspective suggests that low visitation by the nonwhite population exists because national parks and the activities they offer do not reflect the interests and identity of minorities. Differences in leisure travel between African Americans and whites have been examined by Perry Carter (2008), who found that African Americans travel farther than whites regardless of income. While whites are more likely to stay in hotels for vacations, African Americans are more likely to stay and visit with friends or relatives. African Americans also travel in larger groups than whites and are less likely to fly but more likely to use buses (often on church trips) or cruise ships during their travel. Likewise, African American populations visit places that have multicultural points of interest, and their contribution to the U.S. tourism economy has increased in the last decade, from $48 billion to $63 billion (Burt 2018).

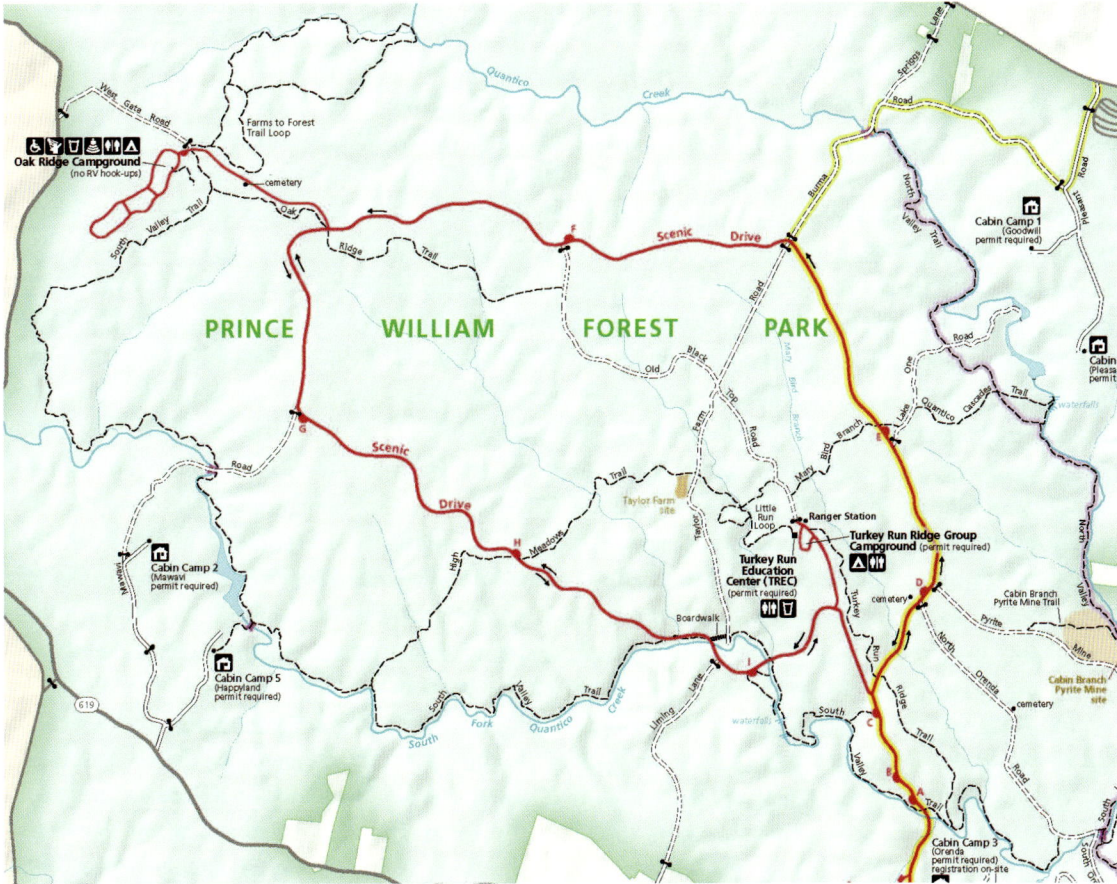

FIG. 8.12   Portion of map of Prince William Forest Park in Virginia in 2021, showing the location of cabin
camps. Camps 1 and 4 (*in the upper right*) were originally reserved for African Americans. NPS.

FIG. 8.13   Lewis Mountain Negro
Area, Shenandoah National
Park, Virginia, in the 1930s.
This segregated facility was
deemed necessary to attract
white visitors to the park.
NPS.

Hispanics do not participate in outdoor physical activity at the same rate as non-Hispanic whites, but they are more likely to picnic and watch sports than others (Stodolska, Shinew, and Li 2010). Beach activities vary by different demographic groups in the United States. Hispanics are noted to be more interested in water sports than other groups (Wolch and Zhang 2004). Hispanics of different origin (such as Mexicans and Cubans) may have considerably different values about nature (Carr and Williams 1993). For example, Hispanics of Central American origin tend to have more homogeneous attitudes toward nature than Mexicans (N. Roberts 2007). Asians likewise cannot be viewed as a homogeneous group, as Asians from different countries have been shown to have different values and interests. Asians who have higher incomes and education are more likely to visit natural areas (Winter, Jeong, and Godbey 2004).

The idea of cultural assimilation holds that as nonwhite ethnic communities take on the characteristics of the cultural majority, they will come to share its values (Floyd 1999; Logan, Alba, and Zhang 2002). This implies that minorities will gradually take on the positive association with outdoor recreation and wilderness shown by many white Americans. Deborah Carr and Daniel Williams (1993) found that Mexicans and Central Americans in the United States had different attitudes toward visiting national forests. Families that had more recently arrived in the United States had different perceptions of and attitudes toward outdoor recreation than those that had been in the country for several generations. For example, those born in a different country were more likely to be part of an organized group than native-born Hispanics. However, this line of research has been criticized for using whiteness as the norm for the nonwhite population and, in fact, preventing their full participation by denying the uniqueness of culturally distinct recreational-based activities (Arai and Kivel 2009).

While each philosophy has some degree of validation, there is no single explanation that can solely explain the complex issue of minority underrepresentation in park visitation. Each hypothesis, therefore, should be examined in light of the local circumstances at individual parks to determine how relevant it might be in explaining the visitation patterns of nonwhite recreationists. It is also likely that many minorities are not represented in the visitor survey data either because of a lack of response from these communities or because the NPS never reached out to them. It is important that the NPS learn from these communities in order to gain a better understanding of their underrepresentation in America's national parks.

Conclusions

Tracking visitor numbers at national parks is not an easy task. We know more about the demographics and movements of grizzly bears and wolf packs than we do about visitors in parks. National park units are complex, and the uniqueness of each park unit means that there is not a single common method available to track visitors (Ziesler and Pettebone 2018). Some parks individually count the visitors entering through their gates, and others use proxies such as vehicle counts to gauge visitation levels. While these data are and can be used to drive several institutional decisions, they also have many limitations.

It is especially difficult to examine domestic and international park visitation. The only data available on these important topics are found through visitor studies conducted by the National Park Service Visitor Services Project, which are only available for a handful of parks and are not conducted with any regularity (Pettebone and Meldrum 2018). The project is hosted by the Social and Economic Sciences Research Center of Washington State University, and its reports use surveys of visitors to derive information about visitor demographics and motives. While these reports are helpful, details of the data get lost in this reporting. For instance, while the origins of visitors are tracked, only the states with the highest number of visitors are included in these reports. States with smaller numbers of visitors are lumped into an "other" category, making it difficult to conduct a statistical analysis of the data for the origin of park visitors by state. The infrequency of these reports also makes it difficult to track changes in visitation on a year-by-year basis by state. Similarly, most social science studies for the NPS are at the park-unit scale, and generalizations based on these sample data can be biased. While the inclusion of participant demographics and characteristics such as race and ethnicity enriches our understanding of park visitation, minority visitors are usually still underrepresented in park surveys, and questions about race and ethnicity were not asked on all Visitor Services Project surveys. Primary data collection is necessary for understanding minority populations' interest in visiting or not visiting national park units.

Moving forward, the advancement of technology and big data may make it easier to track visitors' spatial and temporal patterns and to ensure that their needs are being met. Social media is being looked at as a useful source of park access and visitation data (Hamstead et al. 2018; Kupfer et al. 2021). While the data are not free, they could provide information on exactly where people went within parks and create opportunities to identify the home location of national park visitors down to the neighborhood. Trail counts might be used to gauge the social and ecological costs of backcountry activity in national parks (Pettebone et al. 2019). As new counting methods and technology are made available, it will be important for the NPS to evaluate these innovations to see if they have potential applications in the park visitation data collection system so that a finer level of analysis can be conducted in the ensuing years and decades.

NINE

The Economic Geography of National Parks

The National Park Service runs on money. Land must be bought to create a park, and then roads, along with visitor centers, trails, and bathrooms, must be built and maintained. Staff must be hired to operate and maintain these facilities. Many rural parks also require housing for employees. Most also have one or more water and sewer systems, as well as garbage collection. Boat docks, marinas, and airports are operated in some parks. Historical buildings must be maintained and preserved in harsh and remote locations. Rangers must patrol roads and trails and conduct dangerous search-and-rescue operations.

Parks also generate money. Some collect visitors' money on the way in, but most of the money spent in parks goes to the many businesses operating within parks under contract to the NPS. These businesses provide lodging, meals, campgrounds, tours, gas, or souvenirs to visitors. Some of this money goes to local owners and operators, but often it goes to giant corporations with corporate offices in a distant city. Just outside the park boundaries, gateway communities have long benefited from the presence of park attractions, with places such as Gatlinburg and Pigeon Forge, Tennessee, eclipsing the nearby park entirely.

Parks have long been promoted for their economic benefits, and this remains a popular selling point for the creation of new parks. Again, this money is more likely to go to distant corporations than stay within the community, but the appeal of economic benefits can be very enticing in rural areas. Unfortunately, the flow of money into parks and neighboring businesses has been increasingly interrupted by politics. Federal government shutdowns and even the COVID-19 pandemic have closed businesses within parks or even the entire parks themselves.

Park Budgets

The annual budget of the NPS was about $4.1 billion in 2020, of which about $1.3 billion was for park operations (National Park Service 2020a). This amounts to 0.08 percent of the $4.79 trillion federal budget for 2020, a minuscule level of funding compared to the 60 percent of the federal budget that goes to the Medicare and Medicaid health insurance programs and to Social Security and the 15 percent ($721.5 billion) spent on the military each year.

Each park unit has its own listing in the NPS budget, and the cost of operating parks varies tremendously depending on each park's size, facilities, and location (Rettie 1995) (Fig. 9.1). In 2020 Yellowstone was the single most expensive park, with a budget of $36,531,000, followed by the National Mall and Memorial Parks in Washington, D.C. ($36,351,000), Yosemite ($31,699,000), Golden Gate ($27,016,000), Gateway ($26,316,000), Independence ($25,070,000), Grand Canyon ($22,446,000), Lake Mead ($20,266,000), Great Smoky Mountains ($20,228,000), and Sequoia–Kings Canyon ($17,595,000). Bluestone National Scenic River in West Virginia is currently the least expensive park, at $80,000. Rainbow Bridge and Yucca House, two small southwestern national monuments with no staff and very limited facilities, cost less than $150,000 each. Thaddeus Kosciuszko, America's smallest national park unit, is the fourth least expensive. Many of the other least expensive parks are new units, with

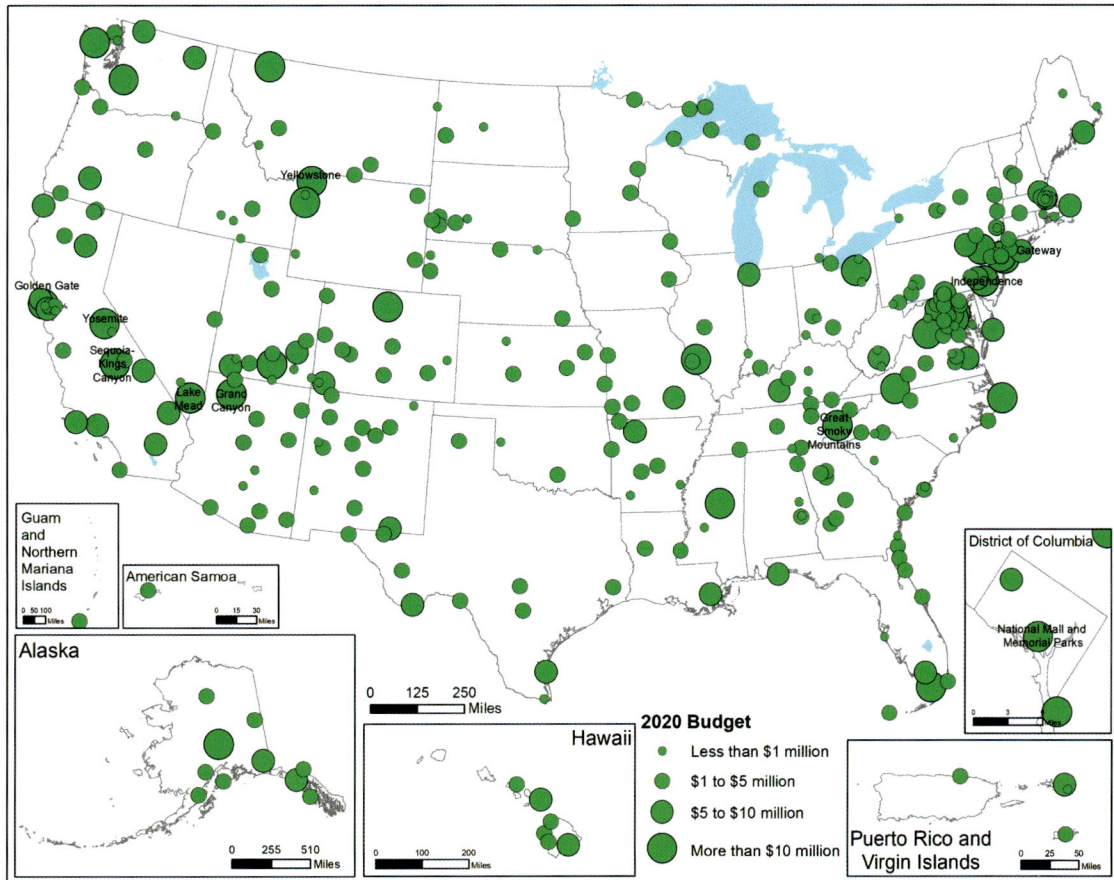

FIG. 9.1   Map of 2020 budgets by park unit.

small budgets to begin the planning process to get the park running. Mill Springs Battlefield, Stonewall, and Medgar and Myrlie Evers Home are among these.

To put these dollar values into perspective, the cost of a single F-35 fighter plane, the latest warplane being built for the air force, navy, and marines, is $135.8 million (Grazier 2020). Yellowstone could be operated for over three years for the cost of one plane, of which the military has 265 in service and plans to buy hundreds more. Yellowstone is open twenty-four hours a day every day of the year, while a typical F-35 is flown less than eight hours a month and is in working order only 28 percent of the time. Likewise, the cost of the navy's newest aircraft carrier, the USS *Gerald Ford* (named after the former president who once worked as a park ranger in Yellowstone), was $13.2 billion (O'Rourke 2013). This is equivalent to operating Yellowstone National Park at current funding levels for over 361 years.

How is a park's budget spent? Park budgets are broken down into several categories, including management and administration, maintenance, facility operations, resource protection, and visitor experience and enjoyment. The last category is the one that most concerns the average visitor; it encompasses fee collection, visitor centers, ranger talks, and overseeing private businesses operating within the parks. Facility operations and maintenance include trash collection, bathroom cleaning, road and trail repair, the operation of wells, water treatment plants, pipelines, sewer treatment, electricity generation, building maintenance, and even airport and marina operation. Management includes not just overseeing the park's budget and the personnel department but also engaging in planning, working with the park's nonprofit partners, and public relations.

Thousands of employees of the National Park Service work at parks, as well as at businesses run by concessionaires and on inholdings. In 2017 the NPS had the full-time equivalent of 15,428 employees. Among park units, the number of FTE (full-time equivalent) employees varies tremendously. Yosemite is the largest, with 634 FTE employees, followed by Yellowstone (528), Grand Canyon (387), Sequoia–Kings Canyon (310), Golden Gate (281), Gateway (279), national capital parks (274), Glacier (264), Grand Teton (247), Rocky Mountain (242), Lake Mead (221), and Mount Rainier (210). Many smaller units have fewer than ten employees, and several units (e.g., Yucca House in Colorado and Yukon–Charley Rivers in Alaska) have no employees at all. Among NPS regions, the Intermountain West has the most employees working in parks (3,955), followed by the Pacific West (3,680). However, the East Coast is also a major cluster of park employees, with 2,485 in the Northeast and 2,160 in the Southeast (not counting those working at headquarters in Washington, D.C.).

A growing expense at many parks is search and rescue (SAR), during which people lost in the backcountry or perhaps stuck on a cliff must be located and rescued by park staff. This can be extremely dangerous for all concerned and often requires expensive boats and helicopters. The cost of SAR has increased tremendously since the early 1990s, when an increase in outdoor recreation and outfitting stores led many more people to buy climbing gear and race off to a wilderness experience they were completely unprepared for. In 2017 park staff had to respond to 3,453 SAR incidents. While canyon and high mountain parks can be predictably dangerous, the most lethal park is a desert lake (National Park Service 2018e). Lake Mead alone accounts for 16 percent of annual SAR missions, with 503 in 2017, or more than one a day. About two people die in the park each month, more than any other park unit, and this total does not include car accidents, drownings, and heart attacks or other natural causes, a frequent occurrence. In contrast, the most dangerous canyon park, the Grand Canyon, suffered only twelve deaths requiring SAR operations that year.

These two parks, along with Yosemite, Rocky Mountain, Sequoia–Kings Canyon, Zion, and the Great Smoky Mountains, all have had one hundred or more incidents and together account for 46.42 percent of the total (National Park Service 2018e). These incidents resulted in 182 fatalities and about 1,500 injuries; they cost $3.4 million. This money must often be taken out of the budget for maintenance, visitor services, or resource protection. More men than women required rescue, with those age twenty to twenty-nine being the most likely to be rescued.

Park budgets are largely dependent on visitation (Rettie 1995). A comparison between each park's budget and the number of visitors shows a correlation of 0.71, which is quite strong. Political support for parks will also likely depend on how widespread visitation is. Having too few visitors may therefore lead to budget cuts or a loss of support. For every park, there is presumably some middle ground or the correct number of visitors, but this is hard to estimate. Further, while park budgets are based on visitation, the cost of operating a park may depend more directly on the size of the park (Rothman and Miller 2013). This can lead to budgets that do not cover the cost of operating the park, which in turn may lead to maintenance being left undone.

The Maintenance Backlog

Unfortunately, most park budgets are not sufficient to keep a park's roads and buildings in adequate shape. The NPS spends about $1.18 billion a year on maintenance, but it is not enough to get the job done. There is a backlog of maintenance to be done, one that was around $2 billion in 1995 (Rettie 1995), but that number had swollen to $11.3 billion in 2017 (National Park Service 2018d). This refers to deferred maintenance or necessary work that has been put off due to a lack of money in the annual budget. About half the amount is for roads, bridges, and tunnels, and the remainder is for needed work on buildings, campgrounds, water systems, sewage treatment plants, marinas, trails, housing, utilities, and other facilities. These facilities have a replacement value of $154.6 billion, making maintenance a bargain. Many of these facilities may be hidden from public view, such as housing and water systems, while campgrounds and road conditions are quite visible.

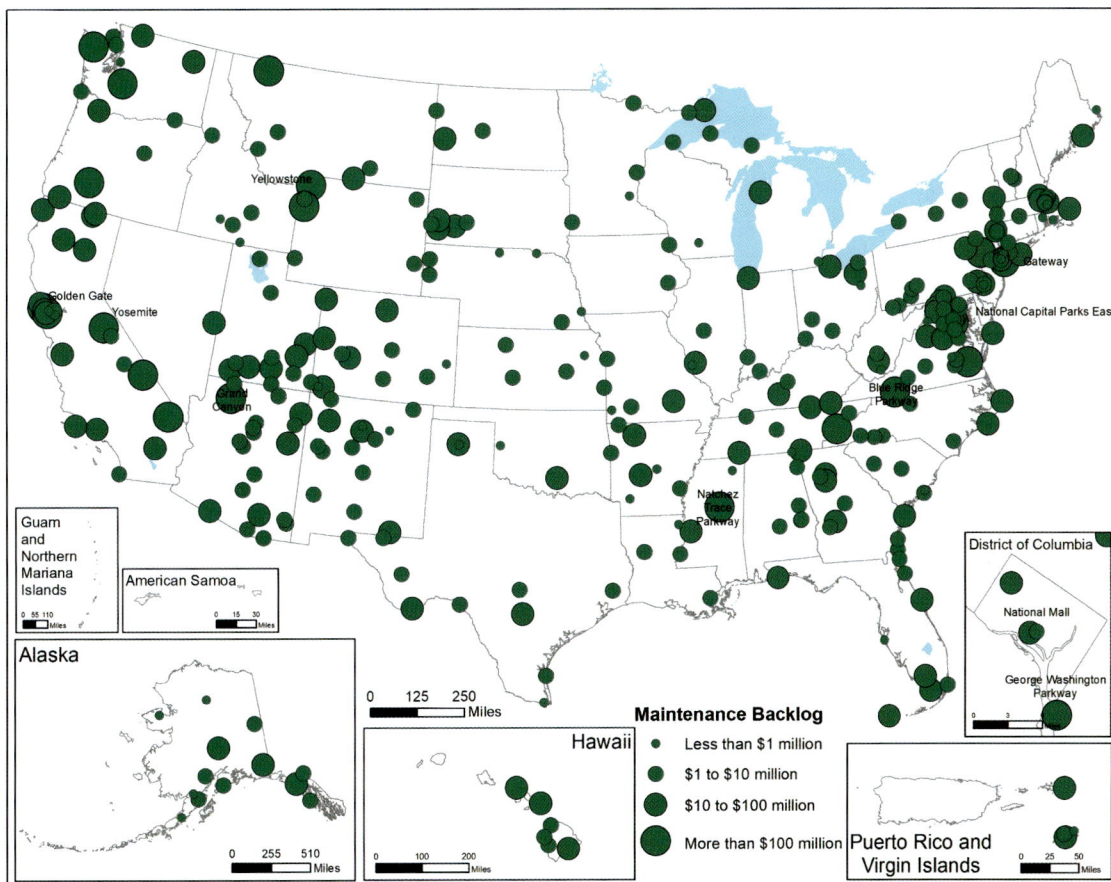

FIG. 9.2   Map of maintenance backlog at national park units, 2018.

Why does this deferred maintenance backlog exist? "Many factors might contribute to growth or reduction in deferred maintenance, including the aging of NPS assets, the availability of funding for maintenance activities, acquisitions of new assets, agency management of the backlog, completion of individual projects, changes in construction and related costs, and changes in measurement and reporting methodologies" (Comay 2017, ii). Many national park structures were built in the 1930s by the CCC or later by Mission 66, but these structures are now at least half a century old. Although many Mission 66 buildings are similar to those built in cities during that time, park buildings often exist in areas with extreme climatic conditions. It may be that Congress often finds it easier to create new units than fund existing ones adequately (Foresta 1984), but it is not clear that new units are to blame for tightening the money available (Rettie 1995; Comay 2017).

In 2018 the units with the largest backlogs

were Gateway, National Mall, Yellowstone, Yosemite, Blue Ridge Parkway, George Washington Parkway, Natchez Trace, Grand Canyon, Golden Gate, and National Capital Parks East (Comay 2017) (Fig. 9.2). Three of these sites are parkways, for which paved roads, bridges, and tunnels account for 85 percent to 93 percent of the total maintenance. Of the top twenty, only the Statue of Liberty and National Mall do not feature roads as a major category (in both cases, piers, floodwalls, and docks are important components), while for the urban recreation areas (Gateway and Golden Gate), roads rank second, well behind buildings.

The Grand Canyon's backlog is $329.4 million, including $119.7 million for water systems, $90 million for paved roads, $31.1 million for sewage, and $28.9 million for trails (Comay 2017). The large amount for water systems is due to the park's dependence on the rickety sixteen-mile Transcanyon Water Distribution Pipeline, which brings water to the

South Rim. This line begins at a spring below the North Rim, runs to the bottom of the canyon and crosses the river on a bridge, then climbs 4,460 feet to the South Rim, where it provides the drinking water for park visitors and employees. The line was built in 1965 and is in poor shape, with numerous leaks and breaks every year, some requiring expensive helicopter use to bring in people and equipment. If the pipe should suffer a massive break that cannot be quickly repaired, the South Rim will have to be closed and all tourists will be forced to leave the park. The $119.7 million only covers deferred maintenance; an entirely new pipeline would cost at least $124 million. It may be the single most important (and expensive) project in the entire national park system.

Everglades has a relatively modest $90 million backlog: $39.5 million for paved roads, $18.5 million for buildings, $13.7 million for marinas and docks, along with needs for parking, dikes, levees and dams, and other amenities. This does not, however, reflect any of the $7.8 billion cost of the Comprehensive Everglades Restoration Plan, passed by Congress in 2000. Only $400 million was ever spent on the plan by Congress, though the state of Florida has shouldered some of the burden (Lodge 2016).

At the other extreme, thirteen units reported no backlog. Several of these parks are new and do not have any facilities or staff, such as Tule Springs Fossil Beds and Stonewall National Monuments. Brices Cross Roads National Battlefield Site has been in the park system since the 1930s but is a roadside park with no facilities beyond a parking lot. Sixty-nine units have maintenance needs of less than $1 million, while 133 require over $10 million.

The reader will note that this backlog is for facilities; the agency does not have an estimate for what is needed for improved visitor programs and resource protection. How larger crowds from more diverse backgrounds will be accommodated has yet to be budgeted, let alone the work necessary to protect endangered species or return animals to areas where they have been eliminated by hunting. And of course, there is no budget to deal with the consequences of climate change; the infrastructure to protect damage from that has yet to be

identified, let alone constructed. Needless to say, had parks been developed differently (or not developed at all), much of this maintenance would not be needed. Wilderness purists may claim this shows that park development was a bad idea, but parks were developed for middle-class tourism centered on automobiles.

In 2020 the Great American Outdoors Act was signed into law. This act will fund the Land and Water Conservation Fund (used to purchase conservation lands) and provide $9.5 billion over five years to help reduce the maintenance backlog. Its effects remain to be seen, but ideally it will help resolve some of the most pressing problems in the park system.

User Fees

Visitors to many parks will often have to pay, whether to enter the park, camp, or take a guided tour or a boat trip, among other activities. Among the latter are special events on the National Mall, which often charge admission. In one case, this was $105 for general admission and $900 for special access (Benton-Short 2016). Fortunately, most park fees are much cheaper.

National park entrance fees were charged as early as 1908 (Summers 2005). Mount Rainier was the first park to charge a fee, set at $6 per person, which sounds inexpensive but when adjusted for inflation is equivalent to over $150 in 2020. (The current park entrance fee is only $30.) In the years that followed, fees were occasionally collected at some parks, with the money collected going to the general treasury rather than staying in the park or even within the NPS budget.

In 1996 a new policy allowed the NPS to charge entrance fees at one hundred park units, with the provision that 80 percent of the money raised would stay in the unit where it was collected (Summers 2005). This was an experiment, and it was found that creating entrance fees did not reduce visitation, while the money collected led to improvements in campgrounds, museums, trails, and educational programs in these parks. Entrance fees have increased since then, with 109 parks having them in 2020. These fees range from three dollars per person

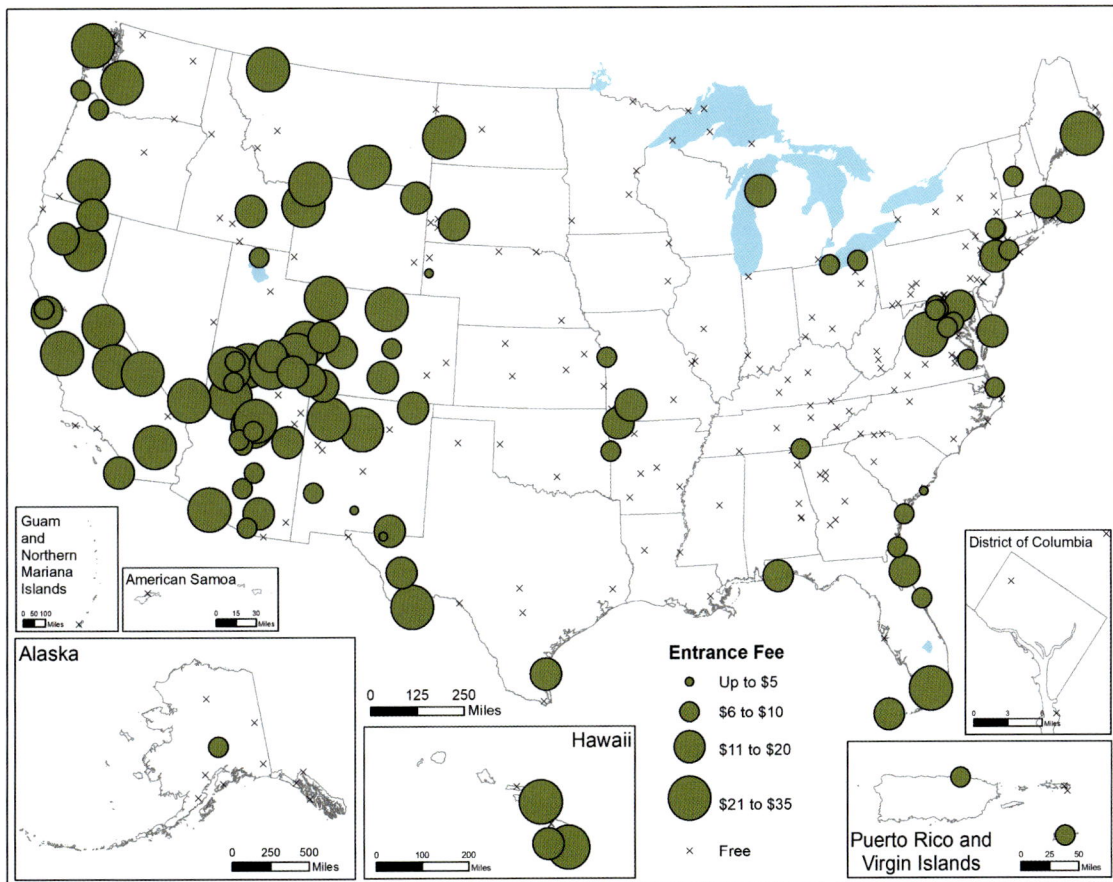

FIG. 9.3   Map of entrance fees by park unit, 2019.

to thirty-five dollars per vehicle. Depending on how people experience a park, these fees may be assessed per vehicle or per person and may be seasonal.

In general, larger units with higher visitation charge the most (Fig. 9.3). Large western parks have the highest fees; the Sierra Nevada and Colorado Plateau stand out not just for their breathtaking scenery but also for their breathtaking entrance fees. The parkways and the Great Smoky Mountains are exceptions in having no entrance fees. (But beginning March 1, 2023, visitors with cars entering the Great Smoky Mountains National Park must purchase a parking tag: it costs $5 daily; $15 for weekly; and $40 for an Annual pass.) Shenandoah does, however, charge a fee, clearly standing out from its free neighbors. Tiny Abraham Lincoln Birthplace is free, as required by the legislation that created it in 1916.

Geography is important in the operation of national park entrance fees. In many smaller units, there is one entrance to the park, and it is easy to collect the fees when visitors arrive, whether at a toll booth on the entrance road or at a desk in the visitor center through which they must walk to see the attraction. To see Montezuma Castle all visitors must walk through the visitor center and then down a short trail, minus fifteen dollars per person. Although Petrified Forest encompasses a large acreage, park regulations state that a person can enter the park only at either of the two road entrances, where toll booths await to collect a twenty-dollar fee from each car. The park is open only during daylight hours, and the entrance gates are shut in the evenings.

The more ways there are to enter a park unit, the more effort is required to collect fees. At Yellowstone, five toll booths where roads enter the park boundary operate all day, every day to collect the thirty-five-dollar entrance fee. To the south of the park is Grand Teton National Park and the John D. Rockefeller, Jr. Memorial

FIG. 9.4   *The Tetons and the Snake River*, by Ansel Adams, 1942. This view of the Grand Tetons, taken from an overlook along Highway 26/89/191, may be the most spectacular view that can be found in the park system without paying an entrance fee. National Archives and Records Administration, NAID 519904.

Parkway, which together also charge a thirty-five-dollar entrance fee at four toll booths. (The northernmost of these is operated with Yellowstone's southern entrance station.) The other three are not at the boundary line but farther into the park, allowing travelers to pass through the park on the main highway or fly into Jackson Hole airport without paying. (Both the road and the airport were built before the area became a park and so remain fee-free.) The Snake River Overlook on U.S. Highway 26/89/191 offers a stunning vista of the Tetons, captured most famously by Ansel Adams in 1942 in his photograph *The Tetons and the Snake River* (Fig. 9.4). This overlook is outside the fee area and is perhaps the most spectacular view to be found in the park system without paying an entrance fee.

Lake Mead National Recreation Area is adjacent to the Las Vegas metro area, with which it is connected by several busy entrances with toll booths charging twenty-five dollars per vehicle. (Those only visiting Hoover Dam or passing through the park on busy U.S. 93 do not pay.) The toll booths at entrances far from Las Vegas are often closed, so people entering there can enjoy the lake for free. This reflects the fact that most visitors arrive from the Las Vegas entrances, but it also suggests political favoritism toward residents in outlying areas. They can get in for free, while Las Vegas residents and tourists pay.

Death Valley National Park is the biggest park in the lower forty-eight states and has at least forty-one road entrances, many of them remote and very rough dirt roads, along with three airports and no restrictions on hikers entering or exiting the park at any location. Most visitors enter along California Highway 190, which passes through the park, but the state highway department did not allow toll booths along this road. With no possible way of stopping every visitor entering the boundaries or even putting up toll booths on the main road, the park instead operates an honor system, with the thirty-dollar fee payable at the visitor center, at several other ranger stations, and at a dozen machines located throughout the park. Those passing through on Highway 190 without stopping or traveling to or from the privately owned Furnace Creek resort do not need to pay. Those in other areas of the park who do not have a permit displayed in their vehicle can be ticketed by rangers. (The fine for this offense is not listed in any park information.) The park is enormous and has few rangers, and the number has been shrinking in recent years (Rothman and Miller 2013), so the chances of getting caught are slim.

Shiloh National Military Park is much smaller but has seven road entrances, all well-used state and county highways, along with Shiloh Methodist Church (for which the battlefield is named) on a privately owned inholding. Entrance fees for those not passing through the park or going to church were to be paid at the visitor center before beginning the park road tour, which crossed or used multiple county roads and passed the church. Many did not bother stopping or did not even know where the visitor center was. Others interpreted the fee as a charge to visit the museum and decided to skip that while touring the rest of the park. The cost of enforcing these visitor fees was not worth the trouble, and they were dropped in 2011 (National Park Service 2011).

A frequent argument is that entry fees discriminate against the poor; however, even a thirty-five-dollar fee is trivial, given the travel costs that visitors from distant cities must pay to visit rural parks. Anyone who has planned a trip to Yellowstone, let alone to an Alaskan park unit, will agree. And $35 for a family for seven days compares very favorably with $159 for one person for one day at Disneyland. There are also options for reducing the cost. The America the Beautiful Annual Pass costs eighty dollars and will get you into any national park (and many other federal lands) for free for one year. It is an incredible bargain for those who intend to visit several parks or to visit their favorite many times. (It will not, however, cover campgrounds and tour fees in parks.) And many visitors don't have to pay anything. Those over sixty-two can get a pass for only twenty dollars (or a lifetime pass for eighty dollars), and those who are active-duty military and their dependents can get a free annual pass, as can those who put in at least 250 hours of volunteer work on federal lands, and fourth graders. The permanently disabled can also get a free lifetime pass.

Economic Activity within Parks

Although parks seem like refuges from the cash-driven outside world, there is considerable economic activity going on within them. Visitors need restaurants, lodging, supplies, gasoline, boat rides, bicycle rentals, tours, and countless other things. Aside from campgrounds, drinking water, and a few tours operated by the NPS, these amenities are almost all provided by private businesses or concessions operated under contract to the NPS.

Concessions vary in size from small, family-owned businesses to international corporations. Major concessionaires include Xanterra Travel Collection (which operates in Crater Lake, Death Valley, Glacier, Grand Canyon, Mount Rushmore, Yellowstone, Rocky Mountain, and Zion, among others), Aramark (Denali, Grand Canyon, Glacier Bay, Mesa Verde, Glen Canyon, Olympic, and others), Delaware North (Grand Canyon, Shenandoah, and Sequoia–Kings Canyon), and Forever Resorts (including Big Bend, Lake Mead, Glen Canyon, Whiskeytown, Grand Canyon, Mammoth Cave, and Bryce Canyon). These companies make over $1 billion each year within national parks and employ twenty-five thousand workers. You may not have seen their names, but if you have eaten a meal or slept in a bed in a park, you have given these corporations your money.

Xanterra is one of several companies owned by the Anschutz Corporation, whose billionaire

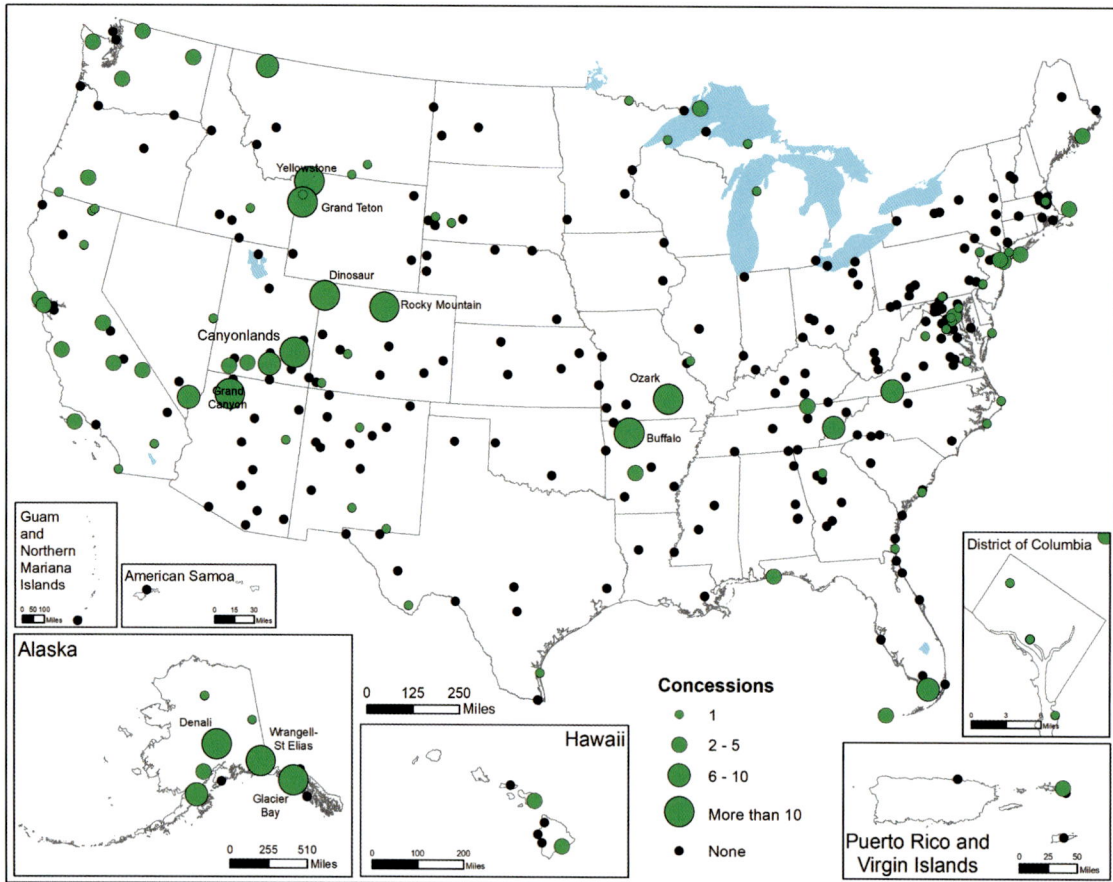

FIG. 9.5   Map of the number of concessions in each national park unit, 2018.

founder, Philip Frederick Anschutz, is considered one of the ten largest landowners in America. In addition to Xanterra's park operations, the company also owns railroads, is heavily involved in agriculture, conducts oil drilling (including some wells just outside Glacier), owns several newspapers and professional sports teams, and runs the Coachella Music Festival. Aramark operates food, janitorial, and uniform services in twenty-two countries, with revenues of almost $16 billion. Delaware North has $3.2 billion in revenues and operates food concessions at professional sports facilities, Disney parks, airports, and NASA's Kennedy Space Center.

There are more than 630 concessionaires in 128 different park units (Fig. 9.5). The number and function of concessionaires vary widely, though any concession or proposed business within a park must be judged as necessary and appropriate for the park mission. Yellow-

stone has the most (forty-seven), with Glacier Bay, Grand Teton, Canyonlands, Grand Canyon, Rocky Mountain, Denali, Ozark, Buffalo, Dinosaur, and Wrangell–St. Elias having more than ten each. In addition to running lodging and food service, these concessionaires provide rafting trips, hiking guides, and other outdoor activities and may have no facilities in the park. This is especially evident in river units such as Buffalo, Ozark, Canyonlands, Grand Canyon, and Dinosaur. Several large cruise lines (Carnival, Holland America, Norwegian, Viking, and Princess) also operate in Glacier Bay.

When Yellowstone was created the intention was that fees paid by concessions within the park would support its upkeep (Ise 1961). The first decades of the park were marked by frequent problems with concessionaires, and Congress was eventually forced to appropriate money for Yellowstone and later parks. Many of the problems stemmed from the unique operating conditions of national parks, which can be remote and expensive to operate and often have short seasons. Many concessionaires did not own the properties they operated (these belonged to the Park Service) (Carr 2007). This meant they could not obtain loans for improvements, as they had no collateral. These companies also operated expensive properties under short-term contracts that might not be renewed, giving them little incentive to invest in improvements.

The NPS has favored concessionaires having monopolies within parks, as one large company could operate more profitably than several small ones. It would also be easier for the NPS to deal with one large company rather than lots of small ones. The NPS sought to merge smaller companies in order to achieve this. Mount Rainier went from having forty-two separate companies operating within it in 1915 to just one in 1918 (Carr 2007). Having monopolies allowed fewer companies to operate more profitably and allowed new lodges to be built in the 1920s and 1930s.

In 1965 many persistent concession problems were finally resolved with a new law that allowed concessionaires to have a "possessory interest" in the properties they operated (Carr 2007). Concessionaires were more willing to spend money on properties, and banks were more willing to make loans for improvements. But even as management improved, so did attempts at closing many concessions in parks. This was partly a reaction to the developments of Mission 66 and an attempt to reverse what was seen as a prodevelopment trend. The Old Faithful Inn and lodgings at Mammoth Cave, Rocky Mountain, Bryce, and Zion were all threatened with closure but survived. Cedar Breaks and Lassen did have lodgings closed; at the latter, visitation fell after park lodging and a restaurant were closed.

Other concessions closed due to changing visitor patterns. When Lake Mead National Recreation Area was created in the 1930s around the nation's largest reservoir it was remote; Las Vegas was a town of less than ten thousand, with few hotels. A lodge was therefore built to house visitors who might be staying at the lake for several weeks (National Park Service 2018c). After World War II camping became a common pastime, while vacations were reduced in length, eliminating much of the reason for the lodge. The lodge also suffered due to the changing relative location of the park as the spectacular growth of nearby Las Vegas resulted in the construction of over 150,000 inexpensive hotel rooms a short distance away. And as a final challenge, the water level in the lake has receded since the 1980s. The marina near the lodge was moved, leaving the formerly lakeside lodge high and dry. The Lake Mead Lodge could not adjust and was shut down in 2009 and later demolished.

Inholdings

Many national park units include private land within them, called inholdings. These may exist for a variety of reasons, even within western parks created from the public domain. The NPS has long had a policy of purchasing these lands, but it has been a slow and contentious process and will not be resolved anytime soon. Often these lands are vacant and will never be developed, but others may include resort areas or other commercial establishments that provide lodging, meals, supplies, and gasoline. Some of the larger facilities, such as the Furnace Creek Inn and Ranch Resort within Death Valley, are small towns. Furnace Creek has two hotels, a post office, a golf course, a general store, restaurants, a gas station, employee housing, a park, campgrounds, and a museum (Fig. 9.6). The resort is located on two sections of private land that originated in patented mining claims in the nineteenth century, when Furnace Creek Ranch was a functioning ranch supporting nearby mining operations. The ranch grew several field crops and many vegetables, along with cattle ranching, before the ranch gradually became a tourist resort in the 1920s. Furnace Creek Inn was developed in the 1920s as a luxury desert resort. Both the inn and the resort were owned by the Pacific Coast Borax Company before passing to the Fred Harvey Company and later Xanterra. Although such resorts may be indistinguishable from a concession on NPS land, they are not under contract with the NPS and not a concessionaire.

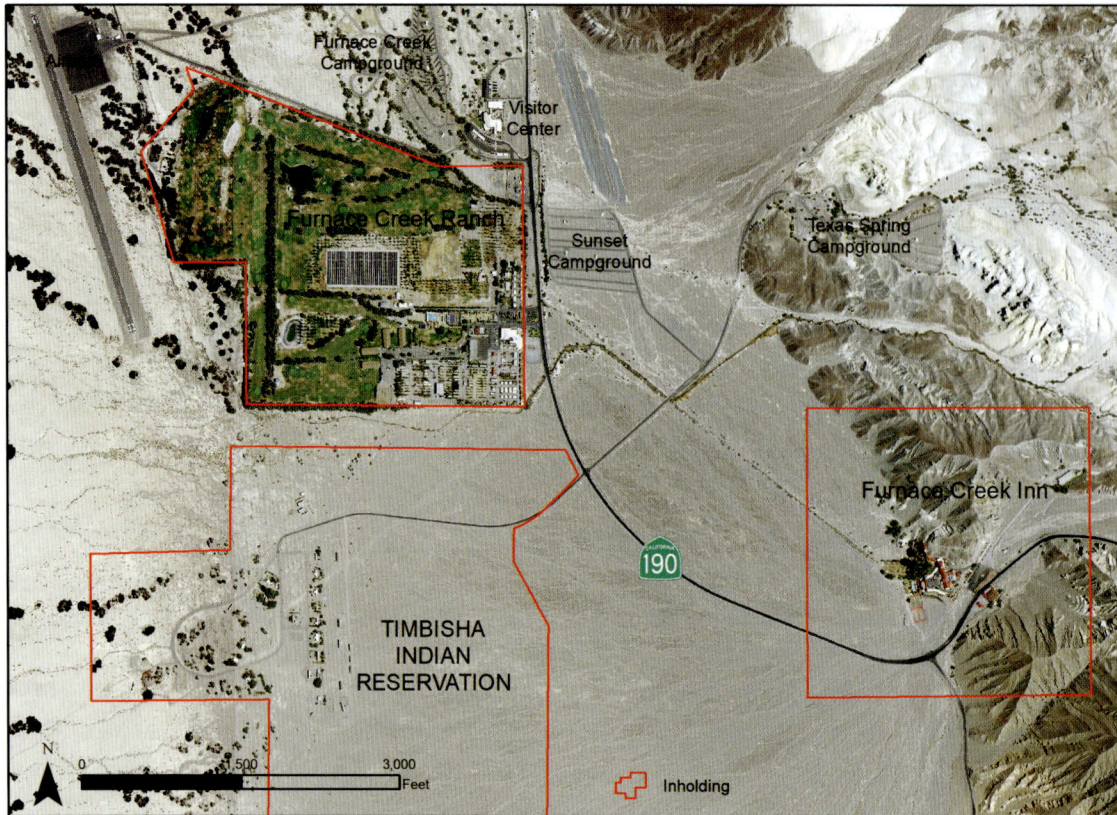

FIG. 9.6   Furnace Creek Inn and Ranch Resort, Death Valley National Park, California, 2019.
These properties form a large inholding with hotels, restaurants, bars, a golf course,
and other amusements.

Figure 9.6 also shows several NPS-owned facilities at Furnace Creek: the 1930s Texas Springs Campground, built by the CCC; the Mission 66 era visitor center and Furnace Creek Campground; the Sunset Campground (a vast parking lot for RVs that arrive in the winter); and an airport (designated LO6 by the Federal Aviation Administration). California Highway 190 runs through this visitor complex, and the Timbisha Indian Reservation is to the south. This combined privately owned, state, tribal, and federal area is supported by local water and sewer systems, with utilities providing electricity and telecommunication lines from outside the park. Visitors will simply see a green oasis within a vast desert park.

Elsewhere in Death Valley, the Stovepipe Wells Resort was once an inholding but was sold to the NPS in the 1970s. This resort also has a general store, a pool, a gas station, a saloon (complete with a government-owned nude painting above the bar), motel units, and a restaurant. It is now operated by a concession-aire, the Death Valley Lodging Company, supervised by park staff. The difference between the two properties can be seen in the price of gas sold at each gas station. On one occasion, regular unleaded at Stovepipe Wells was selling for $4.33 a gallon, while at Furnace Creek (less than thirty miles away) it cost $5.91 a gallon. Higher prices are justified as the cost of doing business in such a location, just as they are by businesses in airports.

Cooperating Associations

Every national park unit is affiliated with a cooperating nonprofit 501(c)(3) association, which typically sells books, maps, and souvenirs in a gift shop in the visitor center, as well as operating several educational programs (Vaughn and Cortner 2013). In the past, these nonprofits were often called natural history associations, but their names have become more varied. Their location within a park's visitor center ensures that most visitors will see the nonprofits when they pass through the building. If

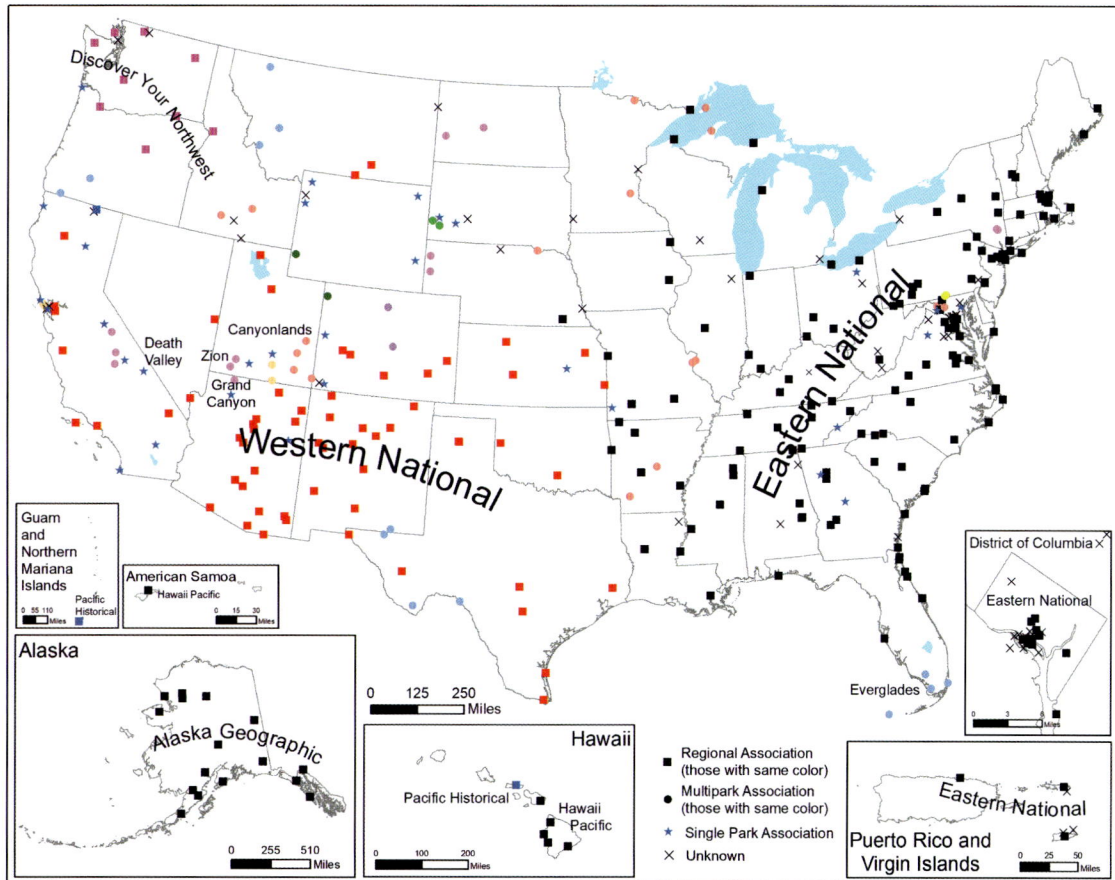

FIG. 9.7   Map of national park cooperating associations, 2021.

you've ever bought a postcard, book, or T-shirt in a visitor center, you did so from one of these associations.

There are currently fifty-three cooperating associations serving national park units (and several more serving national forests, wildlife refuges, and other protected lands) (Fig. 9.7). Some parks have their own associations, while others serve multiple units in the same area (Vaughn and Cortner 2013; Public Lands Alliance 2017). Several are large regional organizations. The biggest is Eastern National, which serves 148 park units in the eastern half of the country. The Western National association has sixty-three units centered in Arizona and New Mexico but extending into adjacent states. Alaska Geographic is associated with the sixteen units in that state. Discover Your Northwest has nine units in Washington, Oregon, and Idaho. The northern Intermountain West is unusual in having no dominant association. Parks in this area, as well as within the ter-

ritories served by large associations, have their own association, as at Death Valley (the Death Valley National History Association) and the Grand Canyon (the Grand Canyon Association), or they have one that serves several adjacent units, as for Zion, Bryce, and Pipe Spring (the Zion National Park Forever Project).

These groups publish books and pamphlets on their parks, host conferences and special events, bring city children for a taste of nature, and contribute money to park visitor service projects, such as improving tourist facilities. If you are a frequent park visitor, you may be a member of one or more of these organizations. In addition to supporting an organization working to promote the national parks, you can receive a discount at any national park visitor center gift shop. The Death Valley Natural History Association offers memberships for as little as thirty-five dollars a year, with a 15 percent discount on anything purchased at the Death Valley visitor center or in most other

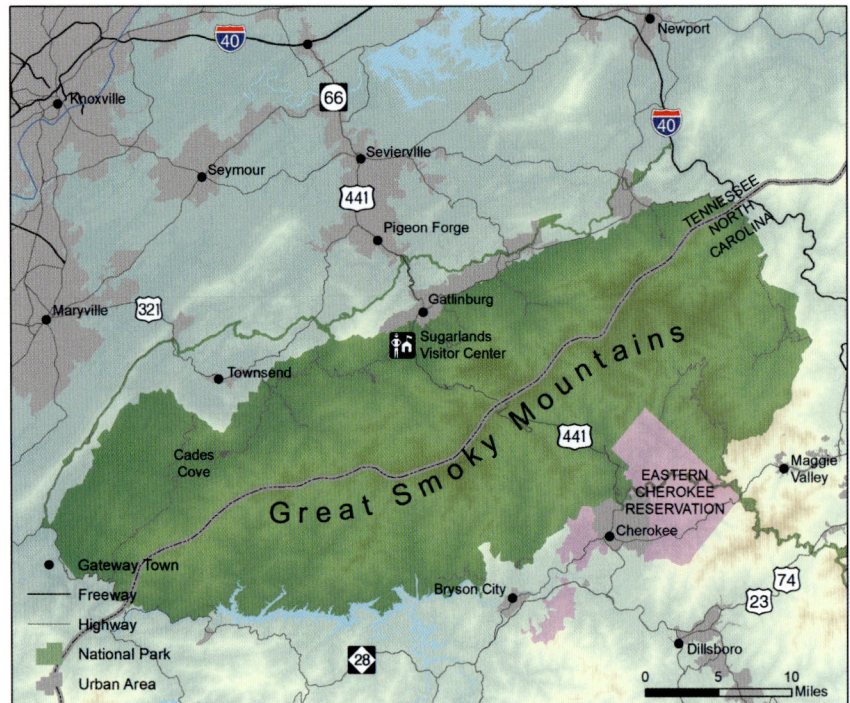

FIG. 9.8   Map of Great Smoky Mountains gateway towns, 2021.

Gateway Communities

The steady stream of tourists to national parks has led to the development of gateway towns, those communities whose economy is largely devoted to providing food, lodging, gas, and souvenirs to park visitors. These may be small towns that took on a new economic purpose after a nearby park was created or brand-new places created to serve tourists. In addition to often being situated within spectacular scenery, they may have their own attractions for visitors (Arflack 2020). Among the best-known examples are Gatlinburg and Pigeon Forge in Tennessee and Bryson City and Cherokee in North Carolina, all of which are just outside Great Smoky Mountains National Park (Fig. 9.8). Other well-known examples are Bar Harbor, just outside Acadia, and Springdale, Utah, at the entrance to Zion.

The Tennessee towns are perhaps the most extreme examples of this phenomenon and have become independent tourist attractions. Many visitors to these towns likely never set foot within the nearby national park. Gatlinburg

parks. Memberships in these associations are a good value.

is adjacent to the park's headquarters at Sugarlands Visitor Center and serves as the major gateway for more than twelve million park visitors each year. The city once was a small logging town with few houses for a century until the establishment of the park in 1934. The town has transformed into a tourist hub with glittering Las Vegas–style, tourist-oriented development, and it has become a real estate mecca (hotels, rental properties, and second homes) (Martin 2007). The town has more than eleven thousand guest rooms and many activities, such as the Ripley's Believe It or Not Museum, an aquarium, a ski resort, the Space Needle observation tower, and the Sky Lift, and it serves as a major vacation destination for families with children who are looking for more than just the natural beauty of the Smoky Mountains. The town is the leading tourism revenue generator in the state.

Pigeon Forge is nestled in the foothills of the Great Smoky Mountains near Gatlinburg and also transformed itself from a small farming community founded in 1817 into a major regional tourist destination. The city took advantage of being between I-40 and the Smokies and boomed with developments such as shopping (more than three hundred factory outlets), motels, and large-scale attractions, which include

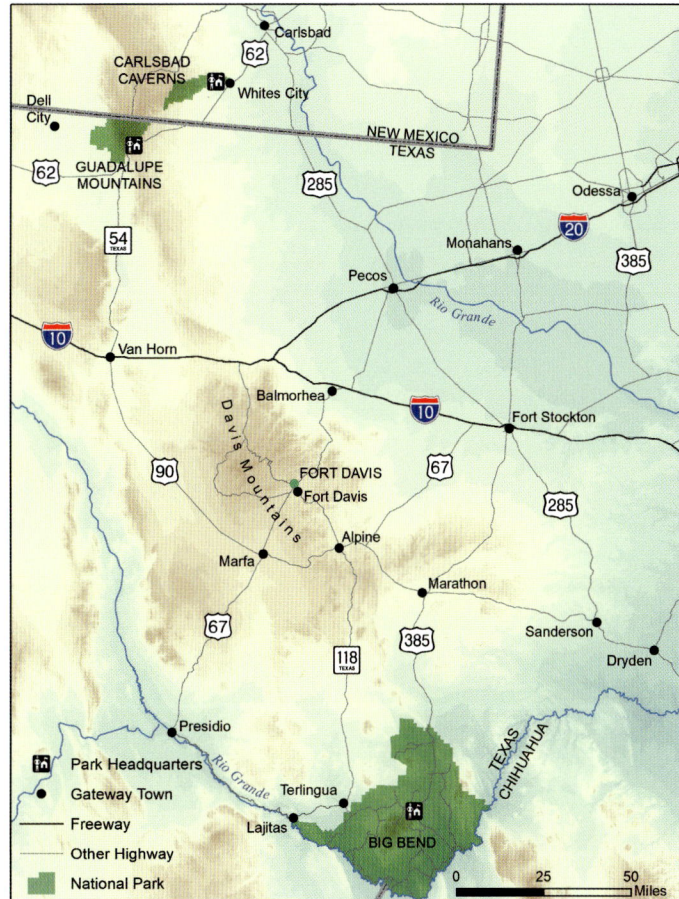

FIG. 9.9   Map of Big Bend, Guadalupe Mountains, and Carlsbad Caverns gateway towns, 2021.

minigolf, arcades, go-karts, and musical revue theaters (Fletchall 2013). The largest attraction, however, is Dollywood, a 140-acre theme park owned by famous country singer and eastern Tennessee native Dolly Parton. Pigeon Forge receives more than nine million tourists annually, and to accommodate this growing influx of tourists, the number of hotel rooms in the city increased from two thousand in the 1980s to more than fourteen thousand in 2013 (Bonimy 2014). The city received record revenues each year from tourism, reaching $1 billion in 2019.

Townsend, Tennessee, and Bryson City, North Carolina, are smaller gateways. Located fifteen miles southwest of Gatlinburg, Townsend is a small gateway town located adjacent to the park boundary and near the third most important entrance of the park at Cades Cove. The town was established in 1921 as a logging town by those involved with the Little River Railroad and Lumber Company. With a population of 496 in 2020, it advertises itself as "The Peaceful Side of the Smokies" and po-

sitions itself as an alternative to its neighboring cities Gatlinburg and Pigeon Forge. The town is built on its natural amenities, which include magnificent views of the mountains, clear rivers, family-owned lodges, scenic trails, and bike paths. It also serves visitors who are interested in visiting Cades Cove, one of the most popular destinations in the park. Cades Cove has been preserved as a "living museum," with a variety of historic buildings and in-depth information about the buildings and the people who built them and once lived there. The city is becoming one of the most popular gateway towns for park lovers who prefer the quietness of town and historical aspects of the park as well as adventurous hikes. In contrast, Bryson City is a small "cabin community," but one with a goal of becoming more competitive to Gatlinburg and Pigeon Forge.

Big Bend is a much more remote park, with several gateway cities and towns scattered over a large area of West Texas (Fig. 9.9). Among them are Alpine, about sixty miles from

FIG. 9.10   Map of Bryce Canyon and Zion gateway towns, 2021.

the park entrance, while Marathon and Marfa are even farther away. Fort Stockton provides the closest access point to Interstate 10, which brings many visitors to the region. At a larger scale, El Paso (about 315 miles to the west) provides services to those driving east, as well as access to the closest airport with scheduled passenger services; several other Texas cities serve travelers headed west. As at Great Smoky Mountains, these gateway communities have served multiple functions over their lifetimes. Marathon was established with the coming of the railroad in 1882, but this declined in importance around the time Big Bend National Park appeared on the map. Tourism is vital to the tiny town. Nearby Alpine is much larger and has a more varied economy, with the nearest hospital and Amtrak train station to the park. Marfa is another railroad town living on tourism and trying to reinvent itself as an arts center. The city has been used as a location for filming Hollywood movies. Among these are *Giant*, which starred James Dean as a rancher turned oil tycoon in 1956, and, more recently, two Oscar-winning films, *No Country for Old Men* and *There Will Be Blood*.

To the north, Carlsbad Caverns National Park is one of the most famous parks in the Southwest (Fig. 9.9). Every year the caverns receive over four hundred thousand visitors, who can visit the caves only with reservations that allow them to enter the cave at a specific time. Park visitors spend only a few hours in

the park before moving on. There is no lodging, so all visitors must spend the night in the city of Carlsbad or the smaller Whites City, located just outside the park entrance. Carlsbad has over twenty-five thousand people and is a thriving agricultural and regional economic center. It can provide any kind of tourist service needed, in addition to conveniently sharing a name with the nearby caverns. Whites City is much closer to the park, but with only seven inhabitants, it hardly qualifies as a city. It provides convenient tourist services, though it mars the park's entrance. Several other park units have had their boundaries expanded along their entrance roads to prevent this sort of development.

Guadalupe Mountains National Park is a short drive from the more famous Carlsbad Caverns but is one of the country's lesser-known parks (Fig. 9.9). The towering mountains offer tremendous opportunities for backcountry exploration but have no roads into them. Visitors either stop briefly at the visitor center along U.S. Highway 62 before getting back in the car or come prepared for a lengthy backcountry sojourn. There is no nearby gateway town; Carlsbad and distant El Paso are the closest places where services can be found. (Dell City is a small farm town to the west with no tourist services.)

Bryce Canyon City is another example (Fig. 9.10). An enterprising local named Ruby Syrett built Ruby's Inn in 1916, which has expanded to include several motels and hotels, campgrounds, restaurants, and shops straddling the entrance road to Bryce Canyon National Park. Tusayan, Arizona, just outside the boundary of the Grand Canyon, was created for similar purposes and has a population of 558. It is on Highway 180, the main entrance to the South Rim of the canyon, and has a variety of motels, restaurants, and souvenir shops, an IMAX movie theater, and an airport big enough to handle a Boeing 747. It is, however, dwarfed in size by Grand Canyon Village, the park's developed area, which contains a similar number of motels, restaurants, and stores, though with the addition of schools and a train station.

Zion is a less remote park with several gateway cities and towns nearby (Fig. 9.10). Springdale began as a Mormon farming community in 1862 and is just outside the main entrance to the park. It has been transformed into a charming and thriving town with sidewalks, public transport, and an eclectic array of cafés and restaurants. Established in 1896 as another farming settlement, Hurricane is twenty-four miles away from Zion and has undergone commercial growth since the 1970s. St. George and Cedar City are farther away but serve as gateway cities along I-15, the main highway between Las Vegas and Salt Lake City that brings most visitors to the area.

Cedar Breaks is a smaller monument open in the summer, but the Brian Head ski resort lies just to the north (Fig. 9.10). Cedar City functions as a gateway to both from the south, while the small highway town of Parowan is the gateway to travelers from Salt Lake City and other northern origins. Kanab is a small town that benefits from tourism at Glen Canyon and the North Rim of the Grand Canyon.

Yellowstone has several gateway towns, among them Cody and Jackson, Wyoming, Gardiner, Montana, and West Yellowstone, Idaho, most of which have been active since the early twentieth century, when they were railroad gateways. Most tourists visiting Yellowstone pass through at least one of these. Moab, Utah, with just over fifty-two hundred people, is a gateway town for Arches and Canyonlands. It was once a small farming and ranching community with a uranium mill just outside town. Summer tourism is now the dominant industry, with

a surge of seasonal residents. Along with other small gateway towns with large numbers of visitors passing through, Moab can provide a wide range of services, making it similar to small towns located at a freeway interchange that is far from any big city. These towns' small size and large number of services, combined with often spectacular natural settings adjacent to a national park unit, make many attractive as retirement communities.

The concept of gateway town is somewhat different in Alaska. Because of the role of air transportation in reaching parks in remote places, these parks tend to have gateways, as well as park headquarters or visitor centers, well removed from the park. The town of Kotzebue is the headquarters location and main gateway for Cape Krusenstern, Noatak, and Kobuk Valley; here visitors can hire an airplane flight into the parks and purchase supplies. But to get to Kotzebue visitors will likely pass through Fairbanks or, more likely, Anchorage. For those traveling from out of state, this city functions as the gateway to most of the state's parks. Denali is one of the few Alaskan parks with a typical gateway community, though in this case many visitors arrive by train. A range of tourist businesses have sprung up in several small gateway towns outside the park entrance, providing transportation to and from the park as well as other services.

Over time, with space-time convergence, gateway communities have tended to move closer to or become concentrated nearer parks. The first gateway community for Yosemite was San Francisco (T. Davis 2016); visitors had to take a steamboat to Stockton and then two days of stagecoach travel to one of several small towns, where they switched to horses for a two-day trip to the valley floor. By that point, many visitors were too sore and exhausted to appreciate the canyon's beauty. A trip to Yosemite was a ten-day trip, with most of it spent in transit. The market for travel led to toll roads with stagecoach services from several towns in the Central Valley, but in 1907 a railroad line allowed those to be bypassed and created the new gateway town of El Portal, still the closest gateway community to Yosemite.

There are clearly many varieties of gateway

towns, ranging from those immediately outside the main entrance whose inhabitants are entirely dependent on park tourism, to large and diverse recreation centers for whom the park may be a diversion for those needing a break from roller coasters, miniature golf, and go-kart racing. Inhabitants of small towns may welcome a new park next door, eager at the opportunity to invigorate their moribund economies with tourism, but things rarely work out that way (Rothman 1998). Outsiders and, eventually, large corporations often end up in control of the local tourism industry, relegating locals to service occupations. Even worse, the town's character may be transformed in ways that locals are not always happy with. Local businesses will likely be driven out and replaced by chains and franchises, and new and wealthier residents may come to live in vast homes that block the view locals took for granted. Locals may find that they can't afford to live in their own town anymore. The economies of gateway communities do not favor locals. And while park visitors are overwhelmingly white, the workers in tourist towns are increasingly nonwhite and in many cases undocumented migrants from south of the border, producing social as well as economic tensions.

Urban versus Rural Parks

Most discussion of the economic impacts of national park units focuses on large rural units. These are often the economic engine of their regions, and the superintendents who run them have considerable power; they operate independently of local governments (Rothman 2004). Urban parks are different; their economic impact is likely quite small within the city, and they are one of an enormous number of governmental entities and countless political actors. They do not dictate policies but must work closely with a larger number of stakeholders.

Urban parks also have issues that rural parks do not have. Small battlefield parks such as Kennesaw Mountain and Guilford Courthouse have become surrounded by homes and shopping centers and are heavily used by residents as city parks. This was not the original intention, and park policies prevent recreational

activities such as sunbathing, throwing Frisbees, and flying kites in these battlefields. Tensions have emerged between the parks and residents over these matters. At De Soto National Memorial the use of the park for an annual festival also produced tensions (Whisnant and Whisnant 2007). While the festival was originally justified to commemorate the sixteenth-century expedition of Hernando de Soto, it evolved into a promotional event with water-skiing exhibitions and other entertainment before the NPS insisted it be moved out of the park. In the 1960s the NPS accepted that evidence for de Soto's landing was absent and that the expedition probably landed elsewhere; the memorial was therefore a commemoration of the expedition rather than its actual starting point. Many locals continued to insist on it as the actual landing point, leading to strained relations between the town and park.

Economic Impacts

The large numbers of tourists who visit parks have an enormous economic impact on the communities and regions around those parks. The NPS regularly quantifies these impacts, most recently in the 2021 National Park Visitation Spending Effects study (National Park Service 2021). This study found that in 2019, 328 million park visitors spent an estimated $21 billion in local gateway regions while visiting NPS lands across the country. These included direct effects, as with spending on gas, food, or lodging, as well as indirect effects and multipliers, which together supported 341,000 jobs.

Total visitor spending was greatest along the Blue Ridge Parkway (over $1.1 billion in 2019), followed closely by the Great Smoky Mountains at $1 billion (Fig. 9.11). Park visitors supported 13,942 local jobs in communities within sixty miles of the park (Chavez 2020). Golden Gate National Recreation Area and the Grand Canyon are next, followed by Grand Teton and Denali, Yosemite, Cape Cod, Yellowstone, and Glen Canyon. The fact that Denali is sixth on the list despite relatively modest visitation is due to the extremely high cost of lodging, food, transportation, and other services when visiting this park. The park has the ninth high-

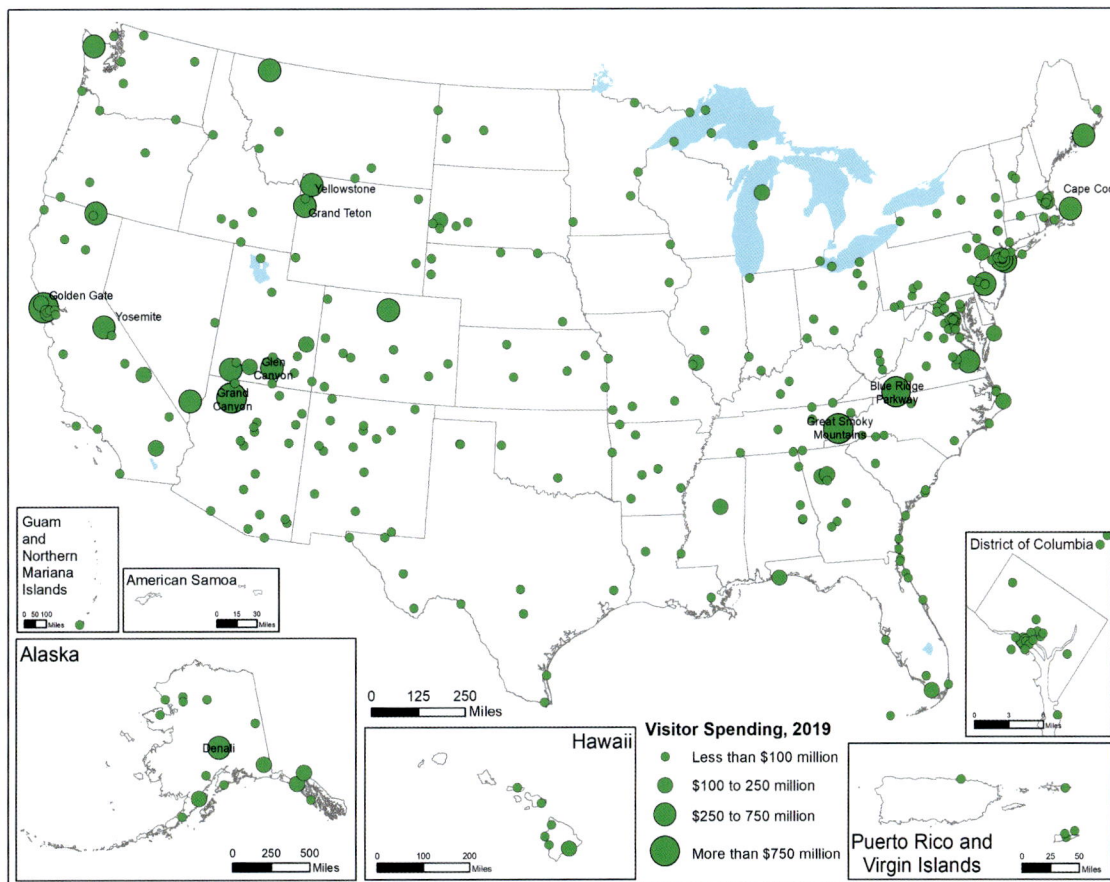

FIG. 9.11   Map of visitor spending at each park, 2019.

est spending per person in the park system; all eight more expensive parks are in Alaska. Bering Land Bridge, one of the most remote and least visited units, is the single most expensive to visit despite having no opportunities within the park to spend money. (Other units in the Brooks Range are similar.) Air transportation to the park, guides, supplies, and other services en route to the park are costly.

At the other extreme is Carter Woodson National Historic Site, with a mere $46,000 in visitor spending in 2019. The park has low visitation and is in the middle of a large city offering all services visitors need. Several other small, lightly visited, and often obscure units are also among those having the smallest visitor impacts (Thaddeus Kosciuszko, Nicodemus, and Minidoka among them). Alaska's remote Aniakchak National Monument and National Preserve, perhaps the least visited park in the entire system, is also among them.

Economic impacts are often used to justify or defend the creation of a new park unit, especially where the creation of a park is seen as threatening rural livelihoods. Alaskan park units created in 1980 were often bitterly opposed by locals, who saw them as a threat to their hunting, fishing, and outdoor lifestyles; in time, they saw opportunities to make money off visitors. And locals do make money in Alaska: aside from Aramark at Denali and Glacier Bay and several cruise ship lines serving Glacier Bay, concessions in Alaska are usually small companies headquartered within the state. For those wanting both to experience nature and to spend locally (and lavishly), Alaskan parks are once again in a different class from the rest of the park system.

Duane Hampton (1981) identified three major arguments used against park proposals or expansions. Among these are economic arguments based on the idea that creating a park will eliminate or reduce economic activities such as mining, farming, ranching, logging, hunting, and oil drilling. Scott Raymond Einberger (2018) refers to this as the "lockout argument": those few who are miners or cattle ranchers or who otherwise make a living off public lands will lose their livelihoods. Local governments may complain that creating a park will reduce privately owned taxable land and greatly reduce their revenues. This is not just an occurrence in the rural West: efforts to expand Fort Raleigh National Historic Site on the North Carolina coast were opposed by Dare County, as 75 percent of the county's land area was already within national park units (Cape Hatteras National Seashore, Wright Brothers National Memorial, and Fort Raleigh) (Binkley and Davis 2003). Since 1976 the Department of the Interior has provided direct payments to states that contain public lands to compensate for this reduction in tax base (Department of the Interior 2021). These "payments in lieu of taxes" provided $59,000 a year to Dare County in 2020, a minuscule portion of the $515 million the government pays out to counties throughout the country each year under this program.

Although not counted as part of park economic impacts, reservoir-based national recreation areas account for about 12 percent of U.S. hydroelectric production. Hoover Dam in Lake Mead National Recreation Area generates about $63 million worth of electricity each year. These facilities are run by the Bureau of Reclamation, and power sales are allocated by several government agencies, such as the Western Areas Power Administration, with the NPS having no part in the process. (It doesn't even get any of the electricity.) The water behind the dam is also owned by the states of Nevada, Arizona, and California and by Mexico, to which it is carefully portioned out according to a legal framework that began in 1922. Los Angeles, Phoenix, Las Vegas, Tucson, and the Imperial Valley all depend on this water, but their economies are not counted as part of the economic impact of the park.

Shutdowns

In the 1950s the well-known author and editor Bernard DeVoto (1897–1955) wrote several widely read essays about the deteriorating condition of national parks during the postwar visitation boom (Muller 2005). His essay "Let's Close the National Parks" is the best remembered (DeVoto 1953). With Congress not giving enough money to adequately run the parks, he suggested that the park system should be reduced temporarily to a level Congress would support. Yellowstone, Yosemite, Rocky Mountain, and the Grand Canyon should be closed and their staff and funds redistributed to other parks. If nothing were to improve, he suggested closing Zion, Big Bend, Great Smoky Mountains, Shenandoah, Everglades, and Gettysburg, after which he felt the public outcry would eventually push Congress into action.

While Congress did not take these drastic steps at the time (and did eventually fund Mission 66 to address the problem), all or most national parks have in fact been closed due to disputes over the federal budget between Congress and the president. There have been ten government shutdowns since 1980, several of them lasting at least several weeks. A 2013 shutdown required closure of national parks, leading to the public backlash forecast by DeVoto when visitors found themselves locked out of their favorite public lands. In response, the NPS developed a contingency plan to keep them open during future shutdowns. This was tested during the December 2018 to January 2019 shutdown, which was not only the longest but also the most harmful to the national park system.

Smaller units, such as historic sites, battlefields, and small (day-use) parks, were usually closed during this shutdown. Units such as Thaddeus Kosciuszko, Belmont-Paul, Carter Woodson, African Burial Ground, George Washington Carver, Abraham Lincoln Birthplace, Guilford Courthouse, and Fort McHenry were among these. Many larger units remained open but with no visitor services. Visitor centers were closed, trash was not collected, restrooms were not cleaned, and entrance and campground fees were not collected. Among these were Devils Tower, Grand Teton, Yellowstone, Glacier, Natchez Trace, Shiloh, Vicksburg, Denali, Gates of the Arctic, Glacier Bay, Mammoth Cave, Gettysburg, Johnstown Flood, Yosemite, National Mall, Joshua Tree, Death Valley, Everglades, Biscayne Bay, Big Cypress, Dry Tortugas, and Channel Islands. Within days

FIG. 9.12 Map of Arizona park status during the 2019 federal government shutdown. Some parks stayed open with state support, others were open with no services, and some units were closed.

park trashcans were filled, restrooms became filthy, and law enforcement problems became an issue at several parks.

Several parks received state support to continue providing services. Utah paid for services at Bryce, Arches, and Zion, and New York helped keep the Statue of Liberty open. At some parks, such as Joshua Tree, locals worked to keep the parks clean. But despite funding available from states and the hard work of volunteers, conditions worsened as the shutdown progressed, and the NPS was forced to close Sequoia–Kings Canyon and Joshua Tree before long.

The range of responses to the shutdown can be seen in Arizona (Fig. 9.12), where the smaller units, along with Petrified Forest, were closed. Sunset Crater and larger monuments and recreation areas (Chiricahua, Glen Canyon, Lake Mead, and Organ Pipe) were open with no services, though most back roads were closed. Only the Grand Canyon received state support to remain open.

The shutdown has had enormous economic consequences for parks and the communities that rely on them. During an average day, park visitors create $57.5 million in economic activity. That income was lost, along with tremendous damage done to some parks that will take additional funding and considerable time to fix.

Conclusions

We like to think of parks as wilderness, where the dollars that our lives revolve around have no significance. The reality is quite different. Money is necessary for parks, and there is never enough of it, despite the massive sums of money spent in and around parks. The NPS has so far avoided any major budget fiascos, such as California's high-speed rail system soaring from an initial cost of $33 billion to nearly $100 billion before the system was canceled (Nagourney 2018). With an annual budget of only $3.2 billion, the NPS cannot experience this sort of fiasco. Perhaps the greatest financial waste the NPS has been involved with is the Comprehensive Everglades Restoration Plan. This was passed with great fanfare in 2000 and called for $8.2 billion to be spent over thirty years by the Army Corps of Engineers and the state of Florida. So far, little has been accomplished; if the plan is ever carried out, both the time needed and the cost will greatly increase. This is not an NPS program, but since the Everglades is often thought to be no larger than the national park boundaries, the NPS might be mistakenly held responsible.

Parks may be welcomed or opposed in part based on the economic rewards promised. The thought of parks as economic prizes sought by rural areas, like military bases, mines, or even prisons, is unsettling but a fact of life in much of the country. There is not even any reason why parks should top this list of prizes; a typical prison employs many more people than the average park and may pay them higher salaries (and prisons won't close during increasingly frequent government shutdowns). Urban residents of distant cities may think it obvious that miners, ranchers, and other rural residents should be happy to have a new park created in their backyard; after all, lots of new tourism jobs will appear. Miners faced with losing a high-paying job

with full benefits for the possibility of seasonal minimum wage work are not so thrilled.

The idea that large multinational corporations operate in many national parks is also unsettling but completely consistent with their dominance in all other aspects of American life. So far, they are American companies; imagine a future when America's parks are operated by foreign corporations. Given trends in other business sectors, this may eventually occur.

The economic geography of national parks depends in large part on what services and products park visitors need and want. This changed over time as camping replaced luxury lodges, gas stations replaced blacksmith shops, and visitors stayed fewer nights on shorter vacations. The Mission 66 program included efforts to concentrate visitors in fewer locations that were to be developed to handle these impacts. This led to the closure of several park lodges, which in turn had impacts on visitors and indirect spending. Future cultural shifts will also likely change preferences for services; increasing numbers of visitors are demanding reliable cell phone and internet coverage within national parks, a trend that will likely produce tremendous harm to parks.

TEN

The Geographic Setting of Parks and Their Challenges

Cumberland Island National Seashore is a quiet, peaceful island along the Georgia coast. It is unique among large park units in that it has a daily attendance limit of three hundred people, meaning that on many days visitors will be outnumbered by the 266 nuclear warheads carried on each of ten ballistic missile submarines stationed at the adjacent Kings Bay naval base. Nuclear weapons and national parks would seem to be utterly incompatible, but there are surprising overlaps. The National Park Service acquired its own nuclear missile in 1999 when Minuteman Missile National Historic Site was created in South Dakota. The park features a missile silo containing a fifty-eight-foot-tall Minuteman II missile, capable of carrying a 1.2-megaton nuclear warhead seven thousand miles. The missile is not armed or fueled, but park managers must comply with the Strategic Arms Reduction Treaty (START II), signed in 1993 between Russia and the United States. The NPS went on to acquire nuclear weapon production facilities at Manhattan Project National Historic Site in 2015. But the service has so far been unable to acquire New Mexico's Trinity Site, where the first atomic bomb exploded in 1945.

We often think of parks as separate from the rest of the world, places where nature is cleaner, where we can leave our worries behind at the entrance sign. But what happens inside a park is part of everything else. While the threat of nuclear annihilation might not be a worry to current visitors, air pollution, congestion, and crime often are. This chapter discusses these and other recent, ongoing, and future challenges to the national park system.

The parks have faced many threats over the years. Remoteness was an early complaint about the parks that has largely disappeared; many are not remote enough today. Resource extraction has been and remains a major threat to many parks. The country's growing population presents some of the most urgent threats to the integrity of the park system, including congestion; incompatible land uses and crowding around park boundaries; air, noise, and light pollution; and global warming. Another group of threats is found within the NPS or other government agencies, including recurring battles over whether the parks should be protected or used for national defense. The NPS was even accused of allowing the Pearl Harbor attack on December 7, 1941! More worrisome are changing cultural values that blur the boundaries between parks and the world outside and could endanger the visitor base parks depend on. They could even cause us to question the fundamental idea of parks altogether.

The Dual Mandate

At the heart of many park controversies are a few sentences within the 1916 law that created the National Park Service. Section 1 of the law created the NPS and specified some information about employees and salaries. The act then required that "The service thus established shall promote and regulate the use of the Federal areas known as national parks, monuments, and reservations hereinafter specified by such means and measures as conform to the fundamental purpose of the said parks, monuments, and reservations, which purpose is to conserve the scenery and the natural and historic objects and the wild life therein and to provide for the enjoyment of the same in such manner and by such means as will leave them unimpaired for the enjoyment of future generations" (Dilsaver 1994a, 46). The wording of this statement directs that the NPS must preserve scenery, wildlife, and history within parks yet also allow the public to enjoy them. This is known as the dual mandate, and it means that the park staff must balance both preservation and development, considered to be the dual mandates for the agency. However innocuous this balancing act may appear in print, it has created many political battles over park management, with frequent accusations that one of the mandates is being emphasized at the expense of the other (Foresta 1984).

In the early years, the NPS was very active in promoting tourism to parks, and the agency was willing to try just about anything to encourage more visitors and better justify its own existence (Sellars 1997). Fish were introduced in parks to attract fishing; predators were killed to encourage moose and deer to proliferate; bears and bison were fed to bring them to the roadsides and lodges, where they could be viewed up close; zoos were established in parks to allow even closer looks at animals; and fires were put out to preserve scenic woodlands (Biel 2006). The NPS even sought to hold the 1932 winter Olympics in Yosemite but ultimately lost to Lake Placid, New York, within the vast Adirondack state park (Sellars 1997).

A scientific approach to the natural resources within parks did not begin until 1930, when biologist George Wright began a study of park wildlife. This study called for the restoration of natural conditions and expanded boundaries to allow complete habitats. Unfortunately, he died in 1936, and so did the scientific view he championed.

In the postwar era, opening parks to tourists and improving the facilities for them became the main NPS goal. The huge and ever-increasing number of visitors required this; Mission 66 was a coordinated attempt to get ahead of this growth. But Mission 66 wasn't just a development project, it was an attempt to better protect park resources. Development would be steered toward a small number of places, which would be "hardened" with boardwalks and asphalt (Sellars 1997). These efforts brought a renewed emphasis on science, including the 1963 Leopold Report and the 1991 Vail Agenda. These again called for ecological management of parks, but this management has made few inroads into how parks are operated. Bear feedings at Yellowstone were gradually ended, some not until the 1970s (Biel 2006). In recent decades, the preservation side has clearly been favored by many observers (Sax 1980; Tweed 2010; Keiter 2013). This has led to battles over whether parks such as Cumberland Island and Apostle Islands should be developed for standard tourism or left largely undeveloped (Bonnicksen and Robinson 1981; Dilsaver 2004).

The needs of the dual mandate can be viewed within broader trends. Environmental values lie on a continuum with anthropocentric (human-centered) views and biocentric (nature-centered) values at opposite ends of the spectrum. The dual mandate places the NPS near the middle of the continuum with its requirement to balance human and natural needs. A more biocentric attitude would reduce human management of and presence in these places, treating them as wilderness off-limits to visitors. Many park issues are driven by a perspective on parks that emphasizes backcountry or wilderness experiences shared by very few visitors (Sax 1980). Traditional mass tourism is shunned, though it remains the cornerstone of the park experience. Tourists are often dismissed with stereotypes of overweight visitors descending in floods of cars or campers and overrunning the natural beauty, in contrast to the authors' more legitimate backcountry experience of the park (Rothman 1998; Tweed 2010). This view has been common since the 1960s, with author Ed Abbey the patron saint of the movement. Abbey worked as a seasonal ranger at Arches, Organ Pipe Cactus, Everglades, and Lassen (Cahalan 2001). He railed against "industrial tourism," which involved paving the parks and allowing as many automobiles as possible into a place such as Arches "National Money-Mint" (Abbey 1968). He was a strong advocate of getting people out of their cars, a common refrain in the decades since. This sentiment has perhaps been most memorably expressed by the comedian George Carlin:

"Everywhere you look there are families with too many vehicles. You see them on the highways with their RVs. But apparently the RVs aren't enough, because behind them they are towing cars, motor boats, go-carts, dune buggies, dirt bikes, ski jets, snowmobiles, parasails, hang gliders. Hot air balloons and two small two-man deep-sea diving bells. The only thing these people lack are lunar excursion modules. Doesn't anyone take a [expletive] walk anymore?" (1996).

So much writing about parks depends on timing. Abbey's 1968 rant against Mission 66 development in parks such as the Grand Canyon seems wildly exaggerated today and ignores that development has been concentrated in a small area. And Abbey was not the originator of these criticisms; John Muir might have been the first when he disparaged tourists to Yosemite in 1870 (T. Davis 2016). Muir had no patience for Native Americans living and pursuing their traditional ways of life in parks, either. Rather, he approved of a very specific type of visitor, one prepared for wilderness but with the ability to truly understand what he (but not she) saw and experienced.

A more anthropocentric view might place larger value in parks as economic resources and encourage development or extractive industries or indifference to the effects of a changing climate. These attitudes tend to be stronger and more common in Alaska, where hunting, fishing, and resource extraction are more important. Creating parks there has therefore been more controversial, with many in the state furious about the 1978 monuments and 1980 parks (Catton 1995, 2010; Norris 1996a, 1996b). These attitudes have slowly become more supportive but can still be surprising to those from the contiguous forty-eight states.

Anthropocentric viewpoints could also place greater emphasis on the needs of visitors. Safety concerns have risen in recent decades, as have Americans' willingness to resort to lawyers to solve real or perceived safety problems. Concern over liability has prompted a growing examination of road-building standards and could lead to park roads being built wider and straighter, with steel or concrete guardrails replacing rustic stone walls or logs and trees and

boulders removed from the side of the road to create wider clear zones (T. Davis 2016). This would eliminate most features that people enjoy about park roads. In the past, visitors could accept or even preferred that being in a remote park meant taking responsibility for your own safety; today this means little to most people, who expect the same level of cell phone service in a park that they find in their home city. It can always be argued that putting up more antennas all over a park will reduce response times in an emergency and improve safety (National Park Service 2014). It is politically difficult for the NPS to resist this view.

Any attempt by park staff to find a balance between these views may be offensive both to those favoring an extreme biocentric view and to those espousing strong anthropocentric views. This is probably a good thing; if the NPS offends both extremes, then it must be following the dual mandate.

External Challenges to the Park System

Regardless of how the NPS applies the dual mandate, it can do little about what goes on beyond a park's boundary. And what goes on outside the boundary has been changing, in part as parks become more accessible to the outside world. This changing geography creates many challenges.

Access

The first challenge, present from the creation of the first parks, was geography and, more specifically, parks' relative location. How would people get to these parks? The early units were hard to reach, limiting visitation and public support, as well as leading to charges of elitism in that only the wealthy had the time and money to visit them. The newly created NPS worked hard to promote tourism by automobile and to make these units more accessible to all Americans. In doing so, agency staff encountered the first battles over the meaning of the dual mandate. NPS staff did not always see a conflict. Building roads in parks was considered a means of protecting them. Creating and sustaining tourism was essential to preventing parks from being mined, logged, dammed, or grazed (T. Davis 2016).

In today's era of mass tourism and easy mobility, it is hard to imagine parks as remote; many problems in fact stem from the opposite. The label "remote" is still attached to many Alaska parks, but remoteness is now usually seen as a good thing, whether because it reduces visitor impacts to a minimum or represents a return to the supposed roots of the parks as wilderness. It is evident now that remoteness had value in part because it allowed places like Yellowstone to be set aside for their scenery, as it was hard to imagine otherwise using such a place (Ise 1961). Fragile cliff dwellings such as those at Navajo National Monument were protected by the fact that the monument was remote and remained so for half a century after it was created (Rothman 1991); a similar story could be told at many western park units not developed until Mission 66. Remoteness, like visitation or development, may be something requiring a careful balance.

Unfortunately, while a remote park may be fun to visit for a few days, it is not always seen as a good place to live, work, and raise a family. Big Bend is considered remote and isolated and something of a hardship for employees with children (Jameson 1996), as is Mesa Verde (D. Smith 2002). Children have largely disappeared from Death Valley, leaving the Mission 66 elementary school nearly empty. Overcoming employee attitudes may be a growing challenge for the parks in the future.

Resource Extraction

National park units contain an enormous amount of land and water no longer open to homesteading, mining, logging, and other uses. Many have coveted these same lands and waters for other uses, including mining, logging, ranching, hunting, drilling, damming, or building residential homes. These activities are a by-product of the changing relative location of parks; as it becomes easier to transport tourists to parks, it also becomes easier to ship out ore, trees, cows, and other resources, often over the same roads.

The General Mining Act of 1872 enshrined the right of prospectors to search for and claim mineral resources on public land (Bureau of Land Management 2020). While Yellowstone was a park two months before the 1872 act went into effect, most western parks had been prospected long before they became parks. An individual filing a mining claim has rights to that property, and these rights do not go away when a park is created. Time-consuming legal action is required to eliminate valid claims in a new park. Even worse, while mining was banned within new park units, Congress passed legislation in the 1930s and 1940s to reopen six units to new mining claims: Denali, Glacier Bay, Death Valley, Organ Pipe Cactus, Coronado, and Crater Lake. Today there are over thirty-seven thousand abandoned mines within the park system, the vast majority of which are in desert parks (Burghardt, Norby, and Pranger 2014). Mojave has over eleven thousand, and Lake Mead and Joshua Tree have over a thousand each. Alaskan parks have 751, and there are 887 in the Northeast (most of them in New River Gorge), 317 in the Southeast, and even 59 in the national capital parks.

Death Valley has the most extensive record of mining in the park system. Gold, silver, copper, lead, tungsten, talc, and borax had been mined in the area since the 1870s (Lingenfelter 1988; Rothman and Miller 2013). Borax mining was a particularly important activity, with the Pacific Coast Borax Company (which once employed Stephen Mather and Horace Albright, the first two directors of the NPS) building roads and developing water resources to support its mining. When borax mining declined, the company converted its properties to accommodations for tourists. To forestall competition, the company persuaded a sympathetic former mining engineer, Herbert Hoover, to use his powers as president to proclaim a national monument in the valley. Mines were started and abandoned repeatedly over the next four decades as the price of different minerals rose and fell: tungsten and antimony were mined during World War II, and lead and talc mining flourished after the war. Several open-pit borax mines were even dug in full view of the main park highway.

By the 1970s mining in the national park system did not enjoy the support of earlier

FIG. 10.1   An abandoned mine in Death Valley. This is part of the mill at Keane Wonder Mine, an early twentieth-century gold mine that remains a popular tourist attraction. Photo by Pierre Camateros, 2007. Wikipedia, Creative Commons Attribution–ShareAlike 3.0 Unported License.

times, and environmental organizations turned their attention to it and demanded changes. The Mining in the Parks Act of 1976 was the beginning of the end of mining in Death Valley and the handful of other park units where it was still permitted. But Death Valley is still littered with over sixteen thousand abandoned mine tunnels and shafts, all of which are hazards to unwary visitors (Fig. 10.1) (Burghardt, Norby, and Pranger 2014). The park contains 44 percent of all abandoned mines in NPS units and 23 percent of all mines rated as high priority, meaning they are serious threats to life. The cost of mitigating them (covering them up or placing gates on tunnel entrances and removing unstable structures and hazardous waste) totals almost $13 million, far more than the park's $9 million annual budget.

While Death Valley's abandoned mines are considered picturesque and are popular tourist attractions, the technology used is a century old and has long since been replaced by fully mechanized strip mining. The days of a lone miner laboriously digging a tunnel into the side of a mountain in search of gold have been replaced by giant machines that will dig through the entire mountain and chemically process it for the tiniest amounts of gold. It may be difficult for visitors to grasp the threat that mining brings when viewing the picturesque remains of mines in parks like Death Valley. Vast open-pit mines in Utah, Montana, and Arizona are examples of what this technology can do today.

Energy generation, whether solar, geothermal, or, especially, hydroelectric, is another long-running threat to parks. The parks have had a particularly long and troubled history with dams and reservoirs. Muir Woods was proclaimed a monument in 1907 to save a grove of redwoods from being cut for a reservoir for the city of San Francisco, while the Hetch Hetchy Valley in Yosemite was subsequently flooded for one (Righter 2005). This precedent would later create tremendous controversies over dams to be built in Dinosaur National Monument and even within the Grand Canyon (Pearson 2002; Righter 2005).

The creation of Lake Mead National Rec-

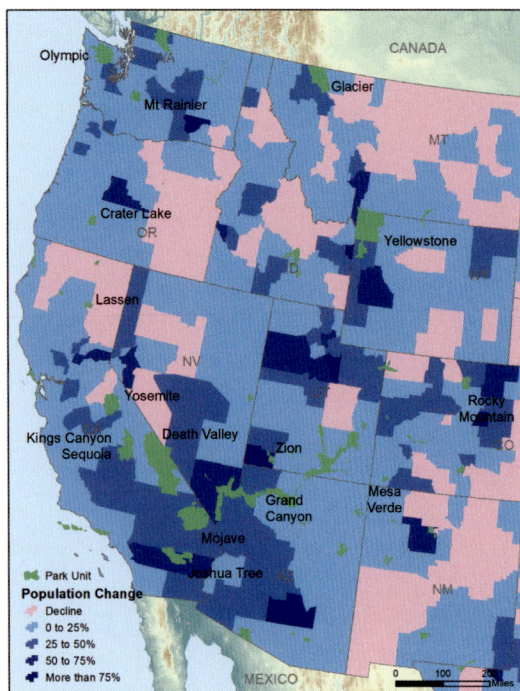

FIG. 10.2   Map of population growth near parks by county, 2000–2016.

reation Area between the 1930s and 1964 set a precedent that reservoirs were not necessarily incompatible with national park units and indeed could provide the feature around which the park unit was created (Dodd 2007). When plans began for a reservoir in Dinosaur National Monument the fact that the NPS had not developed the monument was a handicap to those who opposed the dam: If the NPS had no use for the monument, why preserve it? The NPS even responded by considering the recreational possibilities of the reservoir, with the expectation that the monument could become a national recreation area (Harvey 1994).

Florida's Everglades is a vast freshwater wetland flowing south from Lake Okeechobee, but one suffering from water shortages (McCally 1999; Lodge 2016). Since 1882 parts of the Everglades have been drained and converted to agriculture, providing generations of Americans with cheap sugar and lettuce, while roads and other features have obstructed or stopped the flow of water southward. The national park contains only 25 percent of the Everglades, and the park has been struggling to ensure that enough water flows south to support this ecosystem. The massive Comprehen-

sive Everglades Restoration Plan was launched in 2000 and received widespread media attention, but it faded from public view, with little accomplished.

Even the plants within a park are valuable resources. Saguaro cacti in Arizona's Sonoran Desert grow as much as fifty feet tall and weigh several tons. Yet people steal them. Not casually, the way they might pocket a piece of petrified wood or a fossil from other parks, but with trucks specially equipped with hydraulic lifts to carry the cacti and then reerect them elsewhere. Transplanting any saguaro over four feet tall requires a state permit, but this is not enough to deter people from stealing them from public land. The NPS has had to put trackable microchips within roadside saguaro to dissuade people from stealing them or at least make it possible to convict people if they do (McGivney 2019).

The Effects of Growing Population

When the public domain was first put aside to protect spectacular scenic wonders there were few non-Native people around. Some must have wondered what the point of protecting a place like Yellowstone was, given its great distance from the eastern states where many Americans lived. The NPS sought to move the park system eastward to bring parks closer to the people and increase visitor numbers and support. More people in the parks was the goal, and the agency succeeded in this effort (perhaps too well). But today and for the foreseeable future, people are one of the leading threats to parks. The nation's population increased from 31 million in 1860 to 328 million in 2018 and will likely reach 417 million by 2060 (Colby and Ortman 2015). This means more people visiting parks, but also more cars, more air pollution, more demand on natural resources, and even greater pressure to protect what is left of untouched nature (Fig. 10.2).

Just as the park system was moving toward people, people were moving toward parks, and not just to visit. The Census Bureau measures the center of the nation's population after every census. Since 1790 the population has shifted westward and then southward, reflecting the

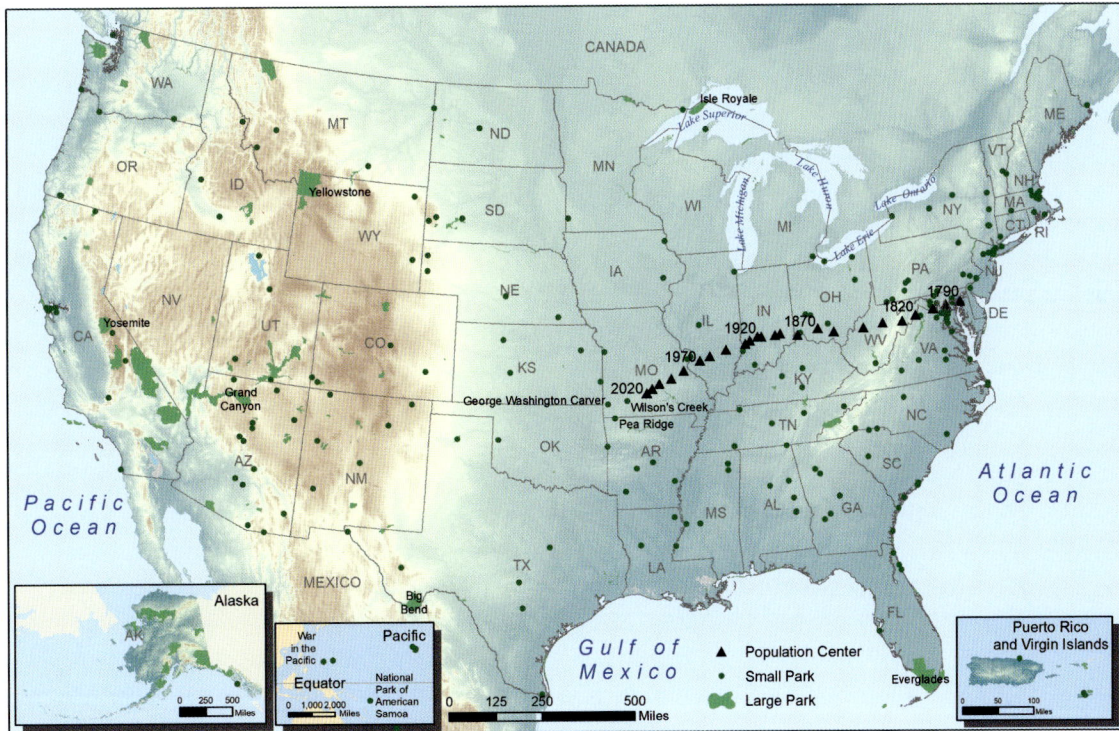

FIG. 10.3   Map of the changing population center of the United States and national parks, 2021.

westward movement of settlement and the growth of large cities (Fig. 10.3). In 2020 the population center was in Missouri, about fifty-six miles northeast of Wilson's Creek National Battlefield. This small park is the most centrally located with respect to the country's population and will probably remain so for several decades before being overtaken by either Pea Ridge National Military Park or George Washington Carver National Monument. That the first two of these central park units are Civil War battlefields and the third commemorates a man born a slave is fitting, as the park system is grappling with issues still unresolved from those times.

One factor driving the changing population center was suggested over sixty years ago by the geographer Edward Ullman (1954), who predicted that amenities, or pleasant living conditions, would become a major determinant of population change in the United States. At that time, California and Florida were considered high in amenities such as sunshine, warm winters, and beaches. (Smog, congestion, and a high cost of living were not yet concerns.) The scenic beauty of national parks has since become a ma-

jor amenity in the rural West. If you can afford to, why not live within sight of the Grand Tetons or Mount Rainier or next to the Grand Canyon? Many people have chosen to do so.

Saguaro was once far outside Tucson; today it provides an urban growth boundary, with homes just forty feet from the park's boundary. Many people commute (at high speeds) to work on scenic roads passing through the park. Commuting and other shortcuts through a rural park also occur at Lake Mead and the Natchez Trace Parkway at Tupelo. The NPS has no power over land uses outside its borders unless it buys the land. This can be expensive and politically difficult, especially when that land has been developed and commands high prices. At Lake Mead the NPS has repeatedly tried to buy a hotel and casino on an inholding within the park boundaries, while at Cuyahoga National Park it did purchase the Richfield Coliseum, former home of the Cleveland professional basketball team. The arena was demolished, and the land was added to the park.

Some parks have become surrounded by urbanization and have essentially become city parks, with a range of problems ranging from urban crime to people who are unaware that park rules are often different from those of city parks (Baker 1995; Rothman 2004; Whisnant and Whisnant 2007). Sunbathing and throwing Frisbees are not permitted in a battlefield park, no matter how inviting a grassy meadow might be, while annual civic pageants with waterskiing shows are fundamentally incompatible with preservation. Even rural parks can be impacted by adjacent lands; on two separate occasions, years apart, park employees at Casa Grande Ruins National Monument were hit and killed by stray bullets fired by children in adjacent farmland (Clemensen 1992). Parks do not have bulletproof borders.

Population has increased greatly around many national park units, often at a much higher rate than the national average (Davis and Hansen 2011). One study examined land use changes around fifty-seven park units from 1940 to 2010. The researchers constructed a typology of different situations, with many western parks classified as "wildland protected," indicating that surrounding lands were largely in the public domain, which ensures continued protection for these parks. Others were "wildland developable," meaning that much of the surrounding land was in private ownership and could be developed. Several parks in the Great Plains and Columbia Basin were surrounded by private lands with considerable agricultural activity and classified as "agricultural." Several eastern parks, along with Saguaro outside Tucson, were classified as "exurban." Finally, several coastal California and South Florida units were classified as "urban." Since 1940 most parks have had increases in surrounding populations, and agricultural land has only slightly decreased.

Population growth has also been a problem within parks in the form of hordes of visitors creating congestion and overcrowding. Canyon parks such as Yosemite and Zion struggle with too few parking spaces, as does even the Grand Canyon's South Rim. The number of vehicles entering Arches regularly exceeds the number of parking spaces available. Trails may be destroyed by too many feet, lodging fills up months in advance no matter the cost, and wildlife are never seen where they used to frequent.

With more people comes more trash to pick up and a need for improved (and expensive) water and sewer systems. In 1975 Crater Lake had to be closed because its water supply had become contaminated with sewage (Harmon 2002). Sewage continued to contaminate the lake itself until a new system was built in 1991.

These problems are familiar to many as the "tragedy of the commons," a term used by Garrett Hardin in 1968 for a situation when individuals can increase their own level of happiness at their own discretion, but the cost of this happiness is spread throughout a larger group. Most people have learned about this concept with a story involving villagers letting their cows graze on a pasture belonging to no single person. Each villager may increase their revenues by putting more cows on the pasture; but the more cows grazing, the more the pasture will be consumed by the cows. No authority limits this process, and if the situation continues, the pasture may be denuded, and all the villagers will suffer the results.

Hardin used national parks as an example of this tragedy in the real world: "At present, they are open to all, without limit. The parks themselves are limited in extent—there is only one Yosemite Valley—whereas population seems to grow without limit. The values that visitors seek in the parks are steadily eroded. Plainly, we must soon cease to treat the parks as commons or they will be of no value to anyone" (1968, 1245). He went on to suggest solutions:

> We might sell them off as private property. We might keep them as public property, but allocate the right to enter them. The allocation might be on the basis of wealth, by the use of an auction system. It might be on the basis of merit, as defined by some agreed-upon standards. It might be by a lottery. Or it might be on a first-come, first-served basis, administered to long queues. These, I think, are all the reasonable possibilities. They are all objectionable. But we must choose—or acquiesce in the destruction of the commons that we call our National Parks. (1968, 1245)

Few of these options seem reasonable today, though some people have called for the NPS to directly limit the number of visitors that may

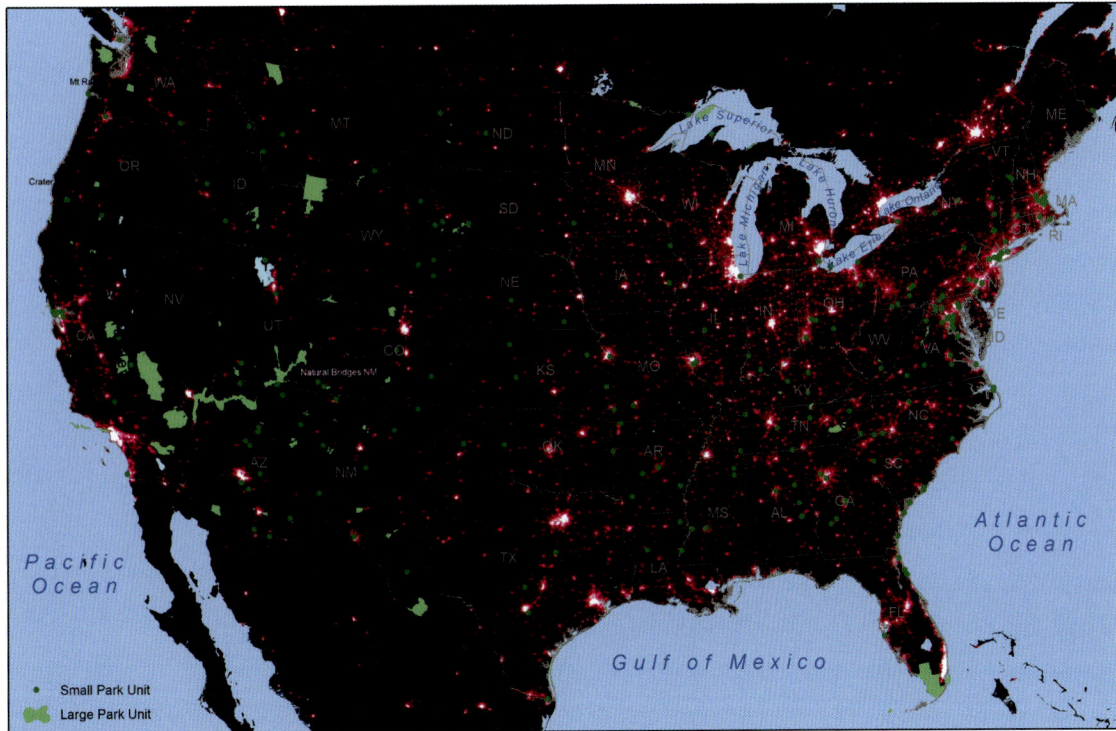

FIG. 10.4   Map showing the night sky darkness of the contiguous United States overlaid with the national park system, 2021. Many units in the interior West still possess spectacular nighttime skies.

enter places such as Yosemite or set entrance fees high to help reduce that number. Lotteries and a first-come, first-served system have been used for backcountry permits but never to gain entrance to a park, and it is doubtful that the public would ever approve of this.

Despite these problems, some parks still have the opposite problem: too few visitors. It should be remembered that park budgets are dependent on visitation, and political support for parks will also depend on how many people show up. If you can find an empty park to visit, you will not only help ensure its survival but also have a park experience that doesn't involve standing in lines or hunting for a parking space. Imagine all the extra time you will have to enjoy the view.

The Parks at Night

The growth of population has brought with it many unintended consequences. One is light pollution, or the absence of darkness at night due to outdoor lights in expanding cities. The brightness of the night sky can be measured (Bortle 2001) and mapped (Cinzano, Fal-

chi, and Elvidge 2001). The darkest skies allow the Milky Way to be seen in full splendor, as well as fainter phenomena such as zodiacal light and the gegenschein. The eastern half of the country has the brightest skies; park visitors will see the glow of city lights on the horizon and a few of the brightest stars. Even in the West, darkness exists in only a few isolated rural areas, such as southern Utah, central Nevada and Oregon, and parts of Montana (Nordgren 2010) (Fig. 10.4). Very few Americans live in such darkness, but several national park units can be found here, and they have used their darkness to attract visitors (Nordgren 2010). In 2007 the International Dark-Sky Association named Natural Bridges as the first International Dark Sky Park (International Dark-Sky Association 2018) (Fig. 10.5). Other park units, including nearby Hovenweep National Monument as well as Chaco Culture, Big Bend, Death Valley, and Grand Canyon, have also received this designation and are popular destinations for astrotourists, a growing movement of those seeking out dark skies.

Dark skies not only preserve the preindus-

FIG. 10.5   Owachomo Bridge at night, with the Milky Way above. Utah's Natural Bridges National Monument offers some of the best night-sky viewing in the entire park system. Photo by Jacob Frank, 2007. NPS.

trial skyscape and provide spectacular sights to urban Americans but also are healthier. A range of evidence suggests that exposure to nighttime lights is associated with greater risks of breast cancer (Kloog et al. 2010). The behavior of nocturnal animals is also affected, as are plants (Correa-Cano et al. 2018). Cacti close their stomata, or surface pores, during the day to reduce water loss; these open at night instead. Bright nighttime lights may reduce cacti's growth and pollination if lights also interfere with nighttime insect behavior.

An international movement seeks to preserve the night sky, including the possibilities of preserves where skies are dark (Charlier and Bourgeois 2013). Population growth limits opportunities for this process in the United States, and it requires a regional approach, with counties and cities passing laws that restrict bright outdoor lights. Perhaps the greatest test of whether these movements are successful will take place in 2061, when Halley's Comet will re-

turn. It was visible to most rural Americans in 1986, but from where in the United States will it be visible during its next appearance? National park units may be among the few places dark enough for you to see it.

The Environment

Many parks have harsh or difficult climates that create environmental conditions few Americans live with. High mountains, the subarctic, scorching hot deserts, active volcanoes, glaciers, and barrier islands are difficult places in which to work, live, and build roads or museums. The National Park of American Samoa was hit by a magnitude 8.1 earthquake, one of the strongest earthquakes ever recorded, on September 29, 2009, with a forty-six-foot-high tsunami destroying the visitor center (189 people in the territory and neighboring countries were killed). War in the Pacific National Historical Park on Guam is on a typhoon-plagued, low-lying, seismically active tropical island, a location that has made for many management challenges: "If a park resource was constructed of metal, it rusted, if it was organic, it had to be mowed or trimmed weekly, if it was steep, it was shaken loose by earthquakes, if it was not firmly embedded in the earth, it was blown away, and if it was anywhere near sea level, it flooded" (Evans-Hatch and Associates 2004, 10a). But the park's fauna presents even worse problems. In the 1940s a ship arriving from New Guinea accidentally introduced brown tree snakes to the island, where there was no natural predator for them. The island quickly became infested with snakes, which decimated the island's bird population, after which spiders grew unchecked. The snakes remain as a pest in the park, in cities, and even inside people's homes. Airplanes leaving the island must be carefully checked to remove any snakes on board; if they were accidentally brought to an uninfested island, the same process could begin again. Residents on the Hawaiian island of Maui defeated an airport expansion project in part because of fears that an expanded runway would allow direct flights from Guam; Maui and its unique fauna are only one flight away from a nightmare holocaust of snakes.

Snakes are also taking over the Everglades. These are Burmese pythons, first brought to the Miami area in the 1980s and sometimes set free but not becoming a viable population until a 1992 hurricane destroyed a breeding facility. There are likely more than one hundred thousand pythons in the Everglades (Lodge 2016), and the spread of the snakes has led to rapid declines in the numbers of rabbits and other small mammals in the area. The future spread of these snakes is intensely debated, but they will likely move north and invade other southeastern states. Other invasive species are causing tremendous harm to parks (and countless places outside them) (National Park Service 2019). Quagga mussels were accidentally introduced in the Great Lakes from the Ukraine in 1989 and have been spreading throughout the nation's rivers and reservoirs, carried not just by currents but also on recreational boats that were not carefully cleaned. These mussels outcompete native species and clog water intakes. The mussels reached Lake Mead in 2007 and have spread to many other southwestern reservoirs since then. Another pest is tamarisk or salt cedar, a tree native to North Africa and brought to the United States; it has spread throughout the West and crowded out native species. It has no value to wildlife and soaks up limited water supplies in desert areas. Southwestern parks have been fighting to eradicate it for decades, with limited success. Tree removal is a never-ending and expensive proposition in these parks; visitors to desert parks may be surprised by how much energy the NPS puts into cutting trees down in these places.

One of the oldest environmental threats to parks remains fire. The first forest firefighting in a national park took place in Yellowstone in 1886 by the army during its time administering the park (Rothman 2007). The year 1910 was terrible for forest fires throughout the West, and fighting fires became one of the new Forest Service's core practices. The NPS paid little attention to fires at first, but by 1930 putting out any fire immediately was an established goal. This was helped by the work of the CCC in building fire lookouts and trails, as well as personally fighting many fires.

The positive effects of fire in burning off undergrowth and preventing larger fires began to be slowly understood (Rothman 2007). Deliberate or prescribed burns began to be used in the Everglades in 1958 with positive results; they were used in Yellowstone from 1968. Today they are common. Unfortunately, events can overtake fire management policy. The summer of 1988 was the driest on record at Yellowstone, and when natural fires broke out, they quickly spread beyond human control and swept across most of the park. Public outrage at the near destruction of America's first national park resulted in a system-wide ban on prescribed burns (Rothman 2007; Marcus et al. 2012). This ban was later reversed, but unfortunately, in 2000 a prescribed burn at Bandelier National Monument went out of control and burned into the town and nuclear weapons labs at Los Alamos.

Fires can be a hazard even in desert parks. In 2020 a grass fire started by lightning from a summer thunderstorm burned forty-three thousand acres of Joshua tree woodland in Mojave

FIG. 10.6 Whiskeytown National Recreation Area after the Carr Fire, 2018. Most of the park was burned by this fatal and destructive fire, one of the most damaging that has yet hit a park unit. NPS.

National Preserve. This fire destroyed the heart of the largest and densest stand of Joshuas. Given that these plants can take eighty years to grow to maturity, this forest will not recover anytime soon.

Since 2000 western fires have become larger and more frequent due to sustained drought. The Forest Service now spends most of its budget on firefighting, and this will likely become a higher priority for the NPS as well. A 2018 fire burned in the Yosemite area for more than a month and closed the park for ten days. While it was burning, the Carr Fire in northern California burned through most of Whiskeytown National Recreation Area, killing eight people and destroying over sixteen hundred buildings in nearby towns. Portions of the park reopened months later, but considerable work was required to survey the damage and rebuild (Fig. 10.6). The lake and some beaches have slowly reopened, but much of the park's backcountry and many of its trails remain closed.

Climate Change

However park boundaries have been determined, the ones in place today will not serve to protect many resources in the future. Temperatures in parks have been rising to near record levels (Monahan and Fisichelli 2014; Monahan et al. 2016), accompanied by a host of other problems. These climate challenges will become worse throughout the park system as climate changes over the next several centuries in response to rising carbon dioxide levels in the

atmosphere. The NPS, like other federal agencies, has long been active in assessing potential impacts, which will affect precipitation, river flow, the distribution of plant and animal species, and visitation.

Although you have probably been conditioned to think of shrinking glaciers when hearing about climate change and national parks, the greatest impacts are likely to be in the southwestern states, home to the greatest collection of parks in the system. Temperatures have risen here faster than in any other part of the country, and droughts have been frequent and severe. It is likely that Death Valley will set a record high temperature exceeding the 134° F recorded in 1913 in the near future and even higher temperatures throughout the rest of the century (though it will never get hot enough to fry an egg on the sidewalk, a messy activity rangers strongly discourage). A prolonged drought has reduced flows on the Colorado River for several decades, with Lake Mead and Glen Canyon breaking records for low lake levels.

While reservoirs are shrinking, coastal parks face the challenge of a rising sea. Retreating glaciers and polar ice packs will raise the oceans by several feet over the century. Much of the nation's Atlantic and Gulf coastlines are made up of barrier islands within national seashores, which will be transformed as the shoreline retreats. Units in coastal cities will also be threatened. The city with the most low-lying park units is Washington, D.C.: the Potomac River near the Lincoln Memorial and the Tidal Basin at the Jefferson and Martin Luther King, Jr. Memorials are at sea level. A slight rise in water level could flood much of the National Mall. Floodwalls or levees will be necessary to protect these and other units, though that will change the view of these monuments.

Other areas will see changes. Hawaii will also be subject to warmer and drier weather (Monahan and Fisichelli 2014), and northern Alaska is another area where temperature increases have been well above the national average. Glaciers are in rapid retreat throughout most of the park system. Northeastern parks are warmer but also wetter. The climate on the Yellowstone Plateau will become warmer, which in the short run could result in heavier winter

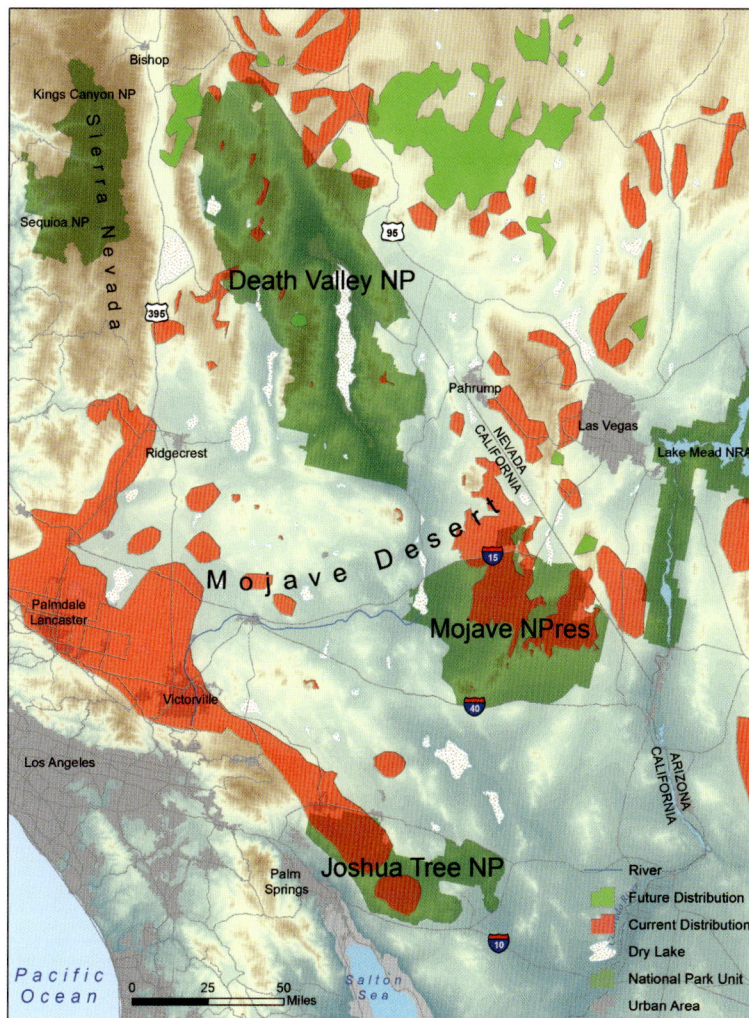

FIG. 10.7 Current and future Joshua tree distribution. In the future, warmer and drier conditions will reduce the populations of Joshua trees within its namesake park, but new habitat may appear farther north in Nevada.

snow at higher elevations before snowfall declines later in the century (Marcus et al. 2012). Less snow means less runoff into the area's rivers and streams.

Climate change will even extend underground. The climate within caves depends on their altitude and latitude, with the Southwest and central Texas having the warmest caves (Moore and Sullivan 1978). As annual temperature goes up, it will eventually rise underground as well. The warmer a limestone cave is, the bigger its formations will likely be due to higher CO_2 content in warmer soils. The formations in southwestern caves such as Carlsbad Caverns will grow faster in the future, assuming they do not dry out due to drought. But at greater depths, cave temperatures reflect past climates rather than today's temperatures; their temperatures are relics of climates that existed hundreds or even thousands of years ago.

These changes are important because several parks have or were even created around plants with a restricted geographic distribution. Among these is the Joshua tree in the Mojave Desert. Several studies predict that current Joshua tree habitat in the Mojave Desert will become unsustainable for the tree in the future due to a warming climate (Cole et al. 2011; Sweet et al. 2019). The northwestern corner of Joshua Tree National Park is the only location in the park that will likely be able to support them by the end of the century. The trees will slowly spread northward into southern Nevada and Utah, as well as higher elevations in northwestern Arizona (Fig. 10.7).

What will this mean for the future of Joshua Tree National Park? The park is a popular rock-climbing destination, and this will presumably remain the case (though perhaps with a shorter winter season). Mojave National Preserve will likewise see its Joshua population shrink. The 2020 fire that burned out much of the densest stand is helping this process along; Joshuas can take up to eighty years to grow to maturity, but due to warmer and drier conditions, the Cima Dome forest may never return. Will Grand Canyon–Parashant National Monument become a new Joshua Tree National Park? Will a new park be created out of national forest lands in Utah or Arizona? Will Joshua trees become concentrated in the Nellis Air Force Base range, off-limits to the public? Similar changes may take place to the distribution of saguaro cactus, though the habitat around Saguaro National Park is predicted to be stable (Albuquerque et al. 2018). But its range may shrink elsewhere, making the park even more valuable as a habitat for it.

An even more limited plant is the Haleakala silversword, which grows near the top of that volcano in Hawaii. After surviving a brush with extinction in the twentieth century it is now thriving. But if Hawaii warms, the plants would have to move uphill or northward to maintain their growing conditions; unfortunately, neither option is possible on an isolated mountaintop in the middle of the Pacific Ocean.

It is not only plants and animals that are relocating due to climate change. Visitation at most park units is related to their temperature (Fisichelli et al. 2015). Most parks are experiencing earlier springs; those in the northern Rockies and the Southeast are among the exceptions (Monahan et al. 2016). Lauren Buckley and Madison Foushee (2012) identified shifts in visitation at twenty-seven park units since 1979. Seven of these units show a shift in temperature and attendance, with peak attendance shifting an average of 4.6 days earlier in the summer. There are many structural factors in park visitation (the length of school years, holiday weekends, and economic conditions among them), but this suggests that climate change is already affecting parks.

The Political Context

The parks are subject to the same range of outside political influences as any other government agency. These include turf wars between the NPS and other agencies over individual units or missions (Hampton 1981). Many of these have been with the Forest Service, created in 1905 within the Department of Agriculture. The Forest Service was not opposed to the existence of parks so much as it felt it should be in control of them. For that reason, it opposed the creation of the NPS in 1916, and the two agencies were frequently involved in disputes over new parks (Foresta 1984; Sellars 2007). The agency also feared losing land to the new agency, which in fact it often did. In Arizona alone, the Grand Canyon, Chiricahua, Coronado, Saguaro, Tonto, Montezuma Castle, Sunset Crater, Walnut Canyon, and Wupatki were all created partially or entirely from national forests. The War Department, which oversaw many battlefields, was also concerned about losing its sites (Sellars 2007). When the NPS was finally created after several years of debate the agency received only the national parks and those national monuments administered by the Department of the Interior; the Forest Service and War Department had successfully defended their territories until 1933, when President Roosevelt took them out of the monument business.

As public institutions, national parks are not immune to political interference. Those outside the system may seek to change policies for their benefit, and those inside the government may manipulate the system for the benefit of themselves or their constituents. Unfortunately, there have also been cases of misconduct and incompetence within the agency and supporting organizations (Berkowitz 2011; Danno 2012). One of the worst incidents in recent years was when a Department of the Interior official attempted to rewrite the NPS management guidelines for parks to remove environmental protections and further open the parks to snowmobiles (Schnayerson 2006). NPS director Fran Mainella offered no defense of the parks after this move and later resigned after letting a wealthy property owner cut trees on national parkland to improve the view from his house.

In the early 2000s considerable discontent had built up within the NPS. Budget cuts during the Bush administration forced parks to cut back to the bare minimum, and park employees were told to use the phrase "service level adjustments" instead of "budget cuts" (J. Wilson 2004). Morale within the NPS suffered as it became clear that those who spoke out would be punished, and there was a real threat that thousands of park workers could lose their jobs to private contractors (J. Wilson 2004; Schnayerson 2006). Nine out of ten NPS employees felt that decision-making was based on politics rather than science.

The Bush administration cuts coincided with Republican efforts to eliminate parks and reduce environmental protections to those that remained, highlighted by an unpopular decision to destroy Yellowstone's winter peace and quiet with snowmobiles. Snowmobile and other winter activities were strongly backed by local businesses outside the park that stood to profit (Yochim 2009). In the 1960s NPS saw snowmobiles as a lesser evil compared to demands to plow Yellowstone's roads all winter; the NPS accommodated these new winter visitors by building the Old Faithful Snow Lodge in 1971. Snowmobiling grew in the 1980s and forced the end of dogsledding. The dual hierarchy worked to great advantage for local businesses, who could use their influence to force the NPS to accommodate ever larger snowmobile numbers. No studies had been done about snowmobiles and their environmental effects until after the escalating numbers forced the agency to act; in 2000 snowmobiles were banned beginning in 2003, but President Bush overturned this. During February 2002 NPS workers at the west entrance were forced to wear gas masks to protect themselves from snowmobile exhaust fumes. In December 2003 a court decision reversed this presidential policy of unlimited snowmobile access, but in February 2004 the ban was again put on hold.

The Military versus the Parks

There are several federal agencies that administer large landholdings in the American West, including not only the NPS but also the U.S. Forest Service and Bureau of Land Management. Another agency is the Department of Defense. The early history of the national park system was closely intertwined with the military. The army provided the first personnel to manage western parks and created the idea of

the park ranger. The army was assigned to protect several early park units before the NPS was created in 1916. Yellowstone and Yosemite were among these; they were patrolled by African American cavalry called Buffalo Soldiers, for whom a national park unit was created in Ohio in 2013. Many national park units, including some of the most famous and revered, are former battlefields, forts, or war memorials, and many were once under the direct administration of the War Department. The War Department also administered several early national monuments before they were transferred to the NPS in 1933.

But this relationship became less friendly over time. During World War II the military took many forts back. Fort Pulaski was a nineteenth-century fort built to guard the port of Charleston; due to the island's military value, it was closed to the public and reoccupied by the army from March 1942 to 1947 (Meadors 2003). Cabrillo National Monument was likewise considered essential for the defense of San Diego, and observation bunkers were dug into the hillside to watch for Japanese ships (Lehmann 1987). Fort Hunt was another defunct army fort that became part of the George Washington Memorial Parkway in Virginia. During the war, it was reoccupied by the army and used to house and interrogate newly captured German prisoners of war (Moore 2006; McDonnell 2015). A small section of Hawai'i Volcanoes National Park was transferred to the army for use as a bombing range. Much of the National Mall in Washington, D.C., between the Washington Monument and the Lincoln Memorial was filled with temporary offices during World War I. These remained in use until after World War II, and the last was not removed until 1970.

A notorious case of conflict between the military and the National Park Service involves Red Hill in Haleakalā National Park (Jackson 1972). This volcanic peak on the island of Maui rises to just over ten thousand feet above sea level, the highest point on the island and with spectacular views into the crater-like valley below the summit. This elevation also made it an ideal spot for a radar station. After World War II ended in 1945 there were accusations that the NPS had delayed construction of an army radar on Haleakalā's summit that could have given warning of the Japanese attack on Pearl Harbor on December 7, 1941 (Morgenstern 1947). This facility had allegedly been opposed by the NPS because its architecture would make it an "eyesore." Was the NPS's concern over a building's appearance to blame for the devastating attack on the U.S. Pacific Fleet? The reality was quite different: the NPS had approved the radar site seven months before the attack, but the army did not begin operating it until a month after the fleet was shattered (Jackson 1972). And the army also had a functioning radar station at Opana Point on Oʻahu that did detect the incoming attack, but its operators were told to ignore the incoming planes. A Hawaiian park unit exists today as a monument to that day's failures, but it is the USS *Arizona* Memorial, not Haleakalā. Although the false accusations against the NPS have been repeated in books and movies, the NPS seems to have had the last word: the army's Opana Point radar station became a national historic landmark due to its role in enabling the Japanese attack.

After the war conflicts between the military and the NPS continued. Geography is responsible for much of this, as both the NPS and the military favor coastal areas, high mountain peaks, and vast unpopulated valleys. Today all four branches of the military have large bases that often neighbor western park units. New Mexico's Tularosa Basin includes White Sands National Park and from 1942 has also been home to a testing range (Schneider-Hector 1993). This began in 1942 as a bombing range and continued after the war to test rockets. Captured German V-2 missiles were assembled and launched across the basin, sometimes falling inside the park's boundaries. Since 1946 the military has been able to close the main road through the basin when missile testing was in progress and has wanted to be able to close the entire park whenever it pleased. In 1963 the military obtained the right to use the western half of the park as part of the missile range, and this area has been closed to public access ever since. Efforts to establish wilderness in the park have been opposed by the air force, which sees this as a potential limitation on its activities.

At other parks, low-flying military aircraft

are a problem. In 1966 it was noticed that sonic booms produced by military jets were causing damage to the fragile cliff ruins at Canyon de Chelly (Brugge and Wilson 1976). The air force eventually prohibited supersonic flights over national parks. Death Valley is surrounded by three large military bases belonging to the army, navy, and air force. The explosions of nearly a thousand nuclear bombs in the nearby Nevada Test Site rattled windows in the park; low-flying air force and navy jets continue to shatter the silence with deafening noise and sonic booms.

But cultural values are diverse, and for many, these low-flying planes have become an attraction. Rainbow Canyon was added to the park in 1994 and was exempted from new restrictions on military planes. Navy planes from the nearby China Lake Naval Air Station frequently use it as part of a low-level training route, called the "Jedi Transition." Visitors at the canyon overlook may frequently see F-18s or other jets flying through the narrow canyon hundreds of feet below them. (Many have posted videos on YouTube.) But don't get too close; in 2019 a navy jet crash in the canyon injured seven visitors standing on the canyon rim (Melley 2019).

The Cultural Context

Navajo National Monument lies within the vast Navajo Nation on the Colorado Plateau but is culturally not a part of that realm. Instead, it preserves spectacular Anasazi cliff ruins and may be the single best site to see these structures (and certainly without the crowds of Mesa Verde). But while many traveled long distances to visit the park, the Navajo who lived around it were not interested. In Navajo culture, the living generally avoid places associated with the dead, and the dead are not discussed; visiting and talking about the ruins violated this belief. Although many Navajo worked at the park, they were uncomfortable playing tourist. But toward the end of the twentieth century, attitudes among some Navajo had changed, and they began to appear among the visitors to the monument (Rothman 1991).

This story highlights an important set of

factors affecting the national park system and its future. The system developed in a culture where the wonders of the natural world and later the nation's history and outdoor recreation were important and in a society where shared ownership of resources was a noble goal. Values aren't static, and changing cultural values could undermine the existence of the park system. A common tactic once used against parks was elitism, or the idea that parks are antidemocratic because they would appeal only to the rich (Hampton 1981). This idea faded as parks became middle-class American vacation destinations due to the spread of roads, automobile ownership, and rising incomes.

But attitudes are still changing. The nation's population has been going through a "big sort," in which people are more likely to live and socialize only among like-minded people (B. Bishop 2009). Many rural parks are in overwhelmingly Republican areas, while park visitors from distant metropolitan areas are more likely to vote Democratic (Bloch et al. 2018). Such a differential between those groups cannot be healthy for the park system. It is likely that these populations have widely differing environmental values, creating conflict over the existence and use of parks that could last generations.

Economic inequality is also on the rise in the United States, and the number of Americans who can travel across the continent for a lengthy visit has decreased. This inequality is especially acute between rural populations living near many parks and urban populations who come to visit. One future for the parks is to once again become the playgrounds of the rich. This is also not likely to be healthy for the future of the park system.

Religion

The First Amendment to the Constitution guarantees the right of Americans to live without having another's religion imposed on them. Government agencies such as the NPS cannot favor one religion over another or promote the views of any religion. For this reason, forcing the Grand Canyon visitor center bookstore to sell an antiscience book written by a fundamentalist group was disturbing to many park visitors (Edds 2004). NPS guidelines exclude religious sites from being considered for inclusion as park units, though there are many churches in the system. Saint Paul's Church National Historic Site is an inactive church, while Tumacácori National Historical Park in Arizona and Salinas Pueblo Missions National Monument in New Mexico preserve four abandoned Spanish missions. The four Catholic churches that make up San Antonio Missions National Historical Park are still active. In addition, parks such as Yosemite, Yellowstone, Grand Teton, Grand Canyon, and the Great Smoky Mountains all have active churches within them. Shiloh not only contains an active church but also took its name from it. Roger Williams National Memorial stands alone in honoring an advocate of the separation of church and state, a fundamental basis for American culture and the prosperity the country has enjoyed.

So far religion has fortunately been a minor issue for the American national park system, especially compared to the tremendous and irreversible damage done to the heritage of countries such as Afghanistan due to religious extremism. Should political restrictions on advocating religion be reduced, it is very likely that the NPS will be dragged into attempts to promote a religious agenda. This may not just be about books sold or proselytizing in parks; it could even restrict visitation. In 2002 the Mormon Church attempted to gain access to the Martin's Cove historic site in Wyoming by pushing for legislation that would require the U.S. government to sell the land. Wyoming residents resisted, and the legislation was defeated. The church then leased public land for a museum next to the site, but a separate lawsuit was required to ensure that all visitors would be able to visit the site, which is on public land, without having to pass through the church's museum (American Civil Liberties Union 2006; Hein 2014). No such attempts to control visitors to national park units have yet been made, but small monuments or historic sites could be vulnerable to these kinds of attacks.

The boundaries of the park system must differentiate between secular and sacred spaces; no matter how spectacular and inspirational a view may be, it exists in the secular or profane landscape for most Americans. But parks are sometimes religious in nature, though in ways most Americans have trouble seeing. Native Americans were removed from parks, but many still have a spiritual connection to the landscape that has nothing to do with ownership or the presence of churches or other religious structures. These ties are rooted in a deeper concept of what is sacred (Eliade 1987; Worster 1992). Several parks include the origin place of several tribes, among them Sipapu, Spirit Mountain, and the Heart of the Monster, along with many other sites of tremendous importance to them. For that reason, Native Americans have been offended by people climbing Devils Tower, as the mountain is sacred to them (and known by other names). In 1995 the NPS started a voluntary climbing ban during the month of June, with signs asking people not to climb it then out of respect for Native American beliefs (Taylor and Geffen 2004). The park cannot enact a mandatory ban, as it would run afoul of the Constitution's ban on the government favoring a religion, in this case, Native Americans valuing Devils Tower as a sacred space. Unfortunately, the ban is rarely observed; those seeking adventure in parks can be remarkably insensitive to Native American concerns. Since the 1970s the Navajo have fought a similar legal battle over Rainbow Bridge, which they hold to be sacred. They particularly objected to the public approaching the bridge and walking underneath it, considered a disrespectful act. In the 1990s NPS management began to reflect greater sensitivity to their concerns and installed a sign asking people to voluntarily refrain from walking under the bridge (Sproul 2001). Given that parklands were formerly occupied by Native American or Native Alaskan groups who are increasingly vocal about reasserting their cultural ties to these lands, these sorts of situations will continue to develop.

Allowing religion into the parks could even mean the elimination of environmental protections and even the parks themselves. This is especially the case in Utah, home to many spectacular parks as well as the Mormon Church, which has a strong antipreservation outlook. This view can also be found in Washington, D.C., where a briefly serving secretary of the interior in the early 1980s often expressed his view that preservation and conservation were simply unnecessary due to the impending return of Jesus. In those less polarized times, this was controversial; in the twenty-first century, it may become government policy.

Authenticity

Parks have distinct boundaries you can't help but notice when you enter them. In some cases, the landscape changes completely, marking the park as a place where there are different rules and we have different expectations. What you see there is real, not computer generated or made of plastic. National parks are not theme parks. While the scenery of the Big Thunder Mountain rollercoaster at Disneyland was based on that of Bryce Canyon and the corresponding ride at Disney World resembles Arches, and Disney World even has a lodge inspired by Yellowstone (complete with an artificial geyser), we know these are fake. National parks represent the real thing. Or do they? What if the boundaries between the park system and everything else were blurred, and we couldn't tell the difference?

The Fiery Furnace Viewpoint parking lot in Arches National Park is a typical NPS parking lot, a one-way loop with eighteen parking spots, clean and well organized, with information signs and restrooms. As access to the Fiery Furnace is limited to ranger-led walks, a rustic log fence restricts movement out of the parking lot. A close look reveals that fence to be artificial, a fiberglass replica of a log fence, which one might expect to find in Disney World but never in a national park (Fig. 10.8)! Whatever the economic, environmental, or other reasons for installing it, the fence serves to erode the differences between a real park and a theme park.

If you know where to look, you can see other blurred boundaries between the real and fake. In Yellowstone, the Pillar of Hercules is a rock hoodoo along the road south of Mammoth Hot Springs. The highway cuts between the pillar and a cliff face, evidence of the skillful roadbuilding by the NPS. It is more skillful than you think: the pillar was moved in 1899 to make room for the road and moved again in 1933 when the road was rebuilt (T. Davis 2016).

FIG. 10.8 A fiberglass fence masquerading as a rustic wood fence at Fiery Furnace Viewpoint at Arches National Park, Utah, in 2007.

Since then it has rested on a concrete pedestal, with great care taken to make it look completely natural.

It is not just rocks that are adjusted for the benefit of tourists. Cades Cove is a peaceful pioneer agricultural landscape in the Great Smoky Mountains. It is a quaint relic of an extinct Appalachian pioneer culture, a time capsule of frontier America. Except that it isn't: the residents of the cove had electricity, indoor plumbing, telephones, and cars and were little different from the surrounding populations (Young 2006). The NPS decided to ignore this when it went about preserving this landscape. Buildings that were considered too modern, which included most of the homes, were removed, and the wires, plumbing, and other artifacts of the twentieth century were stripped from the remainder. A living history program that focused on the nineteenth century was put into place, and the NPS created a fake cultural landscape that never existed.

What if an entire park is not what it seems? Bent's Old Fort was demolished in 1849; the current structure was built in 1976. How many visitors are aware it is not the original? The Native American mounds at Hopewell Culture were leveled when the site became an army training camp in 1917; those there today are reconstructions (Cockrell 1999). Federal Hall National Me-

morial consists of an 1842 building in New York City but commemorates events that happened in the previous building at that location. When the NPS created Fort Raleigh, the site of the lost Roanoke colony, it included land thought to be where the colony stood. But despite careful archaeological work, the lost colony remains lost. It is likely underwater in Pamlico Sound, outside the park boundaries (Binkley and Davis 2003). These may seem unimportant, but units have been removed from the system for these sorts of issues.

There are even places outside of the national park system that have been confused with it. In 1907 a former Mesa Verde park booster who had a falling-out with others built a replica cliff dwelling across the state in Manitou Springs as a competing tourist attraction. Eventually, this was promoted as an actual historic ruin and is still in business as such (D. Smith 2002). Many who have visited it thought they saw actual cliff ruins and without the need to drive all the way across the state to Mesa Verde. Park purists may condescend to these visitors, but the line between real and fake is not always so clear-cut, even within a national park. The ruins at Mesa Verde were cleaned up and partly reconstructed to make them more appealing to tourists. And if those visitors to the fake cliff ruin leave with a greater understanding of our nation's cultural history and perhaps even a greater desire to visit a national park unit, is that not a good thing?

A more insidious threat to parks is that authenticity is a positional good, meaning that it thrives on scarcity (Potter 2010). If an activity considered "authentic" becomes too commonly pursued, then it will lose its cachet; a new, more "authentic" activity must be found to allow authenticity seekers to feel superior to the uninformed masses. It is also perhaps a reflection of space-time convergence and increased access: nowadays anyone can go to Yellowstone, and doing so no longer has the air of adventure and risk it did a century ago. A more remote or challenging place is required now. Where will people go when the Gates of the Arctic are within easy day-trip distance?

This may be the ultimate cost of Mission 66 and the goal of opening parks up to the masses: going to a park may no longer be good enough for those seeking to demonstrate their environmental virtue or show they are more adventurous than you. Perhaps this may help explain the strong, if not shrill, antitourism statements found in many recent books on parks (as in Tweed 2010; Keiter 2013). Some have moved beyond parks to embrace the more "authentic" nature experiences to be found in backcountry wilderness. Perhaps the greatest threat to the future of the national parks is how you react to the thought of sharing your favorite park with two or three times the number of visitors there today. If your love of this park does not depend on how many other people know about it, then the future is promising. But if the thought of this crowding diminishes your love for the park and makes you want to stay away and go find a more "authentic" or adventurous experience elsewhere, then you are helping destroy the park system.

Conclusions

The park system has survived many challenges over the years and shown an ability to adapt to changing demands and conditions. A few challenges have largely disappeared, while others, such as political use of parks and budget battles, are constant and will never be resolved.

Threats are not even necessarily a bad thing; the existence of parks, as well as forests, wildlife refuges, and wilderness, is often a reaction to threats to natural or historic resources. Geography is an important component to these threats; the relative location of parks amid a growing population accounts for much of this. Space-time convergence has been a threat to parks by introducing more invasive species before ecosystems or managers can adjust to them.

The effects of changing climates and cultural values are ominous developments, as there is no political will to deal with these issues at the national level. Should that belatedly develop, tremendous damage will already have been done. The effects of climate change have a geographic dimension; some units are particularly vulnerable because of their location or elevation. Just as the national park system has shifted eastward and into cities in the past centuries, it will likely shift northward and back to higher elevations in response to a warming and drying climate.

Not all threats are in the future. Several park units exist along the U.S.-Mexico border and are subject to a range of threats not found elsewhere. Organ Pipe Cactus National Monument preserves a spectacular section of the Sonoran Desert along the Arizona-Mexico border. While this was a remote corner of Arizona in the past, the monument is now on the front lines of illegal immigration and smuggling. The park has been heavily impacted, including tremendous amounts of trash left by border crossers, illegal roads and trails, a park ranger killed by drug smugglers in 2002, and a new border wall. Much of the monument was closed to visitors for their protection for several years but has since reopened. This is a threat that has long resisted meaningful action and one that may grow in the future.

Because parks are authentic experiences, they can sometimes be dangerous and frightening places. At Canyon de Chelly you can stand on the edge of a 1,000-foot cliff and look across the chasm below you a quarter mile to the 750-foot tall Spider Rock (Fig. 10.9). For city people, this is a similar experience to standing on the eighty-sixth-floor observation deck of the

FIG. 10.9 Spider Rock, Canyon de Chelly. This sacred 750-foot-tall sandstone spire provides an unforgettable landmark for those not afraid of heights. Photo by Daniel Schwen. Wikipedia, Creative Commons Attribution–ShareAlike 4.0 International License.

Empire State Building looking across at nearby buildings on the Manhattan skyline. But at Canyon de Chelly there are no crowds or city noise and no guardrails or fences. The experience of standing on the cliff's edge may be a bit too authentic for some. Even more unnerving, if not terrifying, can be the pitch darkness you can experience during a cave tour at Carlsbad Caverns when the guide asks everyone to turn off their flashlights. And there's the deafening absolute silence of a windless day in Death Valley, which unnerves many city people.

But the most frightening feature in any park might be found in Hawaii's tiny Kalaupapa National Historical Park. This unit contains some of the tallest cliffs in the world, but that is not the most frightening feature of the park or why it exists. It contains what most Americans would think of as a horrifying nightmare from the distant past, for which reason it is the only park unit where children aren't allowed. Few are up to the challenge of visiting and replacing their fear with understanding.

ELEVEN

National Parks in a Global Context

America's national park system does not stand alone; the Bureau of Land Management, Forest Service, and Fish and Wildlife Service, among other agencies, administer vast swaths of the country's public lands. Tribal, state, and local parks may be neighbors as well, all existing within a patchwork of lands where preservation, conservation, and recreation coexist or compete (Weber 2019).

Given the many different agencies and goals involved in managing these public lands, it is often useful to shift the focus to management or level of protection rather than an agency. Are BLM monuments or national conservation areas operated the same way as national park lands? Comparing agency goals and standards is difficult. They have different missions and may use similar terms in quite distinct ways.

It is even more difficult when we compare systems between countries. Most have some form of a national park system, but some allow hunting or farming or have thousands of people living within them, while others are mainly wildlife reserves. Fortunately, the International Union for Conservation of Nature (IUCN), founded in 1948, has established guidelines that allow management priorities in parks, monuments, forests, recreation areas, and other places to be compared. These guidelines refer to all these kinds of lands simply as protected areas.

Scenery, wildlife, history, and culture do not stop at national borders. Can we have national parks that cross them? The idea was once popular in the United States, and over the years the National Park Service has worked to create international parks with Canada, Mexico,

and even Russia, though with very mixed success. The idea of protected areas that span borders, or transboundary conservation, is growing throughout the world, and there are some outstanding examples.

Not all the world's scenic or historic treasures are found within a country. An international effort has protected many locations within Antarctica, as well as in the world's seas. Efforts are even under way to protect humanity's historical heritage on the moon and Mars and protect scientific treasures on other planets. These are possibilities we must prepare for.

Protected Areas

Delicate Arch is the most famous attraction in Arches National Park (Fig. 11.1). You have probably seen many pictures of this rock, and it appears on Utah license plates. The arch stands dramatically in the foreground, silhouetted against distant and often snow-covered peaks. But the La Sal Mountains in the background are far outside the park, within the Manti–La Sal National Forest. In between the peaks and arch, but out of sight behind nearby ridges, are BLM lands. When you admire this view you must remember that America's national park system is only one of several large federal land systems. BLM lands, including national monuments and national conservation areas, national forests, national wildlife refuges, and other places protect and conserve much of the country. Tribal, state, city, and county parks and even private land also preserve much of the country's scenery, history, and recreation lands. This naturally raises the question of whether there is a substantial difference between a BLM national con-

FIG. 11.1   Delicate Arch at Arches National Park. This iconic park view features mountains within Manti–La Sal National
Forest in the background. Photo by Palacemusic, 2005. Wikipedia, Creative Commons Attribution–ShareAlike 3.0
Unported License, 2.5 Generic License, 2.0 Generic License, and 1.0 Generic License.

servation area, an NPS national monument, and a USFS national monument. Are national parks more protected than those other places?

To answer questions like these it is useful to think of these places as different types of protected areas, a generic term that encompasses not just the nineteen different designations of the national park system but also national forests, grasslands, wildlife refuges, BLM lands, and even city parks. All have been set aside and given some protection. The IUCN developed protected area management categories in 1994 to allow greater communication about these places. The IUCN (2008) has devised seven categories of protected areas based on how they are managed.

Category 1A is a strict nature reserve that protects biodiversity or other natural features, with little or no visitation allowed. These re-

serves are usually very small and are also rare. The IUCN database lists 607 of these sites in the United States, the best examples of which might be research natural areas within national forests or on BLM land to preserve small areas of scientific interest. The park system does not have such places, though there are many locations within parks subject to intensive scientific study, and tourists might be excluded as needed.

Category 1B is a wilderness area. This is similar to but does not have the strict visitor controls of a 1A unit. It will also likely be larger. Both categories should have intact ecosystems, but a 1B park may have experienced minor disruption from which it has since recovered. The 765 units of the U.S. wilderness system fit within this category, and when state and local lands are counted there are 1,325 such places in the country. Wilderness areas differ from most other federal protected lands in that they have no specific agency to manage them; they are administered by the agency that controls the land within them. The NPS has 40 percent of the acreage of federal wilderness areas, the Forest Service has 33 percent, the FWS has 19 percent, and the BLM has the remaining 8 percent.

Category 2 refers to national parks, a large natural area that protects environmental features but will allow visitation and the tourist development necessary for that. This is the rarest IUCN category in the United States, with only forty-one, including the Grand Canyon (though the national park system lists sixty-three places labeled as national parks).

Category 3 is a natural monument, a small natural feature that has been preserved. It is usually small and may be just a cave, a waterfall, a rock, sand dunes, or a similar feature or a small cultural feature. It is usually intended to protect an unusual feature rather than an ecosystem, and visitation may be high. Many U.S. national monuments, such as Devils Tower, fall into this category, but national memorials, national historic sites and historical parks, and national battlefields are also included. The IUCN lists 1,804 of these in the United States, the vast majority of which are managed by state or local governments. State, county, and city parks, historic sites, and beaches are common. Even cemeteries and university lands may fit this category.

Category 4 is a habitat or species management area. This is intended to protect plant or animal species and may be of varying size (usually small) and will likely have active management to protect the species or restore habitat. These areas may have been considerably disrupted and in recovery. There are 755 in the United States, mostly national wildlife refuges.

Category 5 is a protected landscape or seascape, an area that has scenic value and that has had considerable interaction between people and the environment. It will usually be large. This is the most numerous type of protected area in the United States, with 28,413 across the country. These include national recreation areas, lakeshores, seashores, parkways, and pre-

serves, along with state, city, and county parks and recreation areas, wildlife management areas, and areas of critical environmental concern. The IUCN also includes large national monuments such as Dinosaur and Organ Pipe Cactus in this category.

Category 6 refers to areas managed with sustainable use of natural resources. These are generally large and managed for extractive uses. There are 418 of these in the United States, including national forests and grasslands, state forests and game lands, waterfowl production areas, state resource management areas, and fish recovery areas.

Altogether, the IUCN database includes 34,074 protected areas in the United States, of which the federal government is responsible for 5,814, or 17 percent. Almost 42 percent are state or local, and 11 percent are cooperatively managed by different governments. The remainder is managed by nonprofit organizations (18.5 percent) or private owners (7 percent). The IUCN classifies park units slightly differently from the National Park Service, with national parks as one category, small monuments in another, and all other units, whether recreation areas or large monuments, in category 5. These definitions can be used to make sense of the world's protected places.

Protected Areas around the World

Every country in the world has some protected areas, according to the UN Environment Programme World Conservation Monitoring Centre's (2020) Protected Planet database. According to this source, there are 217,155 protected areas around the globe. Russia has the largest area of these places, followed by Canada, the United States, China, Brazil, Australia, India, Argentina, Kazakhstan, and the Congo. These are all large countries, and when the percent of each country's territory that is protected is instead calculated, the top countries are Slovenia (53.6 percent protected), followed by Bhutan, Brunei, Lichtenstein, Seychelles, Poland, Croatia, Zambia, Namibia, and Germany. The average is 16.3 percent, slightly higher than the 13.1 percent of the United States that is protected.

The country with the lowest percentage of

its area protected is Libya (0.85 percent protected), followed by Turkey, Lesotho, Haiti, Angola, Afghanistan, Mauritania, Syria, Yemen, Somalia, and Iraq. Several patterns are quickly apparent from this list. Except for Turkey, these are among the least developed countries in the world, and except for Turkey and Lesotho, they have a legacy of decades of instability and violence. All except Lesotho, Angola, and Haiti are also predominantly Muslim-majority countries. While Morocco, Kuwait, and the United Arab Emirates all have above-average protected percentages (and greater than the United States), and tiny, oil-rich Brunei has the fourth highest percentage protected of any country (47 percent), most Muslim-majority countries rank near the bottom in land protection.

Most countries have some form of category 2 national park, though they are usually limited in number and, as in the United States, are a small subset of protected areas. We can take a brief tour of the world's protected areas, beginning with North American countries adjacent to the United States.

North and South America

Canada created its first national park in 1885 (Boyd and Butler 2009). This was Banff National Park in the Rocky Mountains, perhaps still the most famous in the country. While today it is known for its spectacular Rocky Mountain scenery, it was originally created because it contained the first hot springs found in Canada. Its closest competitor was Hot Springs National Park in Arkansas rather than much closer mountainous parks such as Yellowstone and Yosemite (Sheail 2010). Due to its location on Canada's transcontinental railroad line, it attracted more visitors than the more remote Yellowstone in the nineteenth century.

Canada shares not only the Rocky Mountains with American parks but also the fact that most of the earlier units were in the mountainous West, with the same lack of proximity to the eastern population. As in the United States, the Canadian park system has been extended eastward, though also northward to the Arctic coast, and has had to deal with a range of development pressures, including Native populations living within park areas. Today Canada

has 8,161 protected areas, including 1,810 category 2 parks (Fig. 11.2).

Mexico created its first national park in 1935 and could claim forty by 1940, more than in the United States (Wakild 2009). Today Mexico has 1,150 protected areas distributed throughout the country's varied terrain, including the Sierra Madre, a continuation of the Rocky Mountains (Fig. 11.2). Pico de Orizaba National Park contains Mexico's tallest mountain, the 18,491-foot-tall Pico de Orizaba. This is the third highest peak in North America and home to the largest glaciers in Mexico. American national park visitors will find the towering conical volcano to resemble Mount Rainier (though a mile taller) and a counterpart to similar volcanoes around the world, including Kilimanjaro in Tanzania, Mount Fuji in Japan, and Chimborazo in Ecuador (all within protected areas). Iztaccíhuatl-Popocatépetl National Park, with two slightly lower peaks visible from Mexico City, is another high mountain park. Popocatépetl is active and frequently erupts and so is closed to hiking. Both had glaciers until recently, but those on Popocatépetl have disappeared due to volcanic activity and warmer temperatures.

The Sonoran Desert in northwestern Mexico is home to some of the largest protected areas in the country, including El Pinacate y Gran Desierto de Altar Biosphere Reserve near the head of the Gulf of California, a stunning and unearthly volcanic landscape. On the Baja California peninsula, El Vizcaino Biosphere Reserve contains ten-thousand-year-old cave art, as well as spectacular desert vegetation, including the unusual Boojum tree. It also includes Ojo de Liebre (or Scammon's) Lagoon, where gray whales spend their winters within a barren desert landscape. Most of the islands of the Gulf of California are also biosphere reserves; Isla Angel de la Guarda is known for its enormous number of venomous snakes and absence of people.

The tropical Yucatán Peninsula contains many protected areas on land and sea. Most of the coastline is protected, and many tourists visiting Cancun have been to Tulum, a national park. The Mexican Caribbean Biosphere Reserve includes almost half of the Mesoamerican

FIG. 11.2 Map of North America protected areas, classified by IUCN category, 2021.

Barrier Reef System, the largest reef system in the Western Hemisphere and the second longest in the world. It extends from the northern end of the Yucatán Peninsula to Honduras and is also protected within Belize, Guatemala, and Honduras.

Mexico has established several large marine protected areas in both the Pacific and Caribbean. The Baja California Pacific Islands Biosphere Reserve encompasses numerous small islands with bird colonies as well as empty ocean where fishing is tightly controlled. The Deep Mexican Pacific Biosphere Reserve is the largest protected area in the country. It is entirely underwater, starting at twenty-six hundred feet below the surface and continuing down to the ocean bottom, and includes several canyons whose bottoms lie over four miles beneath the surface. The most distinctive features are hydrothermal vents spewing mineral-rich superheated water from below the ocean

bottom. The tremendous water pressure allows water to remain liquid even at 800° F. Here lifeforms live completely isolated from the surface ecosystem, living on the sulfur-rich vents. But don't make any plans to visit; it is impossible to even enter without a deep-diving submersible. It is perhaps the world's most remote protected area.

Costa Rica has a reputation as a nature lover's tropical paradise and contains 164 protected areas, 23 of which are in category 2 and are national parks. These encompass mountain and lowland tropical forests. In Panama you can visit the tropical forests of Soberanía National Park, once part of the United States due to its location in the former Panama Canal Zone. Darien National Park preserves the Darien Gap, a stretch of pristine forest that has never been penetrated by roads and that prevents you from driving between Central and South America. In ecological terms, this national park is one of the

FIG. 11.3   Map of South America protected areas, 2021.

IUCN Category
- Ia
- Ib
- II
- III
- IV
- V
- VI
- Not Reported

most important in the Western Hemisphere: the existence of the jungle prevents the migration of livestock carrying diseases that have been eradicated in North and Central America.

In South America highlights of national parks are Los Glaciares in Argentina, which contains glaciers even more massive than those in Alaska; the Iguazu Falls, split between parks in Brazil and Argentina; and *tepuis*, or high mesas, in Venezuela (Fig. 11.3). Many parks are found in the Andes, a southern continuation of the heavily protected Rocky Mountains of the United States and Canada and the Sierra Madre of Mexico. But even Cabo de Hornos National Park, the southernmost point in Chile, is not the end of this mountain range; it appears again in Antarctica.

Galápagos National Park is perhaps the most scientifically important in the entire world.

This Ecuadoran park contains almost all the Galápagos Islands, a volcanic archipelago with a stunning range of animal and plant life. It is most famous for having been visited by Charles Darwin in 1835; his observations of the varying bird species of the islands were fundamental to the development of his ideas on evolution, the foundation of modern biology.

Australia, New Zealand, and the Pacific

National parks were a transpacific phenomenon, with the United States, Canada, Australia, and New Zealand early leaders, and these countries have park systems and management philosophies most like those of the United States (Sheail 2010). The Royal National Park of Australia was established outside Sydney in 1879; although it was a relatively small recreational park, it was the first "national park" to

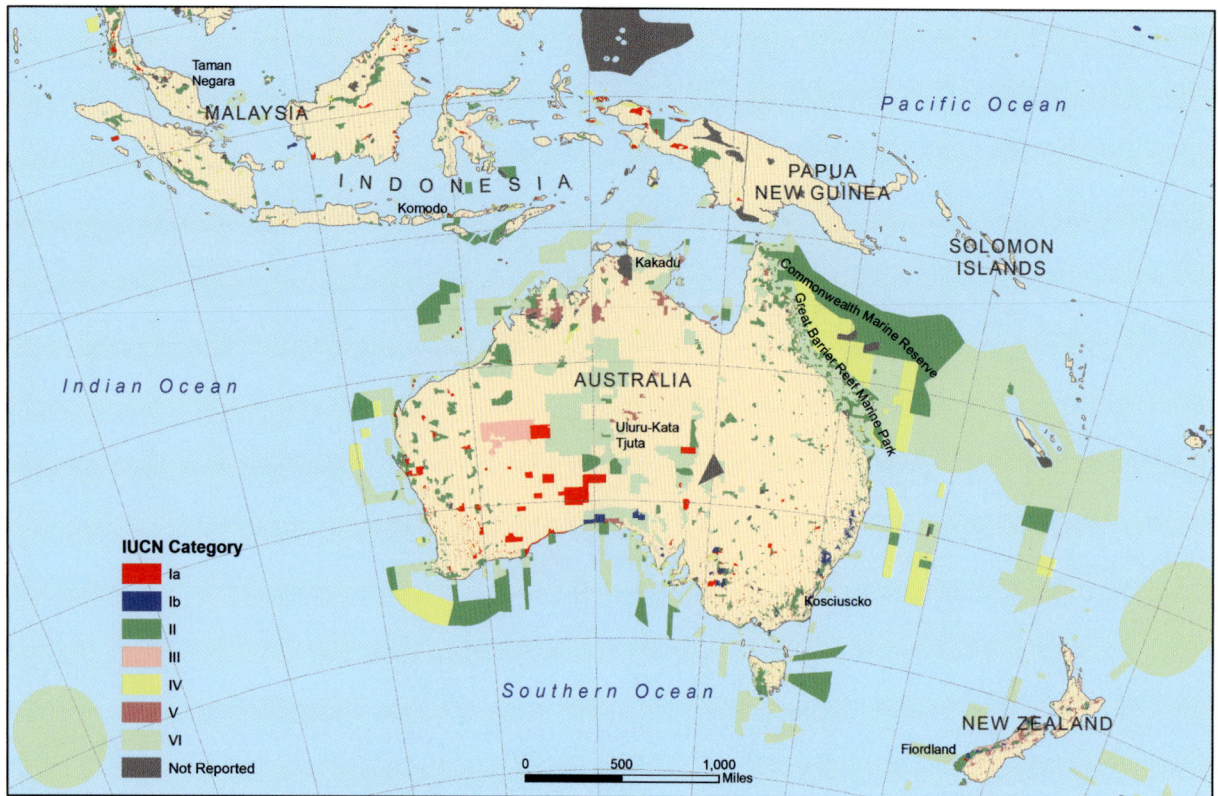

FIG. 11.4   Map of Australia, New Zealand, and Melanesia protected areas, 2021.

be created outside the United States. It did not, however, follow the model of Yellowstone; instead, it imitated the large city parks in vogue in the late nineteenth century (Harper and White 2012).

Much of Australia has been preserved or otherwise protected (Fig. 11.4). Perhaps the best-known place in the country today is the Great Barrier Reef, which contains the planet's longest reef system, stretching over fourteen hundred miles off the northeast coast. It includes over nine hundred islands. The Great Barrier Reef Marine Park was created in 1975 and has a variety of internal zones with different levels of use and protection (IUCN categories 1a and 6). Some are treated as scientific preserves off-limits to visitors, while others are national parks, wildlife refuges, and areas for sustainable fishing.

Kakadu National Park in the Northern Territory is one of the country's largest and contains an array of tropical wildlife (including saltwater crocodiles) and plants (Lawrence 2000). Waterfalls and deep gorges can be found, along with vast floodplains and wetlands. The park

is also a cultural landscape, with some of the country's best rock art. It may in fact be one of the world's oldest cultural landscapes, dating back to the arrival of Aborigines on the continent fifty thousand years ago. Its landscape may be familiar to Americans as having been used for filming several *Crocodile Dundee* movies in the 1980s.

Like the creation of American parks, the process of creating Kakadu took many years, with resistance from local landowners and mining companies (Lawrence 2000). The park includes Aboriginal lands (and the name comes from local Aborigines), though they did not push for the creation of the park, which is mainly oriented around wildlife. The legislation that created it called for joint management between Aborigines and the national government, but this has not been as effective as expected. Promoting tourism has been a goal of park creation, but Aborigines are not enthusiastic about this or even opposed to it. Aborigines do not consider the park to be wilderness; it is a landscape they have always lived in.

Mount Kosciusko, Australia's tallest moun-

FIG. 11.5 **Uluru, a sandstone mountain in central Australia.** Photo by Ogwen, 2004. Wikipedia, Creative Commons Attribution–ShareAlike 3.0 Unported License.

tain, lies within Kosciusko National Park, named for the same Thaddeus Kosciuszko commemorated in Philadelphia's tiny Thaddeus Kosciuszko National Historic Site. Though little known or appreciated within the United States today, Kosciuszko (1746–1817) was a national hero in the United States for his participation in the Continental army during the Revolutionary War; he is also a hero in Poland, Lithuania, and Belarus for his participation in struggles for freedom there. Many other monuments and statues of Kosciuszko can be found throughout Europe. His name was brought to Australia by a Polish explorer in 1840.

One of the most spectacular Australian protected areas is a red sandstone mountain near the "red heart" of the continent (Fig. 11.5). The first European who saw this named it Ayer's Rock; today it is called by its Aborigine name, Uluru. It is far more than a rock; it is a mountain rising 1,142 feet above the surrounding plain. A smaller mountain, previously known as the Ol-

gas and now as Kata-Tjuta, lies sixteen miles away. These mountains were included within an Aboriginal preserve in 1920 but then transferred to a wildlife reserve in 1958. The mountains became a national park in 1977. In 1985 the ownership of the land was given to local Aborigines, but it is still operated as a national park (much like Arizona's Canyon de Chelly). Since 1993 it is known as Uluru–Kata Tjuta National Park. Aborigines consider it sacred and enacted a climbing ban in 2019. Aborigines have a culture stretching back fifty thousand years, the oldest intact civilization on the planet, and it will be worth watching how they apply their knowledge to Uluru.

New Zealand set aside its first national park in 1887 (Sheail 2010). New Zealand's park system is heavily concentrated in the mountains of the South Island. The Southern Alps, glaciers, fjords, waterfalls, and active volcanoes are among the features preserved. The scenery will be familiar to many Americans and Ca-

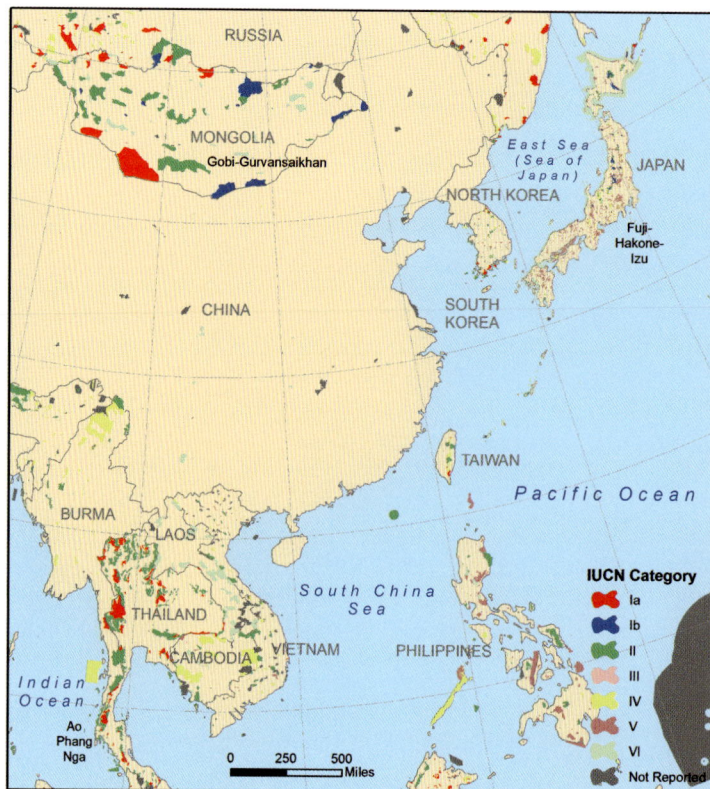

FIG. 11.6   Map of East Asia protected areas, 2021.

nadians, as it resembles that of the Rockies and Cascades (as well as having appeared in many movies). The South Island is at the same latitude as those northern ranges, has similar elevations, and is subject to a similar climate, producing similar landscapes.

Although the Pacific Ocean is home to many small island countries, it has many of the world's largest protected areas in the form of marine reserves. Australia's Great Barrier Reef Marine Park was one of the first (and briefly the largest) such examples. The Phoenix Islands Protected Area is presently one of the largest protected areas anywhere in the world (about the size of California). This includes a vast swath of ocean within the exclusive economic zone of the island country of Kiribati. Seven vast marine national monuments managed by the U.S. Fish and Wildlife Service and the National Oceanic and Atmospheric Administration have been created farther north in the Pacific. Papahānaumokuākea Marine National Monument is 140,000 square miles of ocean and small islands northwest of Niʻihau and Kauaʻi in Hawaii. Rose Atoll Marine National Monument and Pacific Remote Islands National Monument were also proclaimed in 2009. These include many small islands and atolls, managed largely as wildlife refuges. The Mariana Trench Marine National Monument was established in January 2009 and includes almost ninety-seven thousand square miles of ocean (slightly smaller than Wyoming), including a strip almost eleven hundred miles long containing the Mariana Trench, the deepest spot in the world. Only four people have ever visited the bottom of the trench, more than thirty-six thousand feet beneath the surface; by comparison, three times this number have walked on the moon.

East Asia

The countries of East Asia show a variety of contrasting patterns (Fig. 11.6). Perhaps the most striking feature on the map is the absence of any national parks in China. You may find mention on the Internet of hundreds of national parks there, but the IUCN lists only 120 protected areas in the country, none of them category 2. Government propaganda does not make a national park, and it may be that the Chinese government simply does not understand the concept. While many Americans are unhappy with the way the NPS might manage a park, we at least have a voice in its policies and have oversight through a democracy. We can also express our views of the National Park Service without fear of prison.

Japan created its first three national parks in 1934, followed by three more in Taiwan (which it controlled) in 1937 (Sheail 2010). These featured islands, lakes, volcanoes, caves, and high peaks. By 2017 there were thirty-four national parks scattered throughout the country, though they are generally small. Fuji-Hakone-Izu National Park is the most visited in Japan and includes not just Mount Fuji but also a nearby stretch of coastline and a chain of islands. Three hundred thousand people climb Mount Fuji annually, mostly at night (to be at the top at dawn, as at Haleakalā on the island of Maui).

Taiwan eliminated its parks after it became an independent country in 1945, but they were re-created in 1984, along with new ones. Today there are eight. The first national park in the Philippines was created in 1910, but the legal status of its parks and protected areas is confused due to overlapping and conflicting laws and proclamations (World Conservation Union 1992). The IUCN lists thirty-seven category 2 places. Indonesia's Komodo National Park is home to Komodo dragons, the world's largest lizards. Thailand has 238 protected areas, and 147 of them are category 2. Ao Phang Nga National Park was created in 1981 and includes spectacular limestone islands, many in the form of pillars, famous from the 1974 James Bond movie *The Man with the Golden Gun*. Unfortunately, it has been overwhelmed by tourists since then. King George V National Park was created in 1939 in Malaya (now Malaysia), a colony of the Netherlands (Sheail 2010; Kathirithamby-Wells 2012). The park has been known since independence as Taman Negara, which translates as "national park." Far to the north, Mongolia has thirty-one national parks; among these is the Gobi Gurvansaikhan National Park. This includes part of the Gobi Desert, complete with sand dunes, as well as mountain ranges and habitat for camels and snow leopards. Other parks include lakes, the country's highest mountain, and ancient petroglyphs. The country has done an admirable job of pro-

tecting its natural heritage and has developed several areas for tourism.

South Asia

Bangladesh has fifty-one protected areas, which include eighteen category 2 parks, seventeen wildlife sanctuaries, and one marine reserve (Mukul et al. 2008) (Fig. 11.7). Their main function is to serve as wildlife reserves, and all are quite small by U.S. standards. Bhawal National Park was the first of these and was established in 1974 under the Bangladesh Wildlife Preservation Act. More were established after 1980, and since 2008 the number of protected areas in Bangladesh almost doubled, from twenty-six to fifty-one. Given the high population density of the country, these numbers are amazing. The country's best-known and largest natural feature is the Sundarbans freshwater and mangrove forests. Nijhum Dweep National Park was established in 2001 as the nation's largest protected area to preserve these forests and provide a home to the endangered Bengal tiger. Kaptai National Park is the most popular destination for honeymoons and includes varied terrain, with some hills over two hundred feet in elevation (impressive by local standards).

India's first national park was Jim Corbett National Park, named after a British hunter who later led a crusade to preserve the subcontinent's wildlife (World Conservation Union 1992). It contains tigers, elephants, and rhinoceroses, among a wide variety of other tropical species. It was created in 1936, but this is misleading: land has been preserved in India for thousands of years, sometimes in the form of hunting preserves or in tens of thousands of sacred forest groves scattered around the country (M. Lewis 2004).

Most of the country's ninety-three category 2 parks are in the North, in or near the Himalayan foothills, or in the highlands along the southwest coast, areas of relatively low population density along with lush vegetation. Despite being home to many dangerous predators, the country has no history of wildlife extermination, as found in the United States (M. Lewis 2004). While the NPS struggles to reintroduce grizzly bears and wolves in a setting where they pose almost no danger to humans, no such reintro-

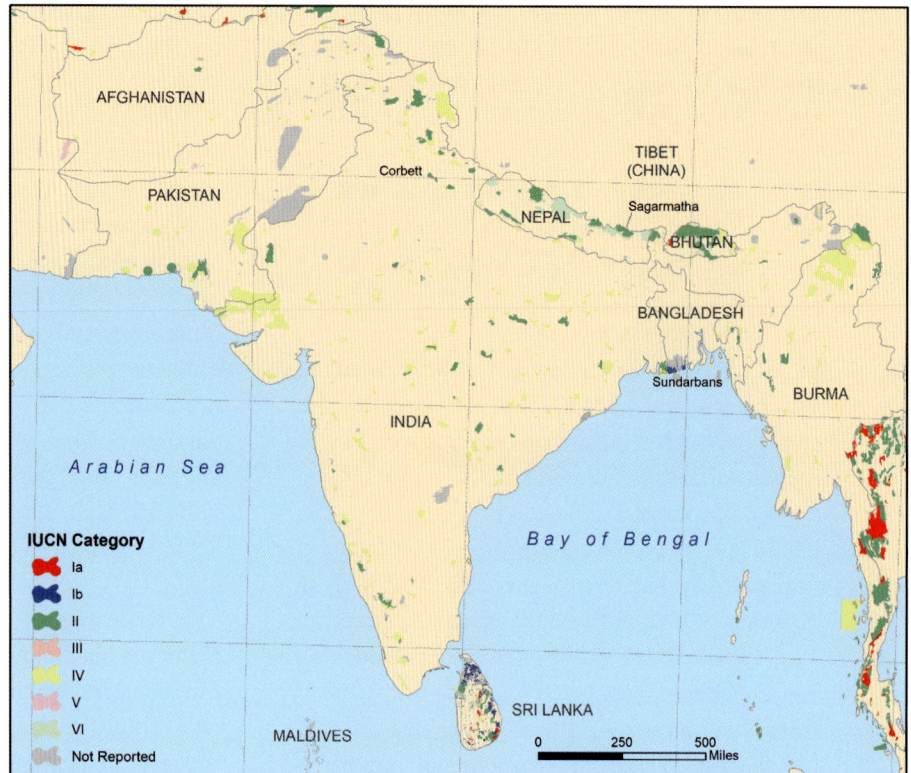

FIG. 11.7   Map of South Asia protected areas, 2021.

ductions are necessary in India despite frequent human deaths.

Pakistan has four hundred protected areas, and among them are twenty-nine national parks, though only five are considered category 2 by the IUCN. These are mostly in the mountainous North of the country. These high-altitude parks include mountain forests, alpine lakes, glaciers, and high peaks, as in the northern Rockies or New Zealand. Central Karakorum National Park is the largest protected area in the country and was established in 1993. K2, the second tallest mountain in the world, is within this park. Farther to the east, Mount Everest, on the Nepal-Tibet border, is within Sagarmatha National Park. Pakistan also has sixty-two category 4 wildlife sanctuaries.

Much of Tibet is over sixteen thousand feet in elevation and is sometimes considered the earth's "third pole." Unfortunately, it is not well protected. Nepal has twenty protected areas designed mainly to conserve the great forest biodiversity in the foothills of the Himalayas. Among these protected areas, ten are designated as national parks, and others are wildlife reserves. Established in 1973, Chitwan National

Park was the first in the country and received World Heritage site status in 1984. A tiny country, Bhutan, bordering India and China has ten protected areas that include four national parks and wildlife or nature preserves. Many of these are located within populated areas that lead to mutual habitat encroachment. The island nation of Sri Lanka has 501 protected areas, and 26 of them are designated as national parks. A permit is required to visit any national park in the country.

Africa

Protected areas in Africa are closely bound up with European attitudes toward nature protection. In the nineteenth century, Britain, France, Belgium, Portugal, and Germany ruled over vast imperial territories around the world (Jones 2012). While they were slow to embrace the national park idea in their own countries, they were willing and eager to create parks in their overseas territories. Belgium's King Albert visited Yellowstone, Yosemite, and the Grand Canyon in 1919 and was inspired to create Virunga and Volcanoes National Parks in what was then the Belgian Congo and now Rwanda and the

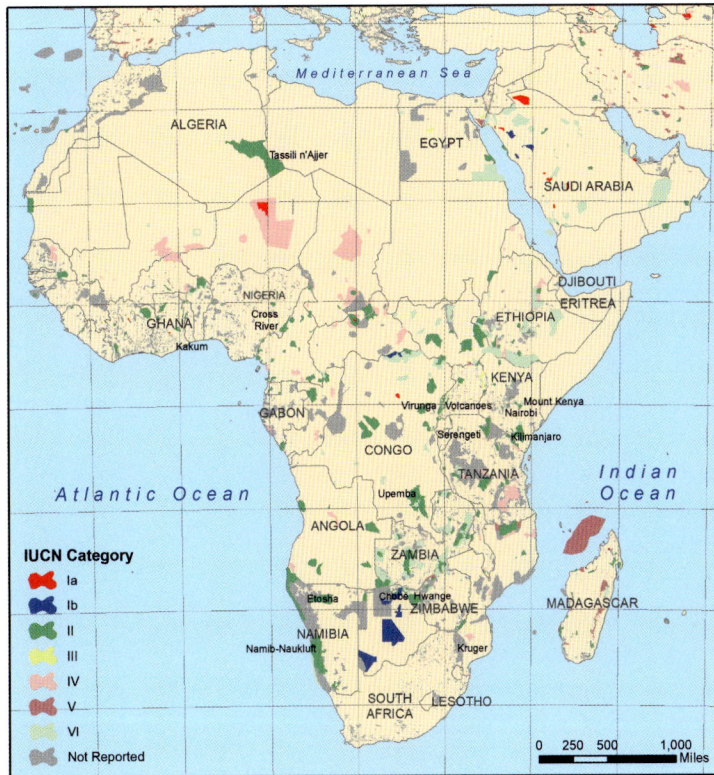

FIG. 11.8   Map of Africa, Middle East, and Western Asia protected areas, 2021.

Democratic Republic of the Congo (Jones 2012). These were among the first places outside the United States to be called national parks.

Parks in Africa were usually created to preserve wildlife, which had become necessary in the late nineteenth century due to massive increases in hunting, especially for ivory (Sheail 2010). In East Africa, Germans created wildlife reserves in Tanganyika (now Tanzania) following the Yellowstone model, justified by their perception that Yellowstone existed to protect the few surviving bison in North America (Gissibl 2012). Germans were fascinated by Africa's wildlife, in part because they saw it as a relic of the distant past and, presumably, like what might have once existed in Europe. Protecting it helped preserve Germany's own past. (The country would not create a national park in its own territory until 1969.) The French approached the matter slightly differently (Ford 2012). In 1921 they created several tourism-oriented national parks in Algeria for the benefit of the many French settlers there but saw no point in creating national parks in other African territories where there were few French (and therefore no tourists). In West Africa and Mad-

agascar, reserves were instead created to protect wildlife. In 1923 a national park was created within France itself, though modeled more on the Swiss scientific reserve concept than on tourism. These parks often required moving large numbers of Africans or Asians out of their homes. But there were more exceptions; in the Belgian Congo and British Malaya, native inhabitants were thought of as part of a park's wildlife and therefore required no relocation (Kathirithamby-Wells 2012).

Today Africa has many national parks and other protected areas (Fig. 11.8), though Eritrea, Lesotho, Djibouti, and Gabon are among the few countries in the world with no IUCN category 2 areas. Most of the larger and best known are in a belt extending from Ethiopia down eastern Africa to South Africa and Namibia; this is an area of highlands, scattered mountains, and tropical vegetation known for its wildlife.

Virunga National Park was created in 1925 to protect mountain gorillas and includes spectacular, glacier-clad volcanic peaks over sixteen thousand feet tall. While many gorillas (and park rangers) have been killed by poachers and militias operating in the area, the park still has an abundance of other wildlife and spectacular scenery. Volcanoes National Park in Rwanda adjoins Virunga and was created the same year; the anthropologist Diane Fossey spent many years here studying gorillas.

Nairobi National Park is on the outskirts of that city in Kenya but has lions, cheetahs, zebras, giraffes, rhinoceroses, leopards, hippos, ostriches, and many other large mammals you would hope to see in an East African park. It is one of Kenya's thirty-six category 2 parks. Its neighbor to the south, Tanzania, has fourteen. Tanzania's Serengeti National Park is a world-famous wildlife preserve, a vast area where countless large mammals may roam undisturbed. Hwange National Park was created in 1928 and is the largest in Zimbabwe. Congo's Upemba National Park was the largest in Africa when it was created in 1939 but has since been shrunk. Kruger National Park was created in South Africa in 1926 and is the country's best-known wildlife park. A map of Kruger shows developed areas, campgrounds, roads, and entrance stations, much like U.S. parks. There are, however, many more roads, reflecting the fact that the scenic resources of a wildlife park move around. Parks in West Africa are generally smaller but are also used to protect wildlife and tropical vegetation. Kakum National Park in Ghana has forest elephants, antelope, monkeys, and a canopy walk allowing visitors to explore the treetops (Appiah-Opoku 2004).

Not all parks in this area were created around wildlife. Tanzania's Kilimanjaro National Park was created in 1973 and includes the glacier-covered heights of the continent's highest mountain, at 19,341 feet slightly lower than Denali, though much more easily climbed. Mount Kenya National Park includes a 16,355-foot glacier-covered peak standing almost exactly on the equator. As Africa straddles the equator, deserts can be found in both the North and South of the continent. Chobe National Park in Botswana and Etosha National Park in Namibia are wildlife parks in the Kalahari Desert. Nambia's Namib-Naukluft National Park lies in the southern Namib Desert, while Algeria's Tassili n'Ajjer National Park is the largest on the continent and covers twenty-eight thousand square miles of the Sahara Desert, but its

centerpiece is a collection of prehistoric cave art dating back to the last ice age. It is also one of very few national parks in North Africa.

The Middle East and Western Asia

The Middle East and North Africa have many stunning historic sites, but national parks are rare. Egypt has only two national parks, one containing spectacular coral reefs in the Red Sea (Fig. 11.8). Iraq also has two national parks, one in the northern mountains and the other in marshlands between the Tigris and Euphrates Rivers. The United Arab Emirates and Israel have only one category 2 protected area each, the former wetlands and the latter the Mount Carmel Nature Reserve. Libya has four category 2 places, and Saudi Arabia has five, one of which is devoted to protecting ibex, a species of mountain goat. Yemen reports one category 2 protected area, the Jabal Bura Valley Forest National Park, containing one of the few forested areas on the Arabian Peninsula (Hall et al. 2008). Among two hundred protected areas in Iran are sixteen national parks, many of them wildlife refuges. The mountainous country has a range of ecosystems ranging from desert to alpine peaks, making for a surprising variety of parks. Golestan, Kavir, and Khar Turan National Parks and Nayband Wildlife Sanctuary are a few of the more famous national parks and nature reserves in the country.

The most recent country to adopt the national park concept is Afghanistan. The country's first national park was created in 2009 with the assistance of National Park Service staff (Revkin 2009). That park, Band-e Amir, includes what is sometimes called the Grand Canyon of Afghanistan, but with the country once again being ruled by brutal Taliban religious extremists, the new park will likely have a short life, as will any tourists who attempt to visit it.

Europe

Europe is not an important part of the world for national parks, with only 7 percent of the world's total (Kupper 2012). Like parks in India and Bangladesh, European parks tend to be small because of high population densities (Fig. 11.9). While Europeans were eager to spread the national park concept around the world,

FIG. 11.9   Map of Europe protected areas, 2021.

they were very reluctant to create them within their own countries. Belgium created several in Africa in the 1920s but did not have a national park of its own until 2006. The primary reason was population: the Yellowstone model of an unpeopled wilderness could rarely be applied here. Sweden became the first European country to have a national park when it created Abisko National Park in the North of the country in 1909 (Sheail 2010). This remains a popular vacation destination, as it is one of the driest and (in summer) sunniest places in the country. Switzerland, with its spectacular glaciated mountains, has only one national park, created in 1914. It was thought of as a scientific reserve rather than a tourist attraction, a different model from that of Yellowstone (Sheail 2010; Kupper 2012). This Swiss model became influential in Europe. The country also has many nature parks, nature experience areas, and a federal inventory of landscapes and natural monuments, one of which includes the iconic Matterhorn. Spain and Italy were the next to create parks, in 1918 and 1922, respectively. Turkey claims forty national parks, but the IUCN does not report any category 2 units. At the other extreme,

Russia has 11,252 protected areas, with 61 category 2 parks.

The Arctic and Antarctica

The Arctic region has largely been ignored during the national park era due to its harsh climate and lack of tourist potential. This began to change toward the end of the twentieth century, especially in the 1970s, when the U.S. national park system had its Alaskan expansion. The world's most northern and largest national park is in North America, but not in Canada or the United States; it is within Denmark's North American territory of Greenland (Fig. 11.10). Northeast Greenland National Park is 375,000 square miles, over eighteen times larger than Wrangell–St. Elias, America's largest park. It and Canada's northern parks are remote and have short tourism seasons, though this is changing as warmer summers make travel and tourism easier.

The Arctic region will become far more accessible and more pleasant to visit due to global warming. This will likely result in several new national park units created around the Arctic to protect this area from development. Will their

FIG. 11.10   Map of Arctic protected areas, 2021.

spatial organization mirror that of high mountains, organized by latitude, with tourist facilities and access points along the southern borders of the parks, and hardier tourists venturing to the Far North? This has been the traditional model of development for the high Arctic (Sugden 1982) and is still how the area is accessed. Or will entirely new facilities, including airports, cruise ship terminals, and luxury resorts, be built in areas now snow-covered wilderness? There are now many corporations that could easily create this infrastructure in return for a monopoly on tourism in a new Arctic park.

The continent of Antarctica contains no countries, so there can be no national parks there. Yet there are many protected areas of different kinds on the continent that preserve its wildlife, history, and scientific sites. The use of Antarctica is governed by the 1961 Antarctic Treaty (Secretariat of the Antarctic Treaty 2020). This has been signed by fifty-four countries and affects all land and waters south of sixty degrees south latitude. The treaty bans military activities or new national claims but allows protected areas to be designated on the continent (Fig. 11.11). There are 162 specially

protected areas, many on islands on the Antarctic Peninsula, the southern end of the mountain chain Americans know as the Rocky Mountains.

Antarctic protected areas include places important for penguin breeding, seal habitat, areas where plant life may be found, fossil deposits, active volcanoes, and the McMurdo Dry Valleys, the largest area of the continent without ice and summer snow. A further seventy-six sites have been designated as historic sites and monuments. These include a flag erected at the South Pole in 1965, relics and ruins of early polar explorers, their graves, and a variety of markers and plaques commemorating early exploration and scientific expeditions. Of these, Shackleton's Hut may be the best known; the expedition Ernest Shackleton led spent the winter of 1908 here before unsuccessfully attempting to reach the South Pole. The dry air has preserved the wooden shack exactly as it was.

Most of these areas are small, and none are intended to preserve scenery or large landscapes. Many have never been surveyed or do not have formal boundaries (IUCN 2008). There are no tourist facilities or park rangers on hand. The U.S. McMurdo Station is the larg-

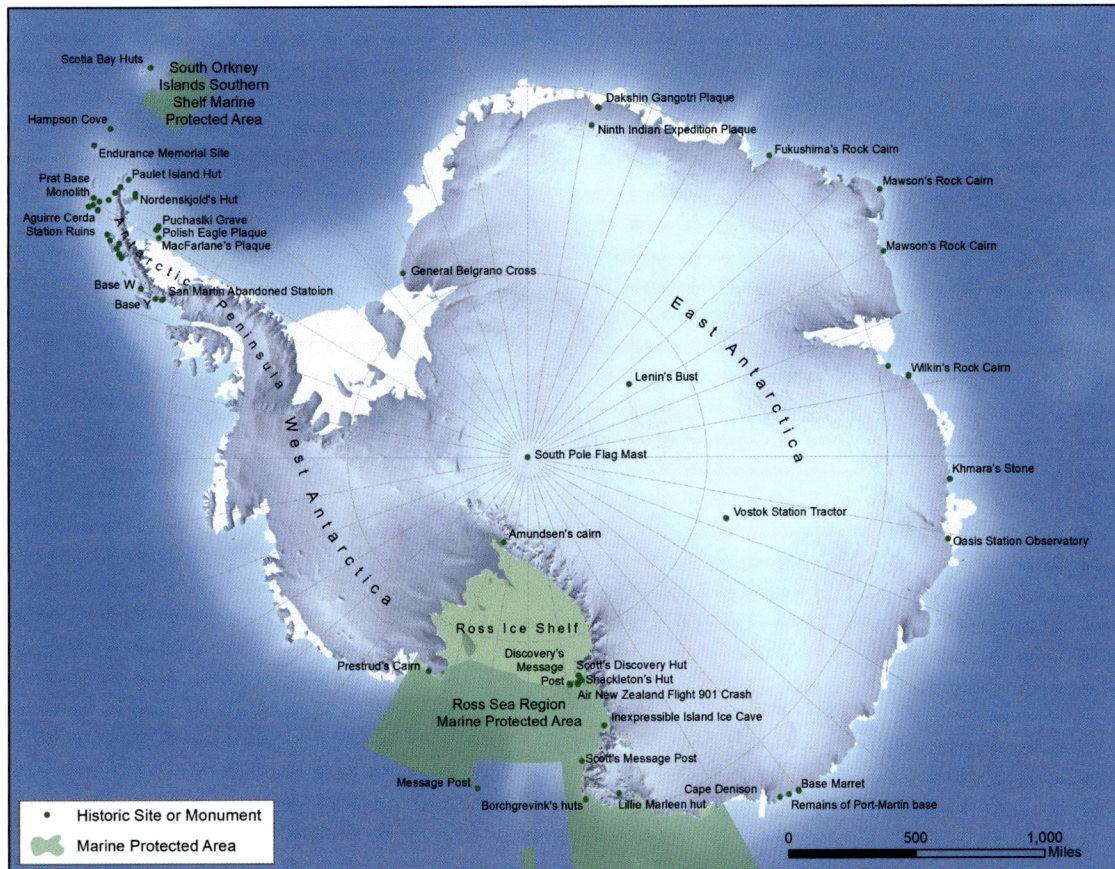

FIG. 11.11   Map of Antarctica protected areas, 2021.

est town and busiest airport in Antarctica, with up to twelve hundred people in the summer. It is not, however, part of Antarctic tourism, as only government personnel live there. Antarctic tourists either fly over the continent from airports in Australia, New Zealand, or South America or visit on cruise ships, making brief landfalls along the coast. Many of these tourists visit only the Antarctic Peninsula, though the Ross Sea is visited by many cruise ships from Australia and New Zealand.

The International Association of Antarctica Tour Operators (2020) includes eighty companies that served twenty-six thousand Antarctic tourists in the summer of 2005–6. These tour operators agree to guidelines designed to minimize impacts, and individual countries may require their companies to meet additional rules. Cruise ships with more than five hundred passengers are not allowed to take passengers ashore; smaller ships allow passengers to go ashore in groups no larger than one hundred, and these will likely only spend a few

hours ashore. King George Island, seventy-five miles off the coast of Antarctica, is a frequent destination and the only airport with scheduled services.

The Antarctic Treaty does not handle tourism well; when it was signed there was none, but overflights, cruise ships, and landings have become common. It is inevitable that this area will be impacted by tourists in the future. How this area will be protected against the inevitable growth of tourism remains to be seen. In 1979 an airliner on a scenic excursion from New Zealand flew into the side of the Mount Erebus volcano, killing all 257 on board. As in the Grand Canyon, flightseeing is an irresistible but fundamentally dangerous temptation; in Antarctica, there is no radar and air traffic control system to help pilots and little chance of rescue for survivors.

International Parks

What about national parks that span national borders? It is evident that mountain ranges,

FIG. 11.12 Coronado National Memorial in southern Arizona. The mountain in the background is in Mexico. Photo by Dave Bly. NPS.

wildlife, and vegetation do not stop at a country's borders, and resolving environmental problems requires the cooperation of all. The possibility of adjacent countries cooperating to create a national park together is an old one, and one that is a growing area among park planners and conservationists.

Canada, Mexico, and the United States all have large and active park systems, and many of them lie along the border. Why not work together for a common cause? NPS staff were enthusiastic about these possibilities in the early twentieth century, and at least thirty-seven initiatives were undertaken to create international or transboundary parks or nature preservation between the United States and Canada or Mexico (Chester 2006; Young, MacEachern, and Dilsaver 2021). Interest in international parks across the U.S.-Mexico boundaries was particularly strong in the 1930s, when sixteen different areas were discussed (Wakild 2009). One of these was to be a multinational site commemorating the 1540 Coronado expedition (Sanchez, Erickson, and Gurule 2001). Coronado National Memorial was created in the Huachuca Mountains in Arizona, with views over the San Pedro valley, the likely route of the expedition (Fig. 11.12). Unfortunately, the Mexican counterpart never materialized, and the U.S. Coronado National Memorial stands alone. In 1935 interest began in a binational park along the Rio Grande, including Big Bend National Park and the Sierra del Carmen in

Mexico (Jameson 1996). The idea died out in the 1950s amid miscommunication and differing priorities.

These U.S.-Mexico parks failed in part because the two countries possessed very different notions of what a national park should be (Wakild 2009). The Mexican parks grew out of the country's revolution and efforts at land reform. The early parks, concentrated around Mexico City, allowed peasants to continue living in and utilizing parklands for their subsistence and were strongly oriented toward resource conservation. Restoring degraded lands was a higher goal than protecting pristine scenery, and middle-class tourism was not a concern. The NPS lacked respect for Mexican park management, while Mexico didn't want to lock up land for the benefit of wealthy tourists.

Canada's national parks are more similar in philosophy to those of the United States, and an international park shared between the United States and Canada can be found. Waterton Lakes National Park was created in 1911 across the border from the much larger Glacier National Park, created the year before (Mortimer-Sandilands 2006). Since 1932 they have been referred to as Waterton-Glacier International Peace Park, but the parks are managed separately. (In fact, the two countries only agreed to the international label because there was to be no joint management.)

One impetus for the international label was that both parks were developed by the Great Northern Railroad, giving them a common development pattern and appearance (Mortimer-Sandilands 2006). Waterton had no railroad connection to the rest of Canada, and bringing in automobile tourists from the United States was the park's initial strategy; this resulted in the Chief Mountain International Highway, completed in 1936. American interest in the park was also very high during Prohibition, with tourists crossing the border in search of alcohol instead of scenery; upon its repeal in 1933, visitation fell off. After Canada declared war on Germany in 1939, vacation travel became more difficult and less common for Canadians, and Americans once again made up many of the park's visitors. This ended in 1942, when Americans had to cut back their own travel during wartime.

Why not a joint park with Russia? In 1991 a bill was entered in Congress that would create the Beringia Heritage International Park with Russia, split between Alaska and Siberia (Blackford 2007). The Bering Land Bridge National Preserve would form the American section of the park, and Russia was to create a park on the Chukotka Peninsula. The bill failed, partly due to local opposition, but the NPS continues a program to raise awareness of the shared heritage in the two countries.

The idea of parks or other protected areas that span national borders now goes by the name of transboundary conservation. The IUCN has three types of transboundary protected areas: class 1, 2, and 3 (IUCN 2011). The first refers to protected areas, the second to conservation land or seascapes, and the third to conservation migration areas. All require active cooperation between countries, but this often remains as difficult as between the United States and its neighbors.

Namibia and South Africa cooperatively manage the |Ai-|Ais/Richtersveld Transfrontier Park (the | symbol represents a clicking sound used in southern African languages) on their mutual border, the centerpiece of which is the spectacular Fish River Canyon, known as the Grand Canyon of Africa. Victoria Falls on the Zambezi River is twice the height of Niagara Falls and is protected by two national parks in Zambia and Zimbabwe. South Africa's Kruger National Park is part of the Great Limpopo Transfrontier Park, shared with Mozambique and Zimbabwe.

An infamous example of a protected area not covered by the IUCN categories is the Korean demilitarized zone (DMZ), separating North and South Korea since 1953. It is 160 miles long and averages 2.5 miles in width, making it about one-third the length of the Blue Ridge Parkway but five times wider. Both sides are heavily fortified, with barbed-wire fences, bunkers, and land mines. Although soldiers can patrol within it and two small villages still exist within it, the area has been almost entirely depopulated since 1953. This is as far from the concept of a national park as can be imagined, but the result of decades of isolation has been the resurgence of wildlife, including tigers, leopards, and bears. It is a de facto wildlife refuge, and South Korea does have several small nature and wildlife reserves within it. When the Korean conflict ends, this refuge will hopefully be preserved; the presence of three million landmines in the DMZ will surely help keep people and development out.

World Heritage Sites

What about cultural sites that transcend national boundaries and can be considered part of the shared heritage of humanity? There is a system for that as well. UNESCO (the United Nations Educational, Scientific and Cultural Organization), a branch of the United Nations, created the World Heritage site designation in 1972, following an American proposal to preserve both natural and cultural heritage (UNESCO 2020b). Individual countries may nominate a site to be a World Heritage site; if approved, it is added to the World Heritage List. Being on the list does not bring any money or create any restrictions on how the site is managed, though if it is damaged, the site may lose its listing. World Heritage status has been sought to promote tourism, because these sites have a higher profile.

There are ten criteria for a site to be considered a World Heritage site. Sites may be natural or cultural in nature; the first six criteria apply to the cultural sites and the remainder to the natural sites. There are currently 1,092 sites in 167 countries (Fig. 11.13). Italy has the most, with fifty-four. The Great Pyramids in Egypt, the Taj Mahal in India, the Easter Island statues, Angkor in Cambodia, and Machu Picchu in Peru are all on the list. Those who are fascinated by Mesa Verde and similar ruins can visit the Dogon villages along the Bandiagara Escarpment in Mali to see similar cliff villages still inhabited by those who built them.

There are twenty-three World Heritage sites in the United States (Fig. 11.14); all but Taos Pueblo, Cahokia, Monticello, and the University of Virginia are included within the national park system (Fig. 11.15). Taos Pueblo is a

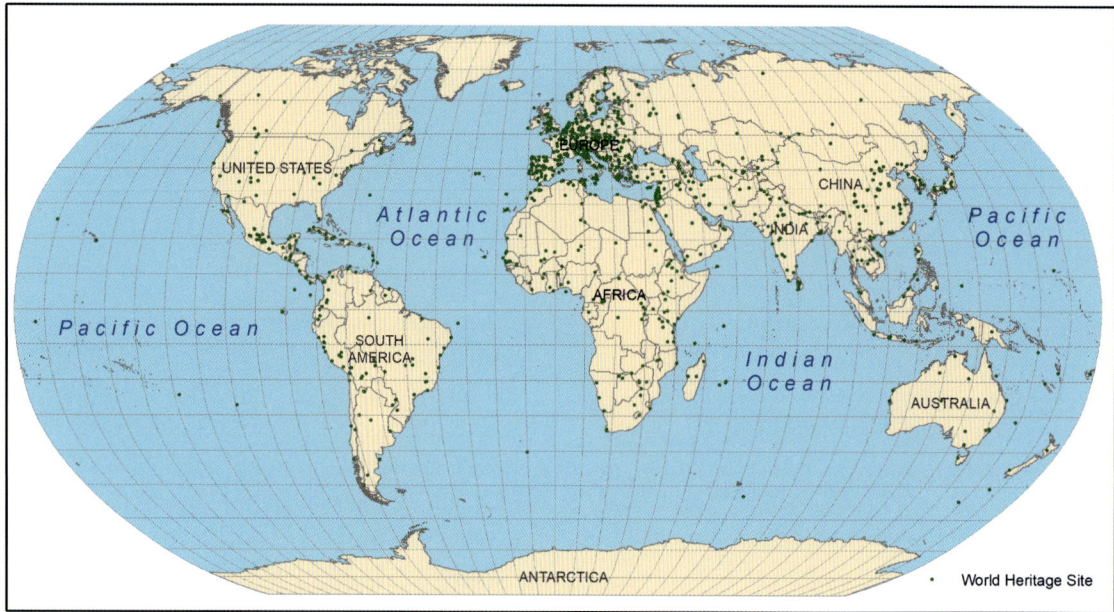

FIG. 11.13   Map of World Heritage sites, 2021. These are heavily concentrated in Europe but found throughout the world.

FIG. 11.14   Map of World Heritage sites in the United States, 2021. Most are part of the national park system.

FIG. 11.15   Pueblo Bonito at Chaco Canyon National Historical Park in New Mexico, a World Heritage site.
The ruins are of an Ancestral Puebloan village abandoned by 1126. NPS.

Native American village close to one thousand years old, while Cahokia, a state park in Illinois, was the largest Native American settlement to have ever existed in the United States. Monticello and the University of Virginia are both the brainchildren of Thomas Jefferson.

Since 1971 UNESCO has also operated the Man and the Biosphere program, which seeks to create an international network of protected ecosystems (UNESCO 2020a). There are 669 biosphere reserves in 120 countries. There are twenty-nine in the United States, some corresponding to specific parks, such as Organ Pipe Cactus, while others are based on larger ecosystems, such as the Mojave and Colorado Deserts in California, which include Death Valley and Joshua Tree National Parks, Anza-Borrego Desert State Park, and the Santa Rosa Mountains Wildlife Management Area. They may even be privately owned, such as the Na-

ture Conservancy's Cascade Head reserve in Oregon.

Parks on Other Worlds

One of the most historic moments in American and all human history is not protected in the U.S. national park system or those of any other country. Apollo 11's landing site on the moon on July 20, 1969, was humanity's first step onto another world. The landing site remains as it was on that day; there are no atmospheric or tectonic forces on the moon to disturb it. No historic site on the earth has been preserved as completely as those on the moon (Spennemann 2006).

In addition to Apollo 11, the artifacts associated with five other manned landings remain on the moon (Fig. 11.16). The descent stages of the Apollo lunar modules, lunar rovers, Amer-

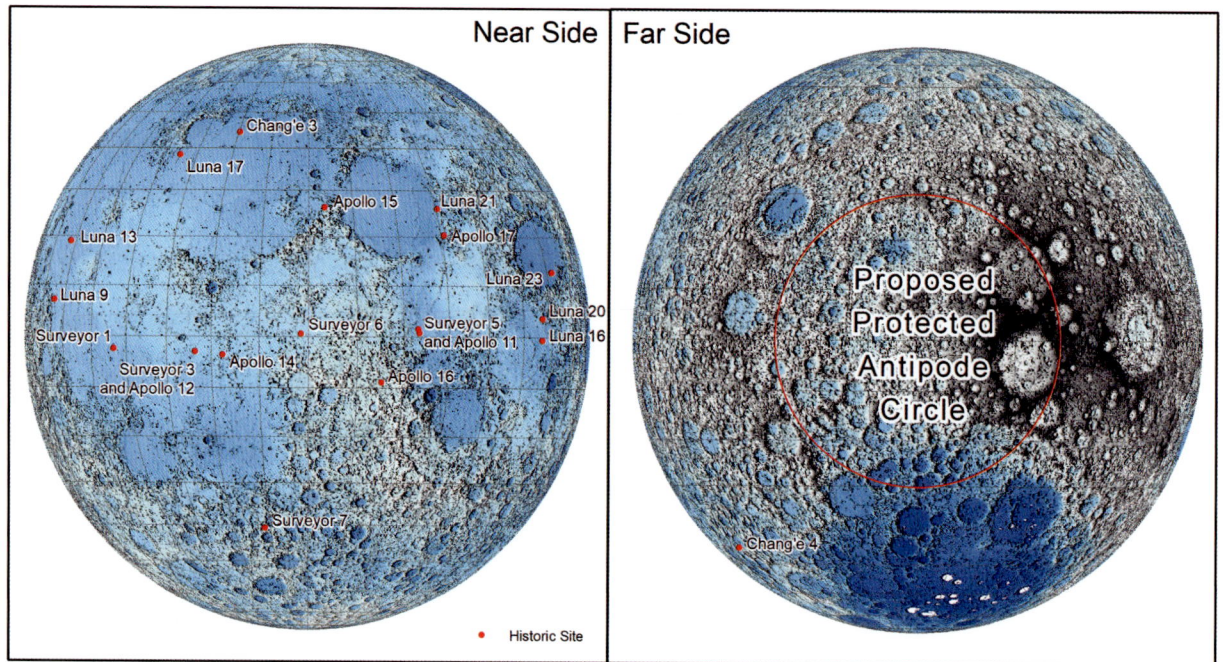

FIG. 11.16   Map of lunar historic sites and a proposed protected area on the far side to be used for scientific studies, 2021.

ican flag, several scientific experiments, and footprints remain at each site (T. Rogers 2004; Spennemann 2004, 2006, 2007; Gorman 2005). At the last three landing sites vehicle tracks extend the human presence as far as three miles from the touchdown site. These sites have been mapped and photographed in detail, and far more is known about them than was known about Yellowstone in 1872. A variety of unmanned landing sites also exist on the moon.

There is no threat to these sites in the foreseeable future, but at some point, humans will return to the moon. The 1967 Outer Space Treaty, to which all spacefaring nations are signatories, prevents a country from claiming territory in space or on a moon or planet but states that objects in space remain the property of and under the jurisdiction of the launch nation (Spennemann 2004). The Apollo 11 lunar module descent stage, American flag, and various pieces of scientific equipment are still U.S. government property.

What will be done with these sites? Possibilities for preserving scenic and historic features on the moon have already been discussed. These possibilities are based on the increase in exploration in recent decades and the likelihood that sites of earlier landings may need

protection. In 2010 California listed artifacts at the Apollo 11 landing site on the state historic site register. This is part of a larger goal of having the site declared a national historic landmark and having the United Nations declare it a World Heritage site (Spennemann 2004). It has been suggested that space heritage sites such as the Apollo 11 landing could be drivers of space tourism (Spennemann 2006). Formally protecting sites may therefore help boost visitation to them, which has been the case with protected areas on the earth. These sites are, however, far more remote than Yellowstone was in 1872 or Antarctica in 1910.

There have been calls to preserve areas on the moon for scientific reasons in addition to protecting humanity's history. Claudio Maccone (2010) suggests creating a scientific preserve on the far side of the moon centered on the crater Dedalus. This area permanently faces away from the earth and is completely shielded from radio transmissions from the surface of the earth or satellites in orbit. This would be the best location for a radio astronomy facility.

What about other planets? Charles Cockell and Gerda Horneck (2004) have outlined a planetary park system for Mars, with six potential parks preserving different aspects of the

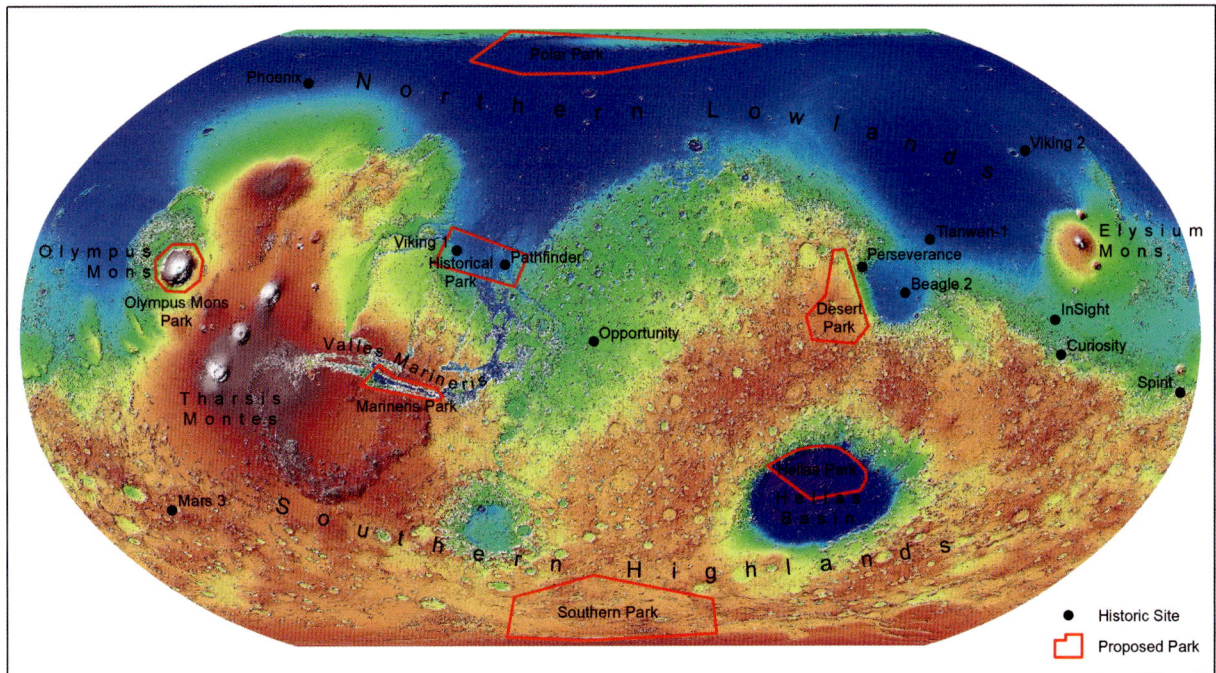

FIG. 11.17   Map of proposed protected areas on Mars, containing major geological features and representative areas of the planet's surface, 2021.

planet's surface (Fig. 11.17). These are Olympus Mons (the largest volcano and the tallest mountain in the solar system), the north polar ice cap, Valles Marineris ("the Grand Canyon of Mars," the biggest canyon in the solar system), and Hellas Basin, representative areas of the planet's surface, along with several spacecraft landing sites to be preserved for their history. Although Mars has been well mapped from space, the proposed park boundaries are quite like those of Yellowstone in 1872: large and generalized due to an absence of information about exactly what is there. Additional exploration and fieldwork will no doubt reveal more about these areas and will identify previously unknown resources, exactly as happened on the earth.

Management of planetary parks and lunar historic sites could be far more challenging than those on the earth. The philosophy of "take only photographs, leave only footprints" that many national park visitors are familiar with would be tremendously destructive on the moon, where the footprints twelve astronauts left during their short stays survive today and are among humanity's most important historical artifacts (Spennemann 2006). The size of these

parks will likely be an issue: Should a historical park preserve the lander, the footprints and tire tracks extending miles around it, or the entire landscape the astronauts or rovers explored? Is a new park model needed to protect these places?

Calls for the preservation of historic sites, scenery, and scientific sites on the moon and elsewhere are based on the same ideas that the national park idea grew out of and embraced: a sense of responsibility to future generations, a land ethic, and a desire to protect beautiful scenery and preserve historic sites (Cockell and Horneck 2004). Unfortunately, these ideas are perhaps not as common or popular as they were in 1872, and vicious nationalism and rampant greed may instead shape (or destroy) humanity's future in space. Despite this, the NPS has already been involved in preserving sites associated with the space program, including NASA's Mission Control in Houston (von Ehrenfried 2018), which is a national historic landmark. The nonprofit organization For All Moonkind is also active in seeking preservation for lunar artifacts.

There will be opposition to these parks and historic sites. At the present time, they are not possible; shortsighted critics can easily ridicule the idea of even discussing them. More meaningful opposition will begin as the reality of designating protected areas on other worlds becomes clear and threats emerge. The U.S. experience shows that economic motivations will lead the opposition, and in fact, a variety of interests are already calling for opening the moon for dubious mining schemes (Schmitt 2006); those still promoting space tourism will not want to see these places put off-limits. For the time being, the rest of the solar system is Earth's backcountry and the ultimate wilderness. But it will not remain that way.

Conclusions

Looking at parks around the world can reveal much about our park system. One of the most significant revelations is something visible in many parks around the world that is very rare in the United States: native peoples. The standard American model for parks is sometimes called the "Yellowstone model," based on preserving an uninhabited wilderness. It is assumed that a park has no residents (other than a handful of employees), and Indigenous peoples have simply disappeared if they even ever lived there (Keller and Turek 1998; Burnham 2000). This assumption does not hold up under the slightest scrutiny. Yosemite was first seen by white visitors during the 1850 Mariposa Indian War when a militia pursuing Native Americans stumbled across the spectacular valley. There, and at Yellowstone a few years later, the removal of native inhabitants was a necessary step to creating a national park. At Yellowstone the NPS went further and created the myth that Native Americans had never lived in the area because they were afraid of geysers; this sort of story spread to other early parks. Aside from Canyon de Chelly and Navajo National Monuments, owned by the Navajo Nation, and the South Unit of Badlands National Park, comanaged with the Oglala Sioux, Native Americans remain only within Death Valley National Park. There the Timbisha Shoshone were granted a reservation in 2000.

The Yellowstone model, or "fortress conservation," in which the outside world is excluded, has been exported throughout much of the world (Stevens 2014). This model has displaced many thousands of people around the world to create parks, with the process of displacement requiring removal from a protected area, restriction on economic activities in and around parks, loss of control over land that has belonged to a group, and loss of access to or destruction of their homeland, religious sites, and even cemeteries. The effects of the Yellowstone model in some areas of the world may be like that created by big dams and reservoirs but rarely studied. It is not just a thing of the distant past; Indigenous people were forced out of Uganda's Bwindi Impenetrable Forest National Park in 1991; the British had allowed them to live in an earlier reserve because they did not hunt the gorillas the reserve was created to protect (Kidd 2014).

The Yellowstone model did not take root everywhere; as noted above, Mexico had a different vision for parks that the NPS could not accept. The importance of Aboriginal management for Australian parks has led to an "Uluru model" for national parks, with Indigenous peoples still living in the park and having at least some management powers; Uluru and Kakadu National Parks in Australia are among the best-known examples (Adams 2014; J. O'Brien 2014). This model will likely become far more influential, perhaps even in the United States.

Parks and protected areas around the world face many of the same challenges and threats as those in the United States, and often far worse. World Heritage sites are at risk for sea-level rise (Reimann et al. 2018). Not only are protected areas in the tiny Indian Ocean nation of the Maldives threatened by sea-level rise, but so is the entire country.

Parks in other countries do not always receive the same level of recognition and respect that Americans give to ours. An extreme example might be Nigeria, where, according to a tourist guidebook, several of the country's national parks are unknown to those who live nearby, who have never even heard of them, and who are unable to give directions to them (Williams 2005). It is also unfortunate that parks in many parts of the world have been involved in conflicts, civil wars, and terrorist activity. Wildlife has been killed for food, historic and cultural attractions have been destroyed, and visitors have been killed. To give a few examples, Cambodia's protected areas were destroyed in the 1970s, and the professional staff that ran them were murdered by the Khmer Rouge (World Conservation Union 1992). Tourists have been savagely murdered in Bwindi Impenetrable Forest National Park in Uganda and massacred at the Temple of Hatshepsut in Egypt. Most of Angola's national parks were destroyed during decades of civil war; only Quicama survives. In 2001 the government of Afghanistan, then under the control of the Taliban, blew up the monumental Buddha statues, a World Heritage site, in Bamiyan. ISIS and other groups have targeted other historic and cultural treasures; the damage done to them is not an accident of war but a deliberate attack based on the requirements of fundamentalist religion to destroy that which does not support it. An international movement was formed to rebuild the statues, but instability in the country has prevented this. It is a tribute to the love of beauty, the natural world, and our shared cultural heritage that many visitors have returned to (some of) these areas. The money they spend has been crucial to paying the salaries of park employees, supporting local communities, and enabling the protection of natural and cultural resources to be restarted.

TWELVE

The Geography of Change

On his first visit to Arizona on May 6, 1903, President Theodore Roosevelt stopped to see the Grand Canyon. He remarked to an assembled audience on the South Rim that they should "leave it as it is. You cannot improve on it. The ages have been at work on it, and man can only mar it." These words have become part of the lore of the national park system and make the preservation of the canyon appear an unqualified success. However, this is only a small part of what he said that day and not completely accurate. His actual remarks about the canyon are a bit different:

> I shall not attempt to describe it because I cannot. . . . I could not choose words that would convey or that could convey to any outside what that canyon is. I want to ask you to do one thing in connection with it in your own interest and in the interest of the country—to keep this great wonder of nature as it now is. . . . I hope you will not have a building of any kind, not a summer cottage, a hotel, or anything else, to mar the wonderful grandeur, the sublimity, the great loneliness and beauty of the canyon. Leave it as it is. You cannot improve on it; not a bit. The ages have been at work on it, and man can only mar it. What you can do is to keep it for your children and your children's children and for all who come after you, as one of the great sights which every American if he can travel at all should see. (Roosevelt 1903, 1)

The reference to buildings, summer cottages, and hotels is usually removed when Roosevelt is quoted, and for obvious reasons: the Grand Canyon Village at the South Rim is a sizable town with several hotels and motels, restaurants, dozens of homes, a shopping center, a school, and many other buildings. The first hotel, El Tovar, appeared the next year at the point where Roosevelt gave his speech. The canyon was not left as it was, and doing so seems to have never been a serious possibility (Nash 2017).

One feature Roosevelt did not name was the recently opened railroad line and train station that brought him conveniently and safely to the canyon rim. We always accept what was there when we arrived yet complain about later changes; Roosevelt was no exception. He made his visit just as this new railroad was unleashing a wave of space-time convergence that would see visitation grow tremendously, leading to the need to build those hotels, restaurants, trails, campgrounds, roads, stores, gas stations, airports, and bus stops on the rim. His visit helped popularize the canyon he preferred to see unchanged and led to its transformation into one of the world's leading tourist attractions.

Horseshoe Bend is a spectacular overlook on the Colorado River within Glen Canyon National Recreation Area (Fig. 12.1). Here the river flows through a sharp bend at the bottom of a one-thousand-foot-deep sandstone canyon. For years few knew about the overlook, and those who did enjoyed it in solitude. Visiting it required pulling off the nearby paved highway into the unmarked dirt parking lot and then hiking a mile through deep sand to the canyon edge. But in 2015 the viewpoint became popular with social media addicts, who began arriving in large numbers to snap selfies. Over two million people visit the overlook each year now. Until recently, there were no railings, and in 2018 a visitor taking a selfie fell seven hundred

FIG. 12.1   Horseshoe Bend on the Colorado River, Glen Canyon National Recreation Area. This has become a popular overlook, leading the NPS to develop the area to better manage crowds. Photo by Brent and Dawn Davis. NPS.

FIG. 12.2   Toroweap Overlook in Grand Canyon National Park. This remote western overlook is undeveloped, but for how long? Photo by Gleb Tarro, 2016. Wikipedia, Creative Commons Attribution–ShareAlike 4.0 International License.

feet to her death. The NPS reacted by installing guardrails, a new path, and a paved parking lot with a ten-dollar entrance fee (Craven 2019). Like the earlier development at the South Rim, this is the latest park location to require hardening to accommodate growing numbers of visitors whose visits were enabled by new technology.

Toroweap (or Tuweep) Overlook offers one of the most spectacular views of the Grand Canyon from a vertical cliff standing three thousand feet above the river (Fig. 12.2). The overlook can only be reached by a long dirt road (often impassable) from the North Rim and has very few visitors. There are no facilities other than a campground. As remote as it is, perhaps one hundred years from now it will be hardened as the South Rim was in 1903 or Horseshoe Bend in 2018.

How will the geography of the national park system change over the remainder of the twenty-first century? Will new kinds of places be added to the system? Will some NPS responsibilities be given to new agencies? Will national monuments administered by other agencies be transferred to the NPS in an echo of the reorganization of 1933? Will parks be developed differently? Who will visit parks in the future, and when? What approaches might the NPS take in its efforts to remain relevant in a changing country? What new threats will emerge? Though the year 2100 appears impossibly distant, it is closer to 2021 than that year is to the reorganization of 1933. There are people alive today who will be visitors to the park system in 2100, and we will use this possibility as a reference point for our discussion of the geography of change in the national park system.

The Changing Geography of the Park System

Statements can be found going back to the creation of the National Park Service in 1916 that the park system includes every worthy site and is complete (Rettie 1995; Sellars 1997), but every year proposals are examined and bills submitted to Congress for new units. An average of 1.9 new park units a year have been added over the past forty years. If that rate continues, there might be 152 more units in the system by 2100, for a total of 576. What might these new units be?

Among the recent or ongoing studies for new park units are President James Polk's home and Fort Pillow in Tennessee, Public School 103 and the President Street railroad station in Baltimore, Camp Granada World War II concentration camp in Colorado, and President George W. Bush's childhood home in Texas. Recent trends in park creation provide some clues as to future additions. Parks have become more representative of the American population over time. African Americans, Indigenous populations (Native Hawaiians, Native Americans, and Native Alaskans), women, Hispanics, and gay people are now represented in the national park system to different extents. The increasing diversity in the park system will surely continue to reflect the nation's demographic composition. One possibility is the disabled. Though the Americans with Disabilities Act of 1990 did not come about due to a dramatic Supreme Court decision, it did create a massive change in the equality of facilities and services offered by businesses and government agencies. Beyond modifying buildings to meet the law, most parks have very low levels of accessibility for people with a mobility disability.

A safe bet for future park units is the creation of more sites for presidents. Nineteen presidents so far have sites associated with them, with the next one in progress. The NPS has been authorized to create a Ronald Reagan boyhood unit in Illinois, but this has stalled because of the unwillingness of the owner to sell Reagan's childhood house to the NPS. Counting Reagan, seven of the thirteen presidents since World War II have park units devoted to them. Perhaps in time, there will be sites devoted to Nixon, Ford, George and George W. Bush, Obama, and Trump. Though perhaps not a particularly accomplished president, Gerald Ford has the distinction of being the only president to have worked for the National Park Service. He was a park ranger at Yellowstone during the summer of 1936.

The list of people who could be honored with a national park unit includes the artists George Catlin, Thomas Moran, and Albert Bierstadt. While these artists are not without criticism for their Eurocentric imagination, their paintings fascinated those who had never visited parks and inspired support for promoting the idea of national parks in the nineteenth century. Thomas Moran's 1872 painting *The Grand Canyon of the Yellowstone* is his most famous, but he also painted spectacular early views of the Grand Canyon (*The Chasm of the Colorado*, 1874) and the Grand Tetons (*The Three Tetons*, 1895). The Yellowstone and Grand Canyon paintings are in the Department of the Interior museum and are seen by few; *The Three Tetons* is in the White House art collection and was displayed in the Oval Office by President Obama. It is a shame that these paintings are not more accessible to the public. The NPS tried to bring *The Grand Canyon of the Yellowstone* home to Yellowstone but failed (Rydell and Culpin 2006). Perhaps someday they will be displayed within a unit devoted to Thomas Moran.

The photographer Ansel Adams (1902–84) is famous for his stunning black-and-white images of the American West. He had photographed several parks in the 1930s, and the NPS commissioned him to photograph the national parks in 1941. He visited many of them, and his images of parks such as Yosemite, Death Valley, Canyon de Chelly, the Grand Canyon, and the future Manzanar National Historic Site are among the most famous of these places. Dorothea Lange (1895–1965) took many photos of Manzanar, as well as iconic and poignant scenes of the Great Depression that are instantly recognizable to all. Perhaps both should have their own parks.

There are many other possibilities. The parks included in the system are just a small set of those that have been proposed over the years. These have included Colorado's Royal Gorge, Virginia's Masanutten Mountain, Arizona's Kofa Mountains and Meteor Crater, glacial grooves on Ohio's Kelly Island, and even the Great Plains (Ise 1961). The site of the first atomic bomb explosion, near White Sands National Park, was proposed as a national monument within a month of the bomb's test in 1945 (Szasz 1984). The NPS investigated the possibility and wrote up a proclamation for President Truman, but he took no action. The NPS acted again in 1946 and 1948 and in the early 1950s. A bill was even submitted in Congress to create a Trinity Atomic National Monument, but the idea faded faster than the radiation at the site. A unit devoted to a nuclear explosion is not everybody's idea of what the national park system is all about, but the NPS has since moved into the atomic age with Manhattan Project and Minute Man Missile historic sites.

There are also many places you have probably never heard of that could be worthy of the national park system. One of the most spectacular petroglyph sites in America is California's Little Petroglyph Canyon. It contains twenty thousand inscriptions, some of which may date back ten thousand years, among the oldest human artifacts in North America. You haven't visited it because it is within a navy bombing range (the China Lake Naval Air Weapons Station) and open to the public on only a few weekends of the year. The world's largest petrified forest is in the United States but not in Petrified Forest National Park or any other NPS unit. It is hundreds of feet underground in Illinois, exposed in the workings of coal mines (Maynard 2012). It may extend over one hundred miles and includes an entire forest that was buried by a flood 307 million years ago. It has been explored by scientists, but the Springfield Petrified Forest will never become a tourist attraction. Instead, it is gradually being destroyed by mining operations. By the time you read this, it may already be gone, burned in a power plant to make electricity for the world above.

In 2019 the remains of the slave ship *Clotilda* were discovered in the vast Mobile Delta of southern Alabama (Delgado 2019). In 1860 this became the last slave ship to reach the United States, more than five decades after the importation of slaves had been outlawed. After the slaves were sold the ship was abandoned and burned to destroy the evidence. Descendants of the slaves and the ship's owners live nearby. What will be done with it? One possibility is to preserve it in place as the centerpiece of the National Slave Ship Memorial, the first unit of the park system created about slavery.

There are many other types of places that have not yet been preserved in the park system and may simply be incompatible with it. Among these are highways; while the Blue Ridge and Natchez Trace Parkways are among the most popular park units, they were created specifically as carefully designed driving environments. Other highways have been examined for inclusion in the system as historic sites due to their role in the development of the American highway system. Among these are the National Road, built in the early nineteenth century from Maryland to Illinois; the Lincoln Highway, which became the first transcontinental highway in 1913; and Route 66, which ran from Chicago to Los Angeles between 1926 and 1985.

In 1994 the NPS completed the National Road Special Resource Study (National Park Service 1994), examining the first of these roads. Four alternatives were suggested, one of which was a national historical park focused mainly on the eastern end. In 1995 the Route 66 Special Resource Study (National Park Service 1995) looked at similar possibilities for this famous highway; there were five options, none of which was a traditional park unit. The size of the road and the fact that portions were still in use would have made it very difficult to manage and interpret. In 2004 the Lincoln Highway Special Resource Study was carried out in response to a 2000 law directing the park service to investigate the road (National Park Service 2004). The remnants of the Lincoln Highway were found to lack historic integrity and did not warrant inclusion in the park system.

Over the last several decades, jaguars have been returning to Arizona. This big cat, distinct from mountain lions found throughout the West, was formerly found in the Southwest but wiped out by hunters in the United States. They survived in Mexico's rugged Sierra Madre and have slowly been moving north again. The return of jaguars was unexpected and viewed as a triumph of nature over hunting, but the species is still heavily threatened. The Forest Service is slowly developing a habitat plan for the jaguar but so far has accomplished little, and much of the proposed habitat could be destroyed by a planned open-pit copper mine. When the Antiquities Act was being written in 1906 jaguars still roamed the Southwest and were seen as pests to be exterminated; today a national monument

could protect the habitat the big cats need to survive in the United States. Few Americans have ever seen one; perhaps by the end of the century they will be thriving. The survival and return of these cats, along with other big carnivores, could be one of the biggest conservation stories of the twenty-first century. North America could be a much wilder place than we are accustomed to.

Changing Names and Designations

Many park names will likely have changed by 2100. Some park names will be in different languages. More parks will have Spanish names; perhaps there will be a park with a Vietnamese name in Alabama, a Gullah one in Georgia, a Yiddish name in Milwaukee, or an Icelandic one in North Dakota. But it is very likely that Native American names will become more common. Ahwahne (Yosemite), Tümpisa (Death Valley), Mi tsi a-da-zi (Yellowstone), Bidáá' Ha'azt'i' Tsékooh (Grand Canyon), Nonnezoshi (Rainbow Bridge), Pahokee (Everglades), Tahoma (Mount Rainier), and Tohopeka (Horseshoe Bend National Military Park) are possibilities. These are all Native American names, not recorded in the distant past but used today by those living around or even in those national parks. Many of these names have been in use for hundreds and perhaps thousands of years. Devils Tower has several Indigenous names (Mato Tipila), and this may become common. In 2016 Bears Ears National Monument was created in Utah (though under administration of the BLM). The proclamation named the monument after twin buttes in Utah that resemble the ears of a bear, known in various languages as Hoon'Naqvut, Shash Jaa', Kwiyagatu Nukavachi, and Ansh An Lashokdiwe, all of which mean Bears Ears. The monument has received strong support from the Ute Mountain Ute Tribe, Navajo Nation, Ute Indian Tribe of the Uintah Ouray, Hopi Nation, and Zuni Tribe. This monument's multilingual name may represent a fundamentally new way of naming parks.

Arizona's Tonto National Monument was created to preserve Native American ruins but was named for its location within the Tonto basin (Dallett 2008). Archaeologists refer to the culture that built the ruins as the Salado, after the nearby Salt River. Visitors may associate the site with the Lone Ranger, whose trusty Native American sidekick was named Tonto. In addition to having no connection to the name, *tonto* is Spanish for "fool" (though the name was allegedly from a Native American language). A 2002 general management plan for the monument called for a name change that would remove the unfortunate connection to the Lone Ranger's sidekick and better match the park's purpose. This idea was dropped when local Native American tribes found suggestions such as Salado Cliff Dwellings National Monument no better or more meaningful than Tonto; their tribal history may conflict with histories developed by archaeologists.

Opportunities to add new place-names to many parks are limited; in 1985 the U.S. Board on Geographic Names, which makes decisions about the names shown on maps made by federal agencies, adopted a policy of no longer accepting new place-names within wilderness areas (Julyan 2000). About half of parklands are within wilderness areas, including almost all the land inside Alaskan parks and Death Valley. This ensures that future visitors to these places can experience an area not just without roads but also without names. Yellowstone and the Grand Canyon are among those with no wilderness and could see a huge increase in place-names in the future.

The designations of many park units will also change. While there are many small battlefields, historical parks, and scenic areas located in the eastern United States, these national park units do not receive the same attention or visitation as parks with the national park moniker (Weber and Sultana 2013). Indeed, there is evidence that the national park designation is a very important part of attracting visitors (Weiler 2006). There is great interest in pushing Congress to redesignate monuments, lake shores, and cultural sites as national parks. The four newest national parks in the system—White Sands, Indiana Dunes, Gateway Arch, and New River Gorge—are all examples of this process. Two of these new parks are in midwestern states that didn't formerly have any national parks.

In western states efforts have been made to

have Cedar Breaks and Colorado National Monuments redesignated as national parks (Lofholm 2014). Dinosaur and Organ Pipe Cactus are by far the largest national monuments that have not already been redesignated; it remains to be seen if their turn will come. Such efforts can be contentious, with the opinions of locals split between the benefits the change would bring and potential impacts on livelihoods or freedoms, though these opinions may be ignored by those in a distant city who are responsible for the decision.

One of the biggest developments in public land management has been the growth of national monuments since 1990 and their management by a variety of agencies, a throwback to conditions before 1933. These new monuments have many parallels with the early national monuments. Most have few or no visitor facilities and will likely not have any soon, and as with the earliest national monuments, there is little protection for their resources. The BLM's Agua Fria National Monument in Arizona was protected to preserve archaeological sites, but in the absence of any staff, vandalism continues. In Arizona, home to many popular NPS national monuments, the existence of these BLM national monuments can create confusion; people who have visited Chiricahua, Saguaro, or Organ Pipe Cactus may expect similar facilities and recreational opportunities from the nearby Ironwood Forest and Sonoran Desert National Monuments. They will be disappointed, assuming they can even figure out how to access the monuments. Agua Fria and Montezuma Castle National Monuments both have exits on I-17, but only the second has facilities. Given this background and the proximity of many of the new national monuments to existing park units and the urban populations they draw visitors from, it seems likely that these BLM monuments will eventually be developed for recreation. Whether they will truly then become a second park system or are transferred to the NPS remains to be seen.

Monuments have also become extremely contentious. Since the 1970s these have been created almost entirely by Democratic presidents and opposed strongly by Republicans. The use of the Antiquities Act has always been political and often associated with controversy, but it has become an increasingly partisan issue that now threatens monuments and the act itself. Creating parks legislatively requires (or at least allows for) a greater degree of bipartisanship and public support. In the hyperpartisan twenty-first century, it may be that the use of the Antiquities Act is no longer the best way to permanently protect an area.

Changing Boundaries

Of all aspects of the park system that will change over the coming decades, it is certain that boundaries will be among them. Boundary changes, whether corrections, expansions, or reductions, have been a constant and are generally a good thing. There have been an average of 4.22 national park unit boundary changes per year since 1872. Many changes no doubt will be small improvements or fix minor issues, but some changes provide considerable opportunities to greatly improve their coverage of the geographic feature they are named for. Many units, perhaps most notably the Grand Canyon and Everglades, do not contain nearly all of what they are named for. The creation of these units with modest boundaries was a great victory at the time, and their continued existence in the face of varied threats is another. Perhaps in the coming years their boundaries will be expanded to better allow the NPS to protect these environments.

Park boundaries may look different in the future. The national park concept was based on the existence of the public domain, with both land and parks under the supervision of the Department of the Interior. Parks soon expanded into land administered by other federal agencies, onto private land through donation and purchase, onto communally owned land (on Indian reservations and in American Samoa, with leases), and even into the sea. Where might they expand in the future? Birmingham Civil Rights National Monument includes four-and-a-half city blocks, streets, a park, and several buildings, but only one is owned by the NPS. The new park does not yet have a management plan, and it is not clear if there will be an attempt by the NPS to acquire other properties, which include an active church. This could pro-

vide a precedent of how urban units might be structured in the future, with limited holdings and a boundary that overlies a larger area.

While park boundaries are important, they have no effect on the electromagnetic spectrum—TV, radio, and cell phone signals freely pass through them. You may find this a nuisance or a great convenience, but it is likely that these transmissions and the antennas necessary to support them will proliferate in the future. The 1964 Wilderness Act excludes roads, but of course cell phones and wireless gadgets did not exist then. The one place in America where you can be free of them is the National Radio Quiet Zone, which spans the Virginia and West Virginia state line. Here transmissions are tightly regulated, with cell phones largely prohibited, to protect the Green Bank radio telescope, which is engaged in astronomical research. Perhaps someday this will be a cell-phone-signal-free national park, perhaps the only one remaining where you can hear birds or the wind.

Changing Spatial Organization

How will the internal spatial organization of parks change in the future? The trend among larger units in recent decades has been to minimize tourist development within parks and rely on gateway communities for tourist services and even park facilities. This may partly reflect limited budgets, but it also reflects transportation, population growth around parks, and cultural values. It is easier to reach a vacation destination than ever before, there are far more services available outside parks than a half century ago, and perhaps there is less expectation of services within parks or a stronger desire to experience less developed parks. There may be substantially less demand for these services in the future, and changes in any of these parameters could alter trends in park development.

But time may work to the advantage of parks. Many older parks have become impressive collections of architectural styles ranging from rustic lodges, more modest but still rustic CCC creations, modernist architecture of the 1950s, and new postmodern styles. This will surely continue; in 2100 architectural styles not yet invented will be part of the historic landscape of parks, and their role in preserving

FIG. 12.3   Landscape Arch in Arches National Park. This arch spans 290 feet but will not exist much longer.
Photo by RichieB, 2012. Wikipedia, Creative Commons Attribution–ShareAlike 2.0 Generic License.

the country's built environment will be much greater. Perhaps solid CCC buildings, hand-built with stone, old-growth lumber, and hand-shaped red tile roofs, will survive next to visitor centers built by robots out of synthetic polymers that are carefully camouflaged to be invisible from hiking trails and overlooks.

Changing spatial organization will also reflect natural events. Though we like to think of parks as unchanging, change can come fast to the park system. Glacier Bay National Park includes Lituya Bay, where on July 9, 1958, a massive rockslide produced a tsunami with a 100-foot-tall wave that swept 1,710 feet up a mountain on the other side of the bay. On October 18, 2015, a massive thunderstorm at the north end of Death Valley washed out miles of roads and damaged much of Scotty's Castle, a desert mansion built in the 1930s. It will likely require at least $75 million to repair the damage to buildings, roads, and utilities, far exceeding the park's $9 million annual budget.

Landscape Arch is the longest of its kind in Arches National Park (Fig. 12.3). The extraor-

dinary slender sandstone span is 290 feet long, one of countless wonders in the park. But it is not permanent; tons of rock broke off and fell from it in 1991 and 1995, and someday soon it will collapse entirely. It and many other arches in the park will no longer exist in 2100, though others will come into being as erosion continues its work on the soft sandstone. Trails and perhaps roads will gradually be realigned to reflect the changing scenery of the park. (The names of these future arches will likely be quite different from those of today, reflecting a more diverse cultural base of names.)

Park cultural landscapes will change more rapidly than their physical landscapes, and not just because of how we will change in the future. In 2021 it was announced that the oldest incontrovertible archaeological evidence of humans in North America was discovered in White Sands National Park (Bennett et al. 2021). The National Park Service is now in possession of the oldest archaeological site in the country and in a park where archaeology was not previously a major theme.

Park cultural landscapes may become more diverse in ways we can barely imagine. Prairie dogs are common inhabitants of parks from the Colorado Plateau to the northern plains. At South Dakota's Wind Cave they can be seen standing in their burrows, chattering away. But their barking is more than instinct; it is a language (Slobodchikoff et al. 2012). What do prairie dogs talk about? They're talking about you. Scientists have discovered that they have a vocabulary allowing them to describe different animals and even people, distinguishing a person's height, appearance, shirt color, and behavior. Perhaps someday we will learn their names for parks like Wind Cave. Imagine the tours or fireside talks they could provide!

Changing Park Transport and Spatial Interaction

The rapid space-time convergence brought about by railroads, highways, and airplanes has had enormous consequences for park visitation and development. But space-time convergence has reached a plateau, with vehicles no faster than fifty years ago and no likelihood of change (Warf 2008). Some parks, such as those in American Samoa, have become more remote over the past half century due to changing air networks and few ways to get there. It is unlikely there will be major reductions in travel time to parks, except for Alaska. Here parks may be transformed from remote wilderness areas to become tomorrow's family vacation driving destination. It is only a matter of time before cruise ships start arriving at Cape Krusenstern, with calls for roads and tour buses to take passengers into the interior.

Within parks, the situation may change tremendously. The private automobile has been the core of mobility options within the park system for the past hundred years. The system has allowed a growing middle class the opportunity to experience parks as a family at their convenience. But the system is showing greater strains. Traffic congestion is a growing frustration in the most popular parks, as there are only so many vehicles that can be accommodated without a massive rebuilding program. Instead, the national parks are hoping to accommodate visitors through alternative transportation like buses and shuttles powered by cleaner sources of energy as part of intelligent transportation systems. National park transportation may be quite different in the future, at least at the most-visited units.

While improved transportation has helped fill parks with visitors and created numerous vacation headaches, it also provides a way to vastly expand America's parklands without spending any money. Space-time divergence is normally thought of as a bad thing, an increase in travel time to get somewhere brought about by slower means of travel, reduced frequency of scheduled services, or longer routes. But it could be a good thing for parks and our experience of them. The writer Ed Abbey provided one of the best descriptions of this, as he pointed out that "distance and space are functions of speed and time" (1968, 63). If cars are banned and people must walk everywhere, then the parks will become vastly larger to them; Arches can be experienced in a day or less depending on how much of the park's fifty-two miles of road one drives. On foot, seeing the same scenery would take many days, if not weeks. "We could, if we wanted to, multiply the area of our national parks tenfold or a hundredfold—simply by banning the private automobile" (Abbey 1968, 63). It seems unlikely that Americans will choose this option to expand the park system, but it remains a possibility. It would surely be the greatest ever expansion in the geography of the park system.

Changing Visitation Patterns

Have you ever heard of Carl Sandburg? Few other Americans have, either. His national historic site in North Carolina receives few visitors, and the majority are elderly locals who were already familiar with his work, giving this park unit what may be one of the smallest (and most rapidly shrinking) constituencies to be found in the system. Parks will not exist without visitors, and the days when ever-larger numbers of people will show up no matter what are gone. Visitation has become uncertain, another aspect of the park environment to be carefully monitored and managed. This raises several big questions: Who will visit parks in the future? When will they visit?

Demographic predictions for the United States show clearly that the population will increase greatly in size but also change in composition. By 2065 non-Hispanic whites will make up about 46 percent of the population, with Hispanics, Asians, and African Americans making up 24, 14, and 13 percent, respectively (Cohn 2015). Non-Hispanic middle-class whites have overwhelmingly made up the greatest numbers of park visitors. As other nonwhite groups make up a larger share of the population, will they increasingly visit parks, or will the percent of the population that visits parks shrink? The underrepresentation of any racial or ethnic groups is at best a puzzle but also a concern for the future. Parks are taxpayer-supported facilities open to all and require the support of as many people as possible to survive. If large segments of the American population choose not to visit parks, the future of the system looks bleak. This is one of the most important questions for the future of the national park system.

Regardless of who chooses to visit parks in the future, it is obvious that the diversification efforts made by the Park Service will continue. Interpretation within national park units will continue to embrace and acknowledge the stories of Indigenous peoples' deeper connections with today's national parks. Similarly, while there is nothing that would prohibit Black Americans from visiting any park, the NPS must address the Anglo-American imagination of African American narratives and their place in America when the park system was created. There is also fear that efforts to increase the diversity of park visitors could lead to a backlash or perhaps a concern among white visitors that parks have become nonwhite places. Given that most non-Hispanic white park visitors are middle to high-middle income, this seems unlikely. It can also be expected that parks will be more accessible to people with a disability, although the actions required and their consequences remain to be seen.

People will likely visit parks at different times of the year than today. Seasonal visitation patterns will change at most parks as temperatures continue to rise and extreme heat events become more common. Visitation will grow more rapidly during current nonpeak seasons, allowing a longer season and potentially reducing congestion in summer parks (Fisichelli et al. 2015). The line dividing winter and summer parks will move northward and to higher elevations (Scott, McBoyle, and Schwartzentruber 2004; Hamilton, Maddison, and Tol 2005). The effect on winter parks is less certain; they might see a decrease in future visitation as cool seasons shrink, but the recent experience of Death Valley indicates that summer heat can also attract visitors. The promise of experiencing record temperatures will undoubtedly attract more visitors in coming years. To date, projected impacts to visitation in response to warming temperatures and extreme heat events have only been studied at the scale of whole park units; no research has examined yet how the spatial or temporal patterns of visitation may change within parks.

Seasonal visitation may also change in western parks due to fire. Devastating forest fires will be a much more common occurrence in this region in the future. September and October are peak fire months in California, and with fires becoming more frequent and widespread, park visitation could greatly diminish in late summer. At Yosemite 75 percent of visitors arrive from May to October, with September the fourth busiest month. Even if park facilities and scenery survive, visitors might adapt by becoming more concentrated in cooler, wetter times of the year, increasing congestion. Yosemite may become a spring park, with dwindling numbers of tourists showing up for a look at the valley's cliffs through smoke and haze in the summer and fall.

The federal government's fiscal year ends on September 30, after which the government (and most parks) will shut down if a new budget has not been passed. Unfortunately, this scenario has become more common in recent years and will probably continue. These shutdowns have so far occurred between early October and late January, a time of the year in which park visitors making reservations far in advance must gamble that parks will be open. Families may choose to make expensive vacation plans at more reliable times of the year or just go to Disneyland instead. Shutdowns could combine with the intensifying forest fires to prevent visitation at many western parks during the second half of the year. Parks that see most of their visitors during cooler months may see huge decreases in visitor numbers due to annual shutdowns, perhaps leading to eventual declines in budgets and park staff.

Changing Economic Geographies

Parks operate on money, and future budgets and needs are impossible to predict. We can only hope that parks are better funded and that the maintenance backlog has disappeared by 2100. But there are several possibilities in which the economic geography of the park system may change substantially. Will Alaskan parks remain distinct from the rest of the system in having limited or no facilities, or will they be developed in the same manner as other units? Services in most parks outside Alaska are dominated by giant corporations, a situation the NPS has worked to achieve and about which the public has shown no concern. Will those corporations attempt to exert more influence over park operations? The legal disputes over park names in Yosemite and Hot Springs suggest that these corporations have already begun to see parks as their property. At the same time, many Americans seem comfortable with increasing commercialization of parks. It may be that if corporations take over parks, it is because the public has worked to erase the difference between parks and the rest of the country, not because corporations have claimed them as their own.

When entrance fees became common this led to the discussion of whether park units could become self-sufficient, earning money from entrance and other fees rather than the general treasury (Summers 2005). This is an old idea, one that Stephen Mather favored in 1916. Among the arguments in favor of this are ensuring fairness (only those who use parks pay for them), providing flexibility for park managers (who can raise or lower fees as needed much more quickly than dealing with the annual congressional budget process), ensuring that park managers are aware of the needs and wants of park visitors, reducing congressional meddling in park budgets, and helping to solve the tragedy of the commons. Parks are not yet self-sufficient, but this is a topic that can

be expected to return under the right political conditions.

Privatization is an even bolder topic, one that would sell or lease government facilities to private operators. Beginning in the early 2000s, several states privatized toll roads and port facilities. Could this be extended to parks? It has been discussed on several occasions, though so far the public has no interest in it. A recent variant would be to convert some NPS roads to toll roads (AECOM 2018). Adding toll lanes to the Baltimore–Washington Parkway has been considered as an option for adding capacity, much as with freeways in other parts of the country. Analysis has shown that this and other parkways in the National Capital Region, as well as the Blue Ridge Parkway, could raise large sums of money that would enable the elimination of the maintenance backlog on these roads.

Talk of privatization will likely appear after federal government shutdowns have closed the parks again. States and local governments will likely step in more often to keep parks running, but at some point, people will start asking: why, if state and local governments must constantly operate parks, do we even need national parks at all? Perhaps privately operated parks are the solution to increasing government dysfunction. It may become a hard argument to resist.

Regardless of interest in privatization, it is likely that concession operations in parks will be reduced simply due to the expanding size of large metropolitan areas and the growth of gateway communities. The demise of a concessionaire-operated motel in once-remote Lake Mead due to the growth of Las Vegas is the most dramatic example of this: the motel was considered crucial in the days of a remote undeveloped desert lake but unnecessary on a lake filled with houseboats a short drive from one of the world's largest collection of hotels (National Park Service 2018c). Growing gateway services will make it harder for competing concessions to operate within parks and easier for the NPS to eliminate them. This is yet another example of how parks are not isolated from what goes on outside their boundaries; a park in northern Alaska will have concession needs different from those of a park adjacent to Las Vegas. The national park system is undergoing a long transition as it becomes more urbanized, and many details remain to be worked out.

Changing Threats and the Importance of Geography

Congressional support is obviously important in creating and funding the park system, but in the 1970s many members of Congress saw the creation of new parks in their districts as an important way of directing resources at their constituents (Foresta 1984). This reached its peak (or nadir) in the National Parks and Recreation Act of 1978. The law created or authorized nine new park units, modified the boundaries of twenty-six units, and renamed or redesignated four more. The new units were a mix of small historic sites and battlefields, as well as Santa Monica Mountains National Recreation Area in Los Angeles. The law marked a major expansion of the park system, though one that is still debated today. The act was seen at the time (and ever since) as an example of pork-barrel politics, with individual members of Congress working to get new units established in their districts, regardless of whether the NPS had any interest in these sites.

Perhaps the most definitive example of this issue is Steamtown National Historic Site. Steamtown was a privately owned railroad museum in Scranton, Pennsylvania, facing bankruptcy. A local congressman directed the NPS to investigate it as a park unit. A historic site was duly created at a cost of $66 million, allegedly despite the NPS having no interest in it. In the 1980s it was not uncommon for NPS employees across the nation to criticize the existence of this unit. Debate over the value of the site's railroad collection continues; money spent on the site has increased, while visitation has dwindled.

Geography is important to support for parks. The U.S. population will become even more concentrated in large cities than it is today. The park system will become more urban as well, not only as cities grow to encompass formerly rural parks but also as historic sites and memorials are created within an increasingly urban nation. But while urban recreation areas have so far been a great visitation

success and urban historic sites are crucial to making the system more representative of the American population, this increasing urbanization also raises questions about how urban populations will value rural areas that fewer people will have any direct connection to.

Death Valley is a spectacular park and the largest in the forty-eight contiguous states, but it also contains one of the world's largest deposits of borax, an important if obscure industrial material. Creating the park has not caused this resource to vanish, and it may return in importance in the future as other sources are depleted. Likewise, Glacier Bay contains a major deposit of nickel and once nearly had a massive nickel mine (Catton 1995) before it was stopped by the Mining in the Parks Act of 1976. Nickel is in heavy demand for the manufacture of stainless steel, and consumption is increasing due to its use in high-performance batteries. The act has not eliminated this valuable resource, and any such act may last only until the next session of Congress. A sizable proportion of both Glacier Bay and your cell phone includes nickel. Which would you rather live without?

But the greater threats are those that unfold slowly in plain sight yet are hard to see. The park system, along with the rest of the world, is slowly warming and often drying out. These changes to the system can't yet be fully seen, but it is already evident that the system will change. Parks in 2100 may be in new locations, with parks created around vegetation moving northward and toward higher elevations. Some national seashores and other coastal parks may be distant memories, curiosities consigned to a growing list of former park units. Other coastal parks may sit behind high floodwalls and levees, as along the National Mall and the Statue of Liberty.

What are the most threatened national park units? Discussion of this topic often revolves around Glacier's shrinking glaciers or photogenic animals in a large park such as Yellowstone. But these parks have room to handle changes and strong national, if not global, support; small units with a single localized resource known mainly to locals are the most likely to disappear. What if an entire park disappeared off the map due to a changing climate? It could happen. De

Soto National Memorial is a small beachfront unit on a low-lying peninsula in Tampa Bay. The entire park is less than three feet above sea level in an area with naturally eroding shorelines, and there is no room for expansion (Whisnant and Whisnant 2007). A rise in water level slight enough that you can wade through it will destroy this park. It surely tops the list of parks that will not exist fifty years from now. Lake Meredith National Recreation Area in Texas is another small unit, but with the opposite problem: it has nearly dried up. Visitation has depended on water levels, and the long-term trend for both is downward. With the lake nearly dry, the NPS has had to resort to showing outdoor movies and putting on fireworks shows to attract people to the park (Garcia 2018). Eventually, the NPS will have to figure out what to do with an empty reservoir. The fact that you've probably never heard of either park should be a source of concern; the disappearance of such parks will not be noticed by many. Only when famous parks are threatened will action result.

What will happen to the Colorado Plateau, the heart of the park system? Anyone who has visited any of the plateau's archaeological sites may have already seen the future in the form of the ancient ruins scattered over the plateau. The Anasazi and other farming societies faced a devastating long-term drought 850 years ago, accompanied by increased erosion in stream channels and the gradual disappearance of the pinyon pine they used for lumber and firewood and as a source of food (Stuart 2000). Their societies did not have the resources to survive these challenges, and as a result, they disappeared; the parks our society created around their ruins are monuments to past climate change. We face similar conditions here in the future (Schwinning et al. 2008), but hopefully we have the means to survive and our parks will not become another set of ruins.

A Global View

The issues discussed here facing America's park system naturally apply elsewhere, though very often other countries do not have similar levels of support for their system or are simply in a struggle for survival. Aside from conflict and the threats of rising seas and drought, pop-ulation growth remains a threat to the world's protected areas. Nigeria's population, estimated at around 190 million in 2017, is expected to grow to as much as 800 million by 2100. The pressure this vast increase in a country about twice the size of California (which has only forty million people) will have on the country's neglected protected areas will be immense. Without either the United States' resources or India's patchwork of protected places set within ancient land uses and traditions, it is hard to see how these places will survive. Nigeria will be a country to watch for the future of protected areas and may well be the first country to let its national parks disappear.

National parks in other countries may also greatly change those in the United States. The grand strategy of the national parks has not changed since 1864: set a large place apart from settlement, mining, or other economic use, remove any inhabitants, and develop it for tourism. This strategy has so far withstood changing economic conditions, space-time convergence, and varying political ideologies in the United States. In other countries this approach is much less common and sometimes impossible. This difference has supported a growing challenge to parks in the United States from their original inhabitants, Native Americans. They were removed from many parks but have not disappeared.

The NPS has worked to conceal this story at times, as at Yellowstone, where the NPS embraced early stories told to tourists that Native Americans had never lived in or even visited the park due to fear of the geothermal features. The truth was quite different (Janetski 2002; MacDonald 2018). Native Americans have been coming here for the past eleven thousand years, and twenty-six tribes today consider the park to be within their cultural realm. While Yellowstone National Park was created in advance of many political boundaries, it came after the Crow Indian Reservation, which was partly within the new park. The reservation was reduced in 1880 to exclude it from the park.

While the NPS has grown more culturally aware of these issues over the years and at Death Valley is working with the resident tribe, there are still restrictions in many parks that are frustrating to Native American groups. It is not certain yet how efforts to reclaim or rename these parks will play out over the remainder of the century, but this is one way in which America's park system will surely come to resemble those in other countries.

Conclusions

This book has provided a geographic overview of America's national park system, including its origins, growth, geographic distribution, visitation, transportation, and economic impacts and how it compares to other systems around the world. Concepts such as location, distance, spatial interaction and organization, and cultural landscapes are essential for understanding what the park system is, how it developed, and what it may become in the future. The rel-

ative location of park units within the country has constantly shifted as populations expanded and changing transportation brought parks closer to more people. This changing relative location very often brought new threats to parks or places that were in turn protected by being brought into the system. The approach used to develop parks follows principles of spatial organization that produce a strikingly similar cultural landscape, regardless of how different the scenery may be from one unit to another. Yet these landscapes are constantly changing in response to changing visitation patterns, new technologies, and social demands. The national park system of 2100 will have a different geography, and were we to know that geography, we would learn much about how the country will have changed in the intervening years.

Abbey, Edward. 1968. *Desert Solitaire: A Season in the Wilderness*. New York: Ballantine Books.

Adams, Michael. 2014. "Pukulpa pitjama Ananguku ngurakulu—Welcome to Anangu Land: World Heritage at Uluru-Kata Tjuta National Park." In *World Heritage Sites and Indigenous People's Rights*, edited by Stefan Disko and Helen Tugendhat, 289–311. Copenhagen: IWGIA.

AECOM. 2018. *Preliminary Planning Estimate of Tolling Potential for Selected NPS Roadways*. https://www.pewtrusts.org/-/media/assets/2019/03/181205_pew-tolling-study_final.pdf.

Albert, Richard C. 1987. *Damming the Delaware: The Rise and Fall of Tocks Island Dam*. University Park: Pennsylvania State University Press.

Albright, Horace M., and Robert Cahn. 1985. *The Birth of the National Park Service: The Founding Years, 1913–33*. Salt Lake City: Howe Brothers.

Albright, Horace M., and Marian Albright Schenck. 1999. *Creating the National Park Service: The Missing Years*. Norman: University of Oklahoma Press.

Albuquerque, Fabio, Blas Bonito, Miguel Ángel Macias Rodriguez, and Caitlin Gray. 2018. "Potential Changes in the Distribution of *Carnegiea gigantea* under Future Scenarios." *PeerJ* 6:e5623.

Algeo, Katie. 2013. "Underground Tourists / Tourists Underground: African American Tourism to Mammoth Cave." *Tourism Geographies* 15 (3): 380–404.

Allaback, Sarah. 2000. *Mission 66 Visitor Centers: The History of a Building Type*. Washington, D.C.: National Park Service.

Ambrose, Skip, and Chris Florian. 2008. *Sound Levels and Audibility of Common Sounds in Frontcountry and Transitional Areas in Grand Canyon National Park, 2007–2008*. https://www.nps.gov/grca/learn/nature/upload/GRCAFrontcountryRep20081112.pdf.

Ament, Rob, Anthony P. Clevenger, Olivia Yu, and Amanda Hardy. 2008. "An Assessment of Road Impacts on Wildlife Populations in U.S. National Parks." *Environmental Management* 42 (3): 480–96.

American Civil Liberties Union. 2006. "ACLU Wins Open Access for All Visitors to Martin's Cove National Historic Site in Wyoming." https://www.aclu.org/news/aclu-wins-open-access-all-visitors-martins-cove-national-historic-site-wyoming.

Amtrak. 2019. "Amtrak Schedules & Timetables." https://www.amtrak.com/train-schedules-timetables.

Anderson, Michael F. 2000. *Policing the Jewel: An Administrative History of Grand Canyon National Park*. Grand Canyon: Grand Canyon Association.

Andrus, Cecil D., and John C. Freemuth. 2006. "President Carter's Coup: An Insider's View of the 1978 Alaska Monument Designations." In *The Antiquities Act: A Century of American Archaeology, Historic Preservation, and Nature Conservation*, edited by David Harmon, Francis P. McManamon, and Dwight T. Pitcaithley, 93–105. Tucson: University of Arizona Press.

Appiah-Opoku, Seth. 2004. "Rethinking Ecotourism: The Case of Kakum National Park in Ghana." *African Geographical Review* 23:49–63.

Apt, Alan, and Kay Turnbaugh. 2015. *Afoot and Afield: Denver, Boulder, Fort Collins, and Rocky Mountain National Park: 184 Spectacular Outings in the Colorado Rockies*. Birmingham: Wilderness Press.

Arai, Susan M., and B. Dana Kivel. 2009. "Critical Race Theory and Social Justice Perspectives on Whiteness, Difference(s) and (Anti)Racism: A Fourth Wave of Race Research in Leisure Studies." *Journal of Leisure Research* 41 (4): 459–72.

Arflack, C. G. 2020. "Selling Eden: Economic Development and Conservation in Mount Desert Island and Acadia National Park." PhD diss., University of Memphis.

Ault, Megan Elaine. 2017. "This Name Is Your Name: Public Landmarks, Private Trademarks, and Our National Parks." *Duke Law Journal* 67:145–87.

Baker, Thomas E. 1995. *Redeemed from Oblivion: An Administrative History of Guilford Courthouse National Military Park*. Washington, D.C.: National Park Service.

Barnes, Francis A. 1987. *Canyon Country Arches & Bridges*. Moab: Canyon Country Publications.

Barringer, Mark Daniel. 2002. *Selling Yellowstone: Capitalism and the Construction of Nature*. Lawrence: University Press of Kansas.

Benn, Douglas I., and David J. A. Evans. 2010. *Glaciers and Glaciation*. 2nd ed. New York: Routledge.

Bennett, Matthew R., David Bustos, Jeffrey S. Pigati, Kathleen B. Springer, Thomas M. Urban, Vance T. Holliday, Sally C. Reynolds, Marcin Budka, Jeffrey S. Honke, Adam M. Hudson, Brendan Fenerty, Clare Connelly, Patrick J. Martinez, Vincent L. Santucci, and Daniel Odess. 2021. "Evidence of Humans in North America during the Last Glacial Maximum." *Science* 373 (6562): 1528–31.

Benton, Lisa. 1998. *The Presidio: From Army Post to National Park*. Boston: Northeastern University Press.

Benton-Short, Lisa. 2016. *The National Mall: No Ordinary Space*. Toronto: University of Toronto Press.

Berg, Scott W. 2007. *Grand Avenues: The Story of the French Visionary Who Designed Washington, D.C.* New York: Pantheon.

Berke, Arnold. 2002. *Mary Colter: Architect of the Southwest*. New York: Princeton Architectural Press.

Berkowitz, Paul. 2011. *The Case of the Indian Trader: Billy Malone and the National Park Service Investigation at Hubbell Trading Post*. Albuquerque: University of New Mexico Press.

Biel, Alice Wondrak. 2006. *Do (Not) Feed the Bears: The Fitful History of Wildlife and Tourists in Yellowstone*. Lawrence: University Press of Kansas.

Binkley, Cameron, and Steven A. Davis. 2003. *Preserving the Mystery: An Administrative History of Fort Raleigh National Historic Site*. Washington, D.C.: National Park Service.

Bishop, Bill. 2009. *The Big Sort: Why the Clustering of Like-Minded America Is Tearing Us Apart*. Boston: Mariner Books.

Bishop, M. Guy. 1998. "Mission 66 in the National Parks of Southern California and the Southwest." *Southern California Quarterly* 80 (3): 293–314.

Blackford, Mansel G. 2007. *Pathways to the Present: U.S. Development and Its Consequences in the Pacific*. Honolulu: University of Hawai'i Press.

Bleakley, Geoffrey T. 2002. *Contested Ground: An Administrative History of Wrangell–St. Elias National Park and Preserve, Alaska, 1978–2001*. Washington, D.C.: National Park Service.

Bloch, M., L. Buchanan, J. Katz, and K. Quealy. 2018. "An Extremely Detailed Map of the 2016 Presidential Election." *New York Times*, July 25, 2018. https://www.nytimes.com/interactive/2018/upshot/election-2016-voting-precinct-maps.html.

Bonimy, Madlyn M. 2014. "Tourism in the City of Pigeon Forge, Tennessee: The Engine for Economic Development." https://patimes.org/tourism-city-pigeon-forge-tennessee-engine-economic-development/.

Bonnicksen, Thomas M., and Thomas S. Robinson. 1981. "Constraints on the Development of National Seashores and Lakeshores: A Political Perspective." *Public Administration Review* 41 (5): 550–57.

Borneman, Walter R. 2003. *Alaska: Saga of a Bold Land*. New York: HarperCollins.

Bortle, John E. 2001. "Introducing the Bortle Dark-Sky Scale." *Sky & Telescope* 101 (2): 126–29.

Boyd, Stephen W., and Richard W. Butler. 2009. "Tourism and the Canadian National Park System: Protection, Use and Balance." In *Tourism and National Parks: International Perspectives on Development, Histories, and Change*, edited by Warwick Frost and C. Michael Hall, 102–13. London: Routledge.

Brantley, Max. 2011. "Hot Springs Wins Trademark Fight with Park Service." *Arkansas Times*, May 26, 2011. https://arktimes.com/arkansas-blog/2011/05/26/hot-springs-wins-trademark-fight-with-park-service-update.

Brown, Frederick L. 2011. *The Center of the World, the Edge of the World: A History of Lava Beds National Monument*. Seattle: National Park Service.

Brown, Ronald C., and Duane A. Smith. 2006. *New Deal Days: The CCC at Mesa Verde*. Durango: Durango Herald Small Press.

Brugge, David M., and Raymond Wilson. 1976. *Administrative History: Canyon de Chelly National Monument, Arizona*. Washington, D.C.: National Park Service.

Buckley, Lauren B., and Madison S. Foushee. 2012. "Footprints of Climate Change in US National Park Visitation." *International Journal of Biometeorology* 56:1173–77.

Bullock, S. D., and S. R. Lawson. 2008. "Managing the 'Commons' on Cadillac Mountain: A Stated Choice Analysis of Acadia National Park Visitors' Preferences." *Leisure Sciences* 30 (1): 71–86.

Bureau of Land Management. 2020. "Mining Claims." https://www.blm.gov/programs/energy-and-minerals/mining-and-minerals/locatable-minerals/mining-claims.

Burghardt, John E., Elizabeth S. Norby, and Harold S. Pranger II. 2013. *Interim Inventory and Assessment of Abandoned Mineral Lands in the National Park System*. Natural Resource Technical Report NPS/NRSS/GRD/NRTR-2013/659. National Park Service. https://permanent.fdlp.gov/gpo35893/NPS_AML-IandA_508_Compliant-2013-0308-FINAL.pdf.

Burnham, Philip. 2000. *Indian Country, God's Country: Native Americans and the National Parks*. Washington, D.C.: Island Press.

Burt, Sharelle. 2018. *Travel Noire. Black American travelers spent $63 billion on tourism in 2018*. Retrieved from https://travelnoire.com/african-americans-tourism-spent-close-to-63-billion.

Burtner, Marcus. 2011. *Crowning the Queen of the Sonoran Desert: Tucson and Saguaro National Park; An Administrative History*. http://npshistory.com/publications/sagu/adhi.pdf.

Butler, William B. 2007. *Railroads in the National Parks*. https://www.nps.gov/parkhistory/online_books/nps/railroads.pdf.

Cahalan, James M. 2001. *Edward Abbey: A Life*. Tucson: University of Arizona Press.

Cannon, Kelly June. 1997. *Administrative History: San*

Juan Island National Historical Park. Seattle: National Park Service.

Carlin, George. 1996. *Back in Town*. Atlantic/WEA.

Carr, Deborah S., and Daniel R. Williams. 1993. "Understanding the Role of Ethnicity in Outdoor Recreation Experiences." *Journal of Leisure Research* 25 (1): 22–38.

Carr, Ethan. 1998. *Wilderness by Design: Landscape Architecture and the National Park Service*. Lincoln: University of Nebraska Press.

———. 2007. *Mission 66: Modernism and the National Park Dilemma*. Amherst: University of Massachusetts Press.

Carter, Perry L. 2008. "Coloured Places and Pigmented Holidays: Racialized Leisure Travel." *Tourism Geographies* 10 (3): 265–84.

Catton, Theodore. 1995. *Land Reborn: A History of Administration and Visitor Use in Glacier Bay National Park and Preserve*. Anchorage, Alas.: National Park Service.

———. 2006. *National Park, City Playground: Mount Rainier in the Twentieth Century*. Seattle: University of Washington Press.

———. 2010. *A Fragile Beauty: An Administrative History of Kenai Fjords National Park*. Seward, Alas.: National Park Service.

Charlier, Bruno, and Nicholas Bourgeois. 2013. "'Half the Park Is after Dark': Dark Sky Parks and Reserves; New Concepts and Tools to Grant Natural Heritage Status." *L'Espace géographique* 42 (3): 186–98.

Chavez, K. 2020. "Great Smoky Mountains National Park Hits Record Number of Visitors, Boosts Local Economy." *Citizen Times*, February 13, 2020. https://www.citizen-times.com/story/news/2020/02/13/great-smoky-mountains-national-park-asheville-economy/4742672002/.

Chester, Charles C. 2006. *Conservation across Borders: Biodiversity in an Interdependent World*. Washington, D.C.: Island Press.

Cinzano, P., F. Falchi, and C. D. Elvidge. 2001. "The First World Atlas of the Artificial Night Sky Brightness." *Monthly Notices of the Royal Astronomical Society* 328:689–707.

Clawson, Marion, and Burnell Held. 1957. *The Federal Lands: Their Use and Management*. Baltimore, Md.: Johns Hopkins University Press.

Clemensen, A. Berle. 1987. *Cattle, Copper, and Cactus: The History of Saguaro National Monument*. Denver: National Park Service.

———. 1992. *Casa Grande Ruins National Monument, Arizona: A Centennial History of the First Prehistoric Reserve 1895–1992*. Denver: National Park Service.

Cockell, Charles, and Gerda Horneck. 2004. "A Planetary Park System for Mars." *Space Policy* 20:291–95.

Cockrell, Ron. 1999. *Amidst Ancient Monuments: The Administrative History of Mound City Group National Monument / Hopewell Culture National Historical Park, Ohio*. Omaha, Neb.: National Park Service.

Cohn, D'Vera. 2015. "Future Immigration Will Change the Face of America by 2065." https://www.pewresearch.org/fact-tank/2015/10/05/future-immigration-will-change-the-face-of-america-by-2065/.

Colby, Sandra L., and Jennifer M. Ortman. 2015. *Projections of the Size and Composition of the U.S. Population: 2014 to 2060*. https://www.census.gov/content/dam/Census/library/publications/2015/demo/p25-1143.pdf.

Cole, Kenneth L., Kirsten Ironside, Jon Eischeid, Gregg Garfin, Phillip B. Duffy, and Chris Toney. 2011. "Past and Ongoing Shifts in Joshua Tree Distribution Support Future Modeled Range Contraction." *Ecological Applications* 21:137–49.

Cole, Terrence M. 1992. "Placenames in Paradise: Robert Marshall and the Naming of the Alaska Wilderness." *Names* 40:99–116.

Comay, Laura B. 2017. *The National Park Service's Maintenance Backlog: Frequently Asked Questions*. https://fas.org/sgp/crs/misc/R44924.pdf.

———. 2022. "National Park System: What Do the Different Park Titles Signify?" CRS Report R41816. https://crsreports.congress.gov/product/pdf/R/R41816#:~:text=Most%20designations%20give%20a%20sense,or%20recreational%20(e.g.%2C%20national%20recreation.

Correa-Cano, Maria Eugenia, Barbara Goettsch, James P. Duffy, Jonathan Bennie, Richard Inger, and Kevin J. Gaston. 2018. "Erosion of Natural Darkness in the Geographic Ranges of Cacti." *Scientific Reports* 8:4347.

Cozine, James J. 2004. *Saving Big Thicket: From Exploration to Preservation, 1685–2003*. Denton: University of North Texas Press.

Craven, Scott. 2019. "Heading to Horseshoe Bend? For the First Time, It's Going to Cost You." *AZCentral*, April 19, 2019. https://www.azcentral.com/story/travel/arizona/2019/04/19/horseshoe-bend-new-parking-fees/3523963002/.

Crowell, John. 2020. "Youth Visitor Trends in Four Pacific Coast U.S. National Parks." *Pacifica: Newsletter of the Association of Pacific Coast Geographers*, Spring–Summer, 1, 7–9.

Culpin, Mary Shivers. 1994. *The History of the Construction of the Road System in Yellowstone National Park, 1872–1966*. Washington, D.C.: National Park Service.

Daigle, John J. 2008. "Transportation Research Trends in National Parks: A Summary and Exploration of Future Trends." *Parks Stewardship Forum* 25 (1). 57–64.

Dallett, Nancy L. 2008. *At the Confluence of Change: A History of Tonto National Monument*. Tucson: Western National Parks Association.

Danner, Lauren. 2017. *Crown Jewel Wilderness: Creating North Cascades National Park*. Pullman: Washington State University Press.

Danno, Robert W. 2012. *Worth Fighting For: A Park*

Ranger's Unexpected Battle against Federal Bureaucrats and Washington Redskins Owner Dan Snyder. Shepardstown, W.Va.: Honor Code Publishing.

Davis, Cory R., and Andrew J. Hansen. 2011. "Trajectories in Land Use Change around U.S. National Parks and Opportunities for Management." *Ecological Applications* 21:3299–3316.

Davis, Janae. 2015. "A Tale of Two Landscapes: Examining Alienation and Non-visitation among Local African American Fishers at Congaree National Park." MS thesis, University of South Carolina.

Davis, Jule Hirschfeld. 2015. "Mount McKinley Will Again Be Called Denali." *New York Times*, August 30, 2015. https://www.nytimes.com/2015/08/31/us/mount-mckinley-will-be-renamed-denali.html.

Davis, Ren, and Helen Davis. 2011. *Our Mark on This Land: A Guide to the Legacy of the Civilian Conservation Corps in America's Parks.* Granville, Ohio: McDonald and Woodward.

Davis, Timothy. 2016. *National Park Roads: A Legacy in the American Landscape.* Charlottesville: University of Virginia Press.

Davis, Timothy, Todd A. Croteau, and Christopher H. Marston, eds. 2004. *America's National Park Roads and Parkways: Drawings from the Historic American Engineering Record.* Baltimore, Md.: Johns Hopkins University Press.

Delgado, James P. 2019. *Archaeological Investigations of 1Ba704, a Nineteenth Century Shipwreck Site in the Mobile River, Baldwin and Mobile Counties, Alabama.* https://ahc.alabama.gov/press/FINAL_1Ba704%20Report_SEARCH_redacted.pdf.

Department of the Interior. 2021. "Payments in Lieu of Taxes." https://www.doi.gov/pilt.

DeVoto, Bernard. 1953. "Let's Close the National Parks." *Harper's* 207 (October): 49–52.

Dilsaver, Lary M. 1994a. *America's National Park System: The Critical Documents.* Lanham, Md.: Rowman & Littlefield.

———. 1994b. "Preservation Choices at Muir Woods." *Geographical Review* 84:290–305.

———. 2004. *Cumberland Island National Seashore: A History of Conservation Conflict.* Charlottesville: University of Virginia Press.

———. 2008. "Not of National Significance: Failed National Park Proposals in California." *California History* 85 (2): 4–23.

———. 2017. *Preserving the Desert: A History of Joshua Tree National Park.* Staunton, Va.: George F. Thompson Publishing.

Dilsaver, Lary M., and William Wyckoff. 2005. "The Political Geography of National Parks." *Pacific Historical Review* 74:237–66.

Dodd, Douglas W. 2007. "Boulder Dam Recreation Area: The Bureau of Reclamation, the National Park Service, and the Origins of the National Recreation Area Concept at Lake Mead, 1929–1936." *Southern California Quarterly* 88:431–73.

Doolittle, William E. 2000. *Cultivated Landscapes of Native North America.* Oxford: Oxford University Press.

Dudley, Nigel, ed. 2008. *Guidelines for Applying Protected Area Management Categories.* https://portals.iucn.org/library/sites/library/files/documents/pag-021.pdf.

Duncan, Dayton, and Ken Burns. 2009. *The National Parks: America's Best Idea.* New York: Alfred A. Knopf.

Dunlap, David W. 2008. "Witnessing a House, and History, on the Move." *New York Times*, June 8, 2008. https://www.nytimes.com/2008/06/08/nyregion/08grange.html.

Dutton, Clarence. 1882. *Tertiary History of the Grand Canyon District.* Washington, D.C.: Government Printing Office.

Eck, Ronald W., and Eugene M. Wilson. 2001. *Transportation Needs of National Parks and Public Lands.* http://onlinepubs.trb.org/onlinepubs/millennium/00128.pdf.

Edds, Kimberly. 2004. "At Grand Canyon Park, a Rift over Creationist Book." *Washington Post*, January 20, 2004. https://www.washingtonpost.com/archive/politics/2004/01/20/at-grand-canyon-park-a-rift-over-creationist-book/362795ed-adf0-4ce9-857e-7852a394f973/?noredirect=on&tm_term=.6b8cde24a115.

Einberger, Scott Raymond. 2018. *With Distance in His Eyes: The Environmental Life and Legacy of Stewart Udall.* Reno: University of Nevada Press.

Eliade, Mircea. 1987. *The Sacred and the Profane: The Nature of Religion.* San Diego: Harcourt Brace Jovanovich.

Elliott, Gary E. 1994. *Senator Alan Bible and the Politics of the New West.* Reno: University of Nevada Press.

Erickson, Beth, Corey W. Johnson, and B. Dana Kivel. 2009. "Rocky Mountain National Park: History and Culture as Factors in African-American Park Visitation." *Journal of Leisure Research* 41 (4): 529–45.

Evans-Hatch and Associates. 2004. *War in the Pacific National Historical Park: An Administrative History.* Washington, D.C.: National Park Service.

Ewert, A. W. 1998. "A Comparison of Urban-Proximate and Urban-Distant Wilderness Users on Selected Variables." *Environmental Management* 22 (6): 927–35.

Federal Aviation Administration. 2019. *Air Tour Management Plan.* https://www.faa.gov/about/office_org/headquarters_offices/arc/programs/air_tour_management_plan/.

Federal Highway Administration. 2008. *Status of the Nation's Highways, Bridges, and Transit: Conditions and Performance.* https://www.transit.dot.gov/research-innovation/status-nations-highways-bridges-and-transit-condition-and-performance.

Fenneman, Nevin M. 1916. "Physiographic Divisions of the United States." *Annals of the Association of American Geographers* 6:19–98.

Fenster, Julie M. 2005. *Race of the Century: The Heroic True Story of the 1908 New York to Paris Auto Race.* New York: Broadway Books.

Finney, Carolyn. 2014. *Black Faces, White Spaces: Reimagining the Relationship of African Americans to the Great Outdoors.* Chapel Hill: University of North Carolina Press.

Fisichelli, Nicholas A., Gregor W. Schuurman, William B. Monahan, and Pamela S. Ziesler. 2015. "Protected Area Tourism in a Changing Climate: Will Visitation at US National Parks Warm Up or Overheat?" *PLOS One* 10 (6): e0128226.

Fletchall, A. 2013. "Making Sense of the Strip: The Postmodern Pastiche of Pigeon Forge, Tennessee." *Southeastern Geographer* 53 (1): 102–22.

Floyd, Myron F. 1999. "Race, Ethnicity and Use of the National Park System." *Social Science Research Review* 1 (2): 1–24.

———. 2001. "Managing National Parks in a Multicultural Society: Searching for Common Ground." *George Wright Forum* 18 (3): 41–51.

Floyd, Myron F., and Monika Stodolska. 2014. "Theoretical Frameworks in Leisure Research on Race and Ethnicity." In *Race, Ethnicity and Leisure*, edited by Monika Stodolska, Kim J. Shinew, Myron F. Floyd, and Gordon J. Walker, 7–20. Champaign, Ill.: Human Kinetics.

Ford, Caroline. 2012. "Imperial Preservation and Landscape Reclamation: National Parks and Natural Reserves in French Colonial Africa." In *Civilizing Nature: National Parks in Global Historical Perspective*, edited by Bernhard Gissibl, Sabine Hohler, and Patrick Kupper, 68–83. New York: Berghahn Books.

Foresta, Ronald A. 1984. *America's National Parks and Their Keepers.* Washington, D.C.: Resources for the Future.

———. 2013. *The Land between the Lakes: A Geography of the Forgotten Future.* Knoxville: University of Tennessee Press.

Foster, Mark S. 1999. "In the Face of 'Jim Crow': Prosperous Blacks and Vacations, Travel and Outdoor Leisure, 1890–1945." *Journal of Negro History* 84 (2): 130–49.

Frost, Warwick, and C. Michael Hall, eds. 2009. *Tourism and National Parks: International Perspectives on Development, Histories, and Change.* London: Routledge.

Garcia, Vanessa. 2018. "Meredith's Future on the Rise: Last Year's Lake Visitation Was Highest Since 2000." *Amarillo Globe-News*, January 29, 2018. https://www.amarillo.com/local-news/news/2018-01-19/meredith-s-future-rise-last-year-s-lake-visitation-was-highest-2000.

Geary, Michael M. 2016. *Sea of Sand: A History of Great Sand Dunes National Park and Preserve.* Norman: University of Oklahoma Press.

Gissibl, Bernhard. 2012. "A Bavarian Serengeti: Space, Race and Time in the Entangled History of Nature Conservation in East Africa and Germany." In *Civilizing Nature: National Parks in Global Historical Perspective*, edited by Bernhard Gissibl, Sabine Hohler, and Patrick Kupper, 102–20. New York: Berghahn Books.

Gomez, Phillip J. 1984. "Fort Bowie National Historic Site: The Evolution of a Unique Western Park." *Journal of Arizona History* 25 (2): 171–90.

Gonzales, Jackie M. M. 2017. "The National Park Service Goes to the Beach." *Forest History Today* 23 (Spring): 19–27.

Good, Albert H. (1938) 1999. *Park and Recreation Structures.* New York: Princeton Architectural Press.

Gorman, Alice. 2005. "The Cultural Landscape of Interplanetary Space." *Journal of Social Archaeology* 5:85–107.

Grattan, Virginia L. 1992. *Mary Colter: Builder upon the Red Earth.* Grand Canyon: Grand Canyon Natural History Association.

Grazier, Dan. 2020. "Selective Arithmetic to Hide the F-35's True Costs." POGO: Project on Government Oversight. October 21. https://www.pogo.org/analysis/2020/10/selective-arithmetic-to-hide-the-f-35s-true-costs/.

Guam International Airport Authority. 2019. *A. P. Won Pat International Airport Guam 2019 Annual Report.* https://www.guamairport.com/corporate/reports/annual-report.

Guyton, Bill. 1998. *Glaciers of California.* Berkeley: University of California Press.

Haines, Aubrey L. 1996. *The Yellowstone Story: A History of Our First National Park.* Rev. ed. Boulder: University Press of Colorado.

Hall, M., A. W. Al-Khulaidi, A. G. Miller, P. Scholte, and A. H. Al-Qadasi. 2008. "Arabia's Last Forests under Threat: Plant Biodiversity and Conservation in the Valley Forest of Jabal Bura (Yemen)." *Edinburgh Journal of Botany* 65 (1): 113–35.

Hamilton, Jacqueline M., David J. Maddison, and Richard S. J. Tol. 2005. "Effects of Climate Change on International Tourism." *Climate Research* 29 (3): 245–54.

Hampton, Duane. 1971. *How the U.S. Cavalry Saved Our National Parks.* Bloomington: Indiana University Press.

———. 1981. "Opposition to National Parks." *Journal of Forest History* 25 (1): 36–45.

Hamstead, Zoe A., David Fisher, Rositsa T. Ilieva, Spencer A. Wood, Timon McPhearson, and Peleg Kremer. 2018. "Geolocated Social Media as a Rapid Indicator of Park Visitation and Equitable Park Access." *Computers, Environment and Urban Systems* 72:38–50.

Hardin, Garrett. 1968. "The Tragedy of the Commons." *Science* 162 (3859): 1243–48.

Harmon, David, Francis P. McManamon, and Dwight T. Pitcaithley, eds. 2006. *The Antiquities Act: A Century of American Archaeology, Historic Preservation, and Nature Conservation*. Tucson: University of Arizona Press.

Harmon, Rick. 2002. *Crater Lake National Park: A History*. Corvallis: Oregon State University Press.

Harper, Melissa, and Richard White. 2012. "How National Were the First National Parks? Comparative Perspectives from the British Settler Societies." In *Civilizing Nature: National Parks in Global Historical Perspective*, edited by Bernhard Gissibl, Sabine Hohler, and Patrick Kupper, 50–67. New York: Berghahn Books.

Harvey, Bruce G., and Deborah Harvey. 2017. *The Ordinary Home of an Extraordinary Man: Administrative History of Harry S Truman National Historic Site*. Omaha, Neb.: National Park Service.

Harvey, Mark W. T. 1994. *A Symbol of Wilderness: Echo Park and the American Conservation Movement*. Seattle: University of Washington Press.

Hartzog, George B., Jr. 1988. *Battling for the National Parks*. Mount Kisco, N.Y.: Moyer Bell Limited.

Hassell, Hank. 1999. *Rainbow Bridge: An Illustrated History*. Logan: Utah State University Press.

Heacox, Kim. 2001. *An American Idea: The Making of the National Parks*. Washington, D.C.: National Geographic Society.

Hebert, Keith S., and Kathryn H. Braund. 2016. *Horseshoe Bend National Military Park Administrative History*. Washington, D.C.: National Park Service.

Hein, Annette. 2014. "The Martin's Cove Controversy." WyoHistory.org: A Project of the Wyoming Historical Society. November 8. https://www.wyohistory.org/encyclopedia/martins-cove-controversy.

Hemmat, Steven A. 1986. "Parks, People and Private Property: The National Park Service and Eminent Domain." *Environmental Law* 16 (4): 935–61.

Hess, Alan. 2004a. *Googie Redux: Ultramodern Roadside Architecture*. San Francisco: Chronicle Books.

———. 2004b. *Ranch House*. New York: Harry N. Abrams, Inc.

Hine, Thomas. 1990. *Populuxe*. New York: Alfred A. Knopf.

Hodges, Tina, and Scott Faulk. 2007. *Alternative Transportation in Parks and Public Lands Program Manual*. https://www.transit.dot.gov/sites/fta.dot.gov/files/docs/ATPPL_Manual_1-9-07.pdf.

Hogenauer, Alan K. 1991a. "Gone, but Not Forgotten: The Delisted Units of the U.S. National Park System." *George Wright Forum* 7:2–19.

———. 1991b. "An Update to 'Gone, but Not Forgotten: The Delisted Units of the U.S. National Park System.'" *George Wright Forum* 8:26–28.

Hubbard, Bill. 2009. *American Boundaries: The Nation, the States, and the Rectangular Survey*. Chicago: University of Chicago Press.

Hunt, Charles B. 1967. *Physiography of the United States*. San Francisco: W. H. Freeman and Company.

———. 1975. *Death Valley: Geology, Ecology, Archaeology*. Berkeley: University of California Press.

Hurley, Andrew. 1995. *Environmental Inequalities: Class, Race, and Industrial Pollution in Gary, Indiana, 1945–1980*. Chapel Hill: University of North Carolina Press.

International Association of Antarctica Tour Operators. 2020. https://iaato.org/home.

International Dark-Sky Association. 2018. "International Dark Sky Parks." http://darksky.org/idsp/parks/.

Ise, John. 1961. *Our National Park Policy: A Critical History*. Baltimore, Md.: Johns Hopkins University Press.

IUCN (International Union for Conservation of Nature). 2011. "Global Transboundary Conservation Network." https://www.tbpa.net/page.php?ndx=6.

Jackson, Frances. 1972. "Military Uses of Haleakala National Park." https://evols.library.manoa.hawaii.edu/server/api/core/bitstreams/474cd6cf-2925-4be1-8b85-a9ee13a9a308/content.

Jacobs, Jeremy P., and Rob Hotakainen. 2020. "Racist Roots, Lack of Diversity Haunt National Parks." *E&E News*, June 25, 2020. https://www.eenews.net/stories/1063447583.

Jakle, John A., Keith A. Sculle, and Jefferson S. Rogers. 1996. *The Motel in America*. Baltimore, Md.: Johns Hopkins University Press.

Jameson, John. 1996. *The Story of Big Bend National Park*. Austin: University of Texas Press.

Janetski, Joel C. 2002. *Indians in Yellowstone National Park*. Rev. ed. Salt Lake City: University of Utah Press.

Johnson, Cassandra Y., J. Bowker, Gary Green, and H. Cordell. 2007. "'Provide It . . . but Will They Come?': A Look at African American and Hispanic Visits to Federal Recreation Areas." *Journal of Forestry* 105 (5): 257–65.

Johnson, Christopher E. 2013. *Nature and History on the Sierra Crest: Devils Postpile and the Mammoth Lakes Sierra*. Seattle: National Park Service.

Jones, Karen. 2012. "Unpacking Yellowstone: The American National Park in Global Perspective." In *Civilizing Nature: National Parks in Global Historical Perspective*, edited by Bernhard Gissibl, Sabine Hohler, and Patrick Kupper, 31–49. New York: Berghahn Books.

Julyan, Robert. 2000. "Protecting the Endangered Blank Spots on Maps: The Wilderness Names Policy of the United States Board on Geographic Names." *Names* 48 (3–4): 217–28.

Kack, David, and Jaydeep Chaudhari. 2009. *Grand Teton National Park Public Transit Business Plan*. Western Transportation Institute, College of Engineering, Montana State University, Bozeman. https://www.nps.gov/grte/learn/management/upload/public-transit-business-plan.pdf.

Kaiser, Harvey H. 1997. *Landmarks in the Landscape: Historic Architecture in the National Parks of the West*. San Francisco: Chronicle Books.

———. 2002. *An Architectural Guidebook to the National Parks: California, Oregon, Washington*. Salt Lake City: Gibbs Smith.

———. 2003. *An Architectural Guidebook to the National*

Parks: Arizona, New Mexico, Texas. Salt Lake City: Gibbs Smith.

Kalt, Brian C. 2005. "The Perfect Crime." *Georgetown Law Journal* 93 (2): 675–88.

Karnow, Stanley. 1990. *In Our Image: America's Empire in the Philippines*. New York: Ballantine Books.

Kathirithamby-Wells, Jeyamalar. 2012. "From Colonial Imposition to National Icon: Malaysia's Taman Negara National Park." In *Civilizing Nature: National Parks in Global Historical Perspective*, edited by Bernhard Gissibl, Sabine Hohler, and Patrick Kupper, 84–101. New York: Berghahn Books.

Keiter, Robert B. 2013. *To Conserve Unimpaired: The Evolution of the National Park Idea*. Washington, D.C.: Island Press.

Keller, Robert H., and Michael F. Turek. 1998. *American Indians and National Parks*. Tucson: University of Arizona Press.

Kidd, Christopher. 2014. "Bwindi Impenetrable National Park: The Case of the Batwa." In *World Heritage Sites and Indigenous People's Rights*, edited by Stefan Disko and Helen Tugendhat, 147–62. Copenhagen: IWGIA.

Kirkconnell, Barbara. 1988. *Catoctin Mountain Park: An Administrative History*. Washington, D.C.: National Park Service.

Kloog, Itai, Richard G. Stevens, Abraham Haim, and Boris A. Portnov. 2010. "Nighttime Light Level Codistributes with Breast Cancer Incidence Worldwide." *Cancer Causes Control* 21:2059–68.

Kraft, Susan, and Gordon Chappell. 1999. "Historic Railroads in the National Park System and Beyond." *CRM: Cultural Resource Management* 22 (10): 4–5.

Krahe, Diane L., and Theodore Catton. 2014. *Walking in Credence: An Administrative History of George Washington Carver National Monument*. Washington, D.C.: National Park Service.

Krutch, Joseph Wood. 1952. *The Desert Year*. New York: William Sloane Associates.

———. 1955. *The Voice of the Desert: A Naturalist's Interpretation*. New York: William Sloane Associates.

———. 1958. *Grand Canyon: Today and All Its Yesterdays*. New York: William Sloane Associates.

———. 1961. *The Forgotten Peninsula: A Naturalist in Baja California*. New York: William Sloane Associates.

Kupfer, John A., Zhenlong Li, Huan Ning, and Xiao Huang. 2021. "Using Mobile Device Data to Track the Effects of the COVID-19 Pandemic on Spatiotemporal Patterns of National Park Visitation." *Sustainability* 13 (16): 9366.

Kupper, Patrick. 2012. "Translating Yellowstone: Early European National Parks, Weltnaturschutz and the Swiss Model." In *Civilizing Nature: National Parks in Global Historical Perspective*, edited by Bernhard Gissibl, Sabine Hohler, and Patrick Kupper, 123–39. New York: Berghahn Books.

Kuznia, Rob. 2012. "LMU Professor Has Traveled the World from A to Z." *Daily Breeze*, July 12, 2002.

https://www.dailybreeze.com/2012/07/12/lmu-professor-has-traveled-the-world-from-a-to-z/.

Landres, Peter B., Richard L. Knight, Steward T. A. Pickett, and M. L. Cadenasso. 1998. "Ecological Effects of Administrative Boundaries." In *Stewardship across Boundaries*, edited by Richard L. Knight and Peter B. Landres, 39–64. Washington, D.C.: Island Press.

Lawrence, David. 2000. *Kakadu: The Making of a National Park*. Melbourne: Melbourne University Press.

Lehmann, Susan. 1987. *An Embarrassment of Riches: The Administrative History of Cabrillo National Monument*. Washington, D.C.: National Park Service.

Lewis, Emanuel Raymond. 1970. *Seacoast Fortifications of the United States: An Introductory History*. Annapolis, Md.: Leeward Publications.

Lewis, Michael L. 2004. *Inventing Global Ecology: Tracking the Biodiversity Ideal in India, 1947–1997*. Athens: Ohio University Press.

Lingenfelter, Richard E. 1988. *Death Valley and the Amargosa: A Land of Illusion*. Berkeley: University of California Press.

Lissoway, Brenna Lauren. 2004. "An Administrative History of Organ Pipe Cactus National Monument: The First Thirty Years, 1937–1967." MA thesis, Arizona State University.

Lodge, Thomas E. 2016. *The Everglades Handbook: Understanding the Ecosystem*. 4th ed. Boca Raton, Fla.: CRC Press.

Lofholm, Nancy. 2014. "Colorado National Monument in New Effort to Become National Park." *Denver Post*, April 1, 2014. https://www.denverpost.com/2014/04/01/colorado-national-monument-in-new-effort-to-become-national-park/.

Logan, John R., Richard D. Alba, and Wenquan Q. Zhang. 2002. "Immigrant Enclaves and Ethnic Communities in New York and Los Angeles." *American Sociological Review* 67 (2): 299–422.

Louter, David. 1998. *Contested Terrain: North Cascades National Park Service Complex, Washington, an Administrative History*. Seattle: National Park Service.

———. 2006. *Windshield Wilderness: Cars, Roads, and Nature in Washington's National Parks*. Seattle: University of Washington Press.

Lubick, George M. 1996. *Petrified Forest National Park: A Wilderness Bound in Time*. Tucson: University of Arizona Press.

Maccone, Claudio. 2010. "PAC: Protected Antipode Circle at the Center of the Farside of the Moon for the Benefit of All Humankind." In *Lunar Settlements*, edited by Haym Benaroya, 291–303. Boca Raton, Fla.: CRC Press.

MacDonald, Douglas H. 2018. *Before Yellowstone: Native American Archaeology in the National Park*. Seattle: University of Washington Press.

Mackintosh, Barry. 1983. *Wolf Trap Farm: An Administrative History*. Washington, D.C.: National Park Service.

———. 1985. *The Historic Sites Survey and National Historic Landmarks Program: A History*. Washington, D.C.: National Park Service.

———. 1991. *The National Parks: Shaping the System*. Washington, D.C.: National Park Service.

———. 2005. *The National Parks: Shaping the System*. Rev. ed. Washington, D.C.: National Park Service.

Mackintosh, Barry, Janet A. McDonnell, and John H. Sprinkle Jr. 2018. *The National Parks: Shaping the System*. 4th ed. Special issue of the *George Wright Forum* 35 (2): 2–132.

Marcus, W. Andrew, James E. Meacham, Ann W. Rodman, and Alethea Y. Steingisser. 2012. *Atlas of Yellowstone*. Berkeley: University of California Press.

Mares, Michael A., ed. 1999. *Encyclopedia of Deserts*. Norman: University of Oklahoma Press.

Martin, C. B. 2007. *Tourism in the Mountain South: A Double-Edged Sword*. Knoxville: University of Tennessee Press.

Maynard, W. Barksdale. 2012. "An Underground Fossil Forest Offers Clues on Climate Change." *New York Times*, April 30, 2012. https://www.nytimes.com/2012/05/01/science/underground-fossil-forest-in-illinois-offers-clues-on-climate-change.html.

McCally, David. 1999. *The Everglades: An Environmental History*. Gainesville: University Press of Florida.

McClelland, Linda Flint. 1998. *Building the National Parks: Historic Landscape Design and Construction*. Baltimore, Md.: Johns Hopkins University Press.

McConnell, Curt. 2000. *Coast to Coast by Automobile: The Pioneering Trips, 1899–1908*. Palo Alto, Calif.: Stanford University Press.

McDonnell, Janet A. 2015. "'Far-Reaching Effects': The United States Military and the National Parks during World War II." *George Wright Forum* 32 (1): 89–110.

McGivney, Annette. 2019. "'Yanked from the Ground': Cactus Theft Is Ravaging the American Desert." *The Guardian*, February 20, 2019. https://www.theguardian.com/environment/2019/feb/20/to-catch-a-cactus-thief-national-parks-fight-a-thorny-problem.

McKoy, Kathleen L. 2000. *Cultures at a Crossroads: An Administrative History of Pipe Spring National Monument*. Denver: National Park Service.

Meadors, J. Faith. 2003. *Fort Pulaski National Monument Administrative History*. Washington, D.C.: National Park Service.

Melley, Brian. 2019. "US Fighter Jet Crashes in Death Valley, 7 Park Visitors Hurt." *AP News*, July 31, 2019. https://www.apnews.com/472730134b914427b08062eddo8ea68d.

Mengak, Kathy. 2012. *Reshaping Our National Parks and Their Guardians: The Legacy of George B. Hartzog, Jr.* Albuquerque: University of New Mexico Press.

Mojave Desert Land Trust. 2019. "Land Purchases Help Piece Together Death Valley National Park." https://www.mdlt.org/land-purchases-help-piece-together-death-valley-national-park/.

Molnia, Bruce F. 2008. *Glaciers of North America: Glaciers of Alaska*. U.S. Geological Survey Professional Paper 1386-K. https://pubs.er.usgs.gov/publication/pp1386K.

Monahan, William B., and Nicholas A. Fisichelli. 2014. "Climate Exposure of US National Parks in a New Era of Change." *PLOS One* 9 (7): e101302.

Monahan, William B., Alyssa Rosemartin, Katherine L. Gerst, Nicholas A. Fisichelli, Toby Ault, Mark D. Schwartz, John E. Gross, and Jake F. Weltzin. 2016. "Climate Change Is Advancing Spring Onset across the U.S. National Park System." *Ecosphere* 7 (10): e01465.

Monmonier, Mark. 2006. *From Squaw Tit to Whorehouse Meadow: How Maps Name, Claim, and Inflame*. Chicago: University of Chicago Press.

Moore, George W., and G. Nicholas Sullivan. 1978. *Speleology: The Study of Caves*. 2nd ed. St. Louis: Cave Books.

Moore, John Hammand. 2006. *The Faustball Tunnel: German POWs in America and Their Great Escape*. Annapolis, Md.: Naval Institute Press.

Morgenstern, George. 1947. *Pearl Harbor: The Story of the Secret War*. New York: Devin-Adair.

Moritsch, Barbara J. 2012. *The Soul of Yosemite: Finding, Defending, and Saving the Valley's Sacred Wild Nature*. Rochester, N.Y.: CJM Books.

Mortimer-Sandilands, Catriona. 2006. "'The Geology Recognizes No Boundaries': Shifting Borders in Waterton Lakes National Park." In *The Borderlands of the American and Canadian Wests*, edited by Sterling Evans, 309–33. Lincoln: University of Nebraska Press.

Mukul, S. A., M. B. Uddin, M. S. Uddin, M. A. S. A. Khan, and B. Marzan. 2008. "Protected Areas of Bangladesh: Current Status and Efficacy for Biodiversity Conservation." *Proceedings of the Pakistan Academy of Sciences* 45 (2): 59–68.

Muller, Edward K., ed. 2005. *DeVoto's West: History, Conservation, and the Public Good*. Athens: Ohio University Press.

Myerson, Harvey. 2001. *Nature's Army: When Soldiers Fought for Yosemite*. Lawrence: University Press of Kansas.

Nagourney, Adam. 2018. "A $100 Billion Train: The Future of California or a Boondoggle?" *New York Times*, July 30, 2018. https://www.nytimes.com/2018/07/30/us/california-high-speed-rail.html.

Nash, Stephen. 2017. *Grand Canyon for Sale: Public Lands versus Private Interests in the Era of Climate Change*. Berkeley: University of California Press.

National Parks Conservation Association. 2017. "Friend of the National Parks Award, 113th Congress." https://www.npca.org/resources/3127-friends-of-the-national-parks-113th-congress.

National Park Service. 1964. *Parks for America: A Survey of Park and Related Resources in the Fifty States, and a Preliminary Plan*. Washington, D.C.: Government Printing Office.

⸻. 1984. *Park Road Standards*. Washington, D.C.: Government Printing Office.

⸻. 1991. *Blilioi (Peleliu) Historical Park Study*. http://www.botany.hawaii.edu/basch/uhnpscesu/htms/peleliu/index.htm.

⸻. 1994. *The National Road Special Resource Study*. Washington, D.C.: National Park Service.

⸻. 1995. *Route 66 Special Resource Study*. Washington, D.C.: National Park Service.

⸻. 2004. *Lincoln Highway Special Resource Study / Environmental Assessment*. Washington, D.C.: National Park Service.

⸻. 2010. *Redwood Official National and State Parks Handbook*. Washington, D.C.: Government Printing Office.

⸻. 2011. "Shiloh Battlefield to Eliminate Entrance Fees." https://www.nps.gov/shil/learn/news/shiloh-battlefield-to-eliminate-entrance-fees.htm.

⸻. 2012. "Unigrid." https://www.nps.gov/parkhistory/online_books/brochures/unigrid/index.htm.

⸻. 2014. "Environmental Assessment: Commnet Cell Service Proposal for Stovepipe Wells." https://parkplanning.nps.gov/document.cfm?documentID=54845.

⸻. 2015. *Law Enforcement Program Reference Manual 9: Law Enforcement, Security, and Emergency Services*. https://www.nps.gov/aboutus/foia/upload/RM-9-redacted-V-2.pdf.

⸻. 2016a. "Change of Jurisdiction—National Park Service Units within the Commonwealth of Kentucky." *Federal Register* 81 (170): 60377.

⸻. 2016b. *The National Parks: Index 2012–2016*. Washington, D.C.: National Park Service.

⸻. 2018a. *Criteria for New National Parks*. https://parkplanning.nps.gov/files/Criteria%20for%20New%20Parklands.pdf.

⸻. 2018b. "Identifying & Reporting Deferred Maintenance." https://www.nps.gov/subjects/infrastructure/identifying-reporting-deferred-maintenance.htm.

⸻. 2018c. "Lake Mead Lodge." https://www.nps.gov/lake/learn/historyculture/lake-mead-lodge.htm.

⸻. 2018d. *NPS Deferred Maintenance by State and Park*. https://www.nps.gov/subjects/infrastructure/identifying-reporting-deferred maintenance.htm

⸻. 2018e. 2017 Annual SAR Dashboard. https://nps.maps.arcgis.com/apps/opsdashboard/index.html#/b526c87ae21f4a669eb6c9238c2c4bcf.

⸻. 2019. "Invasive Species, National Parks, and You." https://www.nps.gov/articles/invasive-species.htm.

⸻. 2020a. *Budget Justifications*. https://www.doi.gov/sites/doi.gov/files/fy2021-nps-justification.pdf.

⸻. 2020b. "Coral Reefs—Oceans, Coasts & Sea-shores." https://www.nps.gov/subjects/oceans/coral-reefs.htm.

⸻. 2020c. "Fossils and Paleontology." https://www.nps.gov/subjects/fossils/index.htm.

⸻. 2020d. *National Park Service Transit Inventory and Performance Report, 2019*. https://www.nps.gov/subjects/transportation/upload/NPS_NTI_2019_Report_508.pdf.

⸻. 2020e. National Park Service Visitor Statistics. https://irma.nps.gov/Stats/Reports/National.

⸻. 2021. "Visitor Spending Effects—Economic Contributions of National Park Visitor Spending." https://www.nps.gov/subjects/socialscience/vse.htm.

⸻. 2023a. National Historic Landmarks. https://www.nps.gov/subjects/nationalhistoriclandmarks/list-of-nhls-by-state.htm.

⸻. 2023b. National Register of Historic Places. https://www.nps.gov/subjects/nationalregister/database-research.htm.

National Park Service Office of Legislative and Congressional Affairs. 2021. "Working with Congress." Legislative and Congressional Affairs. https://www.nps.gov/subjects/legal/index.htm.

National Travel and Tourism Office (NTTO). 2023. Statistics and Research Programs. https://www.trade.gov/travel-and-tourism-research.

Newmark, William D. 1985. "Legal and Biotic Boundaries of Western North American National Parks: A Problem of Congruence." *Biological Conservation* 33 (3): 197–208.

Nordgren, Tyler. 2010. *Stars Above, Earth Below: A Guide to Astronomy in National Parks*. Chichester: Springer.

Norris, Frank. 1996a. *Isolated Paradise: An Administrative History of the Katmai and Aniakchak NPS Units, Alaska*. Anchorage, Alas.: National Park Service.

⸻. 1996b. "A Lone Voice in the Wilderness: The National Park Service in Alaska, 1917–1969." *Environmental History* 1 (4): 66–76.

⸻. 2000. *A Victim of Nature and Bureaucracy: The Sad Short History of Old Kasaan National Monument*. http://npshistory.com/publications/alaska/old-kasaan-2000.pdf.

⸻. 2006. *Crown Jewel of the North: An Administrative History of Denali National Park and Preserve*. Anchorage, Alas.: National Park Service.

O'Brien, Bob R. 1966. "The Future Road System of Yellowstone National Park." *Annals of the Association of American Geographers* 56 (3): 385–407.

O'Brien, Justin. 2014. "No Straight Thing: Experiences of the Mirarr Traditional Owners of Kakadu National Park with the World Heritage Convention." In *World Heritage Sites and Indigenous People's Rights*, edited by Stefan Disko and Helen Tugendhat, 313–40. Copenhagen: IWGIA.

O'Brien, William E. 2015. *Landscapes of Exclusion: State Parks and Jim Crow in the American South*. Amherst: University of Massachusetts Press.

———. 2018. "Racialized Assemblage and State Park Design in the Jim Crow South." In *The American Environment Revisited*, edited by Geoffrey L. Buckley and Yolanda Youngs, 101–20. Lanham, Md.: Rowman and Littlefield.

O'Rourke, Ronald. 2013. *Navy Ford (CVN-78) Class Aircraft Carrier Program: Background and Issues for Congress*. Washington, D.C.: Congressional Research Service. https://apps.dtic.mil/sti/pdfs/ADA501532.pdf.

Pattison, William D. 1964. "The Four Traditions of Geography." *Journal of Geography* 63 (5): 211–16.

Pearson, Bryan F. 2002. *Still the Wild River Runs: Congress, the Sierra Club, and the Fight to Save the Grand Canyon*. Tucson: University of Arizona Press.

Pettebone, David, Ashley D'Antonio, Abigail Sisneros-Kidd, and Christopher Monz. 2019. "Modeling Visitor Use on High Elevation Mountain Trails: An Example from Longs Peak in Rocky Mountain National Park, USA." *Journal of Mountain Science* 16 (12): 2882–93.

Pettebone, David, and Bret Meldrum. 2018. "The Need for a Comprehensive Socioeconomic Research Program for the National Park Service." *George Wright Forum* 35 (1): 22–31.

Pinto, Robin. 2007. *Chiricahua National Monument Historic Designed Landscape*. http://www.npshistory.com/publications/chir/historic-designed-landscape.pdf.

Potter, Andrew. 2010. *The Authenticity Hoax: Why the "Real" Things We Seek Don't Make Us Happy*. New York: Harper.

Protas, Josh A. 2002. *Past Preserved in Stone: A History of Montezuma Castle National Monument*. Tucson: Western National Parks Association.

Public Lands Alliance. 2017. "Reciprocal Discount Program." http://publiclandsalliance.org/membership/rdp.

Public Roads Administration. 1947. *Highway Statistics 1947*. Washington, D.C.: Government Printing Office.

Pyne, Stephen. 1995. *Fire on the Rim: A Firefighter's Season at the Grand Canyon*. Seattle: University of Washington Press.

Raynal, Jeremy M., Arielle S. Levine, and Mia T. Comeros-Raynal. 2016. "American Samoa's Marine Protected Area System: Institutions, Governance, and Scale." *Journal of International Wildlife Law & Policy* 19 (4): 301–16.

Reimann, Lena, Athanasios T. Vafeidis, Sally Brown, Jochen Hinkel, and Richard S. J. Tol. 2018. "Mediterranean UNESCO World Heritage at Risk from Coastal Flooding and Erosion due to Sea-Level Rise." *Nature Communications* 9:4161.

Reps, John W. 1965. *The Making of Urban America: A History of City Planning in the United States*. Princeton, N.J.: Princeton University Press.

———. 1979. *Cities of the American West: A History of Frontier Urban Planning*. Princeton, N.J.: Princeton University Press.

Rettie, Dwight F. 1995. *Our National Park System: Caring for America's Greatest Natural and Historic Treasures*. Urbana: University of Illinois Press.

Retzlaff, Rebecca. 2000. "Buses Replace Private Cars on Zion Park Scenic Drive." *Planning* 66 (7): 28–29.

Revkin, Andrew C. 2009. "Afghanistan: The First National Park." *New York Times*, April 21, 2009. https://www.nytimes.com/2009/04/22/world/asia/22briefs-Park.html.

Richmond, Al. 1995. *Cowboys, Miners, Presidents and Kings: The Story of the Grand Canyon Railway*. Flagstaff, Ariz.: Northland Graphics.

Ridenour, James M. 1994. *The National Parks Compromised: Pork Barrel Politics and America's Treasures*. Merrillville, Ind.: ICS Books.

Righter, Robert W. 1982. *Crucible for Conservation: The Creation of Grand Teton National Park*. Boulder: Colorado Associated University Press.

———. 2005. *The Battle over Hetch Hetchy: America's Most Controversial Dam and the Birth of Modern Environmentalism*. Oxford: Oxford University Press.

Roberts, Nina. 2007. *Visitor/Non-visitor Use Constraints: Exploring Ethnic Minority Experiences and Perspectives*. https://www.academia.edu/33807564/Race_ethnicity_and_outdoor_studies_Trends_challenges_and_forward_momentum_book_chapter.

Roberts, Phil. 2012. *Cody's Cave: National Monuments and the Politics of Public Lands in the 20th Century West*. Laramie, Wyo.: Skyline Western Press.

Rogers, Jeanne. 2007. *Standing Witness: Devils Tower National Monument, a History*. Washington, D.C.: National Park Service.

Rogers, T. F. 2004. "Safeguarding Tranquility Base: Why the Earth's Moon Base Should Become a World Heritage Site." *Space Policy* 20 (1): 5–6.

Rohn, Arthur H., and William M. Ferguson. 2006. *Puebloan Ruins of the Southwest*. Albuquerque: University of New Mexico Press.

Roosevelt, Theodore. 1903. Address of President Roosevelt at Grand Canyon, Arizona, May 6, 1903. Theodore Roosevelt Papers. Library of Congress Manuscript Division. https://www.theodorerooseveltcenter.org/Research/Digital-Library/.

Rose-Redwood, Reuben S. 2008. "Genealogies of the Grid: Revisiting Stanislawski's Search for the Origin of the Grid-Pattern Town." *Geographical Review* 98 (1): 42–58.

Rothman, Hal. 1991. *Navajo National Monument: A Place and Its People*. Washington, D.C.: National Park Service.

———. 1994a. *America's National Monuments: The Politics of Preservation*. Lawrence: University Press of Kansas.

———. 1994b. *Preserving Different Pasts: The American National Monuments*. Urbana: University of Illinois Press.

———. 1998. *Devil's Bargains: Tourism in the Twentieth-*

Century American West. Lawrence: University Press of Kansas.

———. 2004. *The New Urban Park: Golden Gate National Recreation Area and Civic Environmentalism.* Lawrence: University Press of Kansas.

———. 2007. *Blazing Heritage: A History of Wildland Fire in the National Parks.* Oxford: Oxford University Press.

Rothman, Hal, and Char Miller. 2013. *Death Valley National Park: A History.* Reno: University of Nevada Press.

Runte, Alfred. 2010. *National Parks: The American Experience.* 4th ed. Lanham, Md.: Taylor Trade Publishing.

———. 2011. *Trains of Discovery: Railroads and the Legacy of Our National Parks.* New York: Roberts Rinehart.

Russell, Peter. 1992. *Gila Cliff Dwellings National Monument: An Administrative History.* Santa Fe, N.M.: National Park Service.

Rydell, Kiki Leigh, and Mary Shivers Culpin. 2006. *Managing the "Matchless Wonders": A History of Administrative Development in Yellowstone National Park, 1872–1965.* Washington, D.C.: National Park Service.

Sanchez, Joseph P., Bruce A. Erickson, and Jerry L. Gurule. 2001. *Between Two Countries: The Story behind Coronado National Memorial.* Santa Fe, N.M.: National Park Service.

Santucci, Vincent L., and John M. Ghist. 2014. "Fossil Cycad National Monument: A History from Discovery to Deauthorization." In *Proceedings of the 10th Conference on Fossil Resources.* Rapid City, S.D. 74–85.

Sax, Joseph L. 1980. *Mountains without Handrails: Reflections on the National Parks.* Ann Arbor: University of Michigan Press.

Schmidt, Ronald S., and William S. Hooks. 1994. *Whistle over the Mountain: Timber, Tracks and Trails in the Tennessee Smokies.* Yellow Springs, Ohio: Graphicom Press.

Schmitt, Harrison. 2006. *Return to the Moon: Exploration, Enterprise, and Energy in the Human Settlement of Space.* New York: Praxis Publishing.

Schnayerson, Michael. 2006. "Who's Running Our National Parks?" *Vanity Fair,* June 7, 2006. https://www.vanityfair.com/news/2006/06/nationalparks 200606.

Schneider-Hector, Dietmar. 1993. *White Sands: The History of a National Monument.* Albuquerque: University of New Mexico Press.

Schullery, Paul, and Lee Whittlesey. 2003. *Myth and History in the Creation of Yellowstone National Park.* Lincoln: University of Nebraska Press.

Schuppe, Jon. 2015. "Mt. McKinley to Denali: How a Mountain's Renaming Got Tied Up in Politics." *NBCnews.com,* August 31, 2015. https://www.nbcnews.com/news/us-news/mckinley-denali-how-mountains-renaming-got-tied-politics-n418811.

Schwinning, Susanne, Jayne Belnap, David R. Bowling,

and James R. Ehleringer. 2008. "Sensitivity of the Colorado Plateau to Change: Climate, Ecosystems, and Society." *Ecology and Society* 13 (2): 28.

Scott, Daniel, Geoff McBoyle, and Michael Schwartzentruber. 2004. "Climate Change and the Distribution of Climatic Resources for Tourism in North America." *Climate Research* 27 (2): 105–17.

Scott, David. 2013. "Economic Inequality, Poverty, Park and Recreation Delivery." *Journal of Park and Recreation Administration* 31 (4): 1–11.

Scott, David, and Kang Jae Jerry Lee. 2018. "People of Color and Their Constraints to National Park Visitation." *George Wright Forum* 35 (1): 73–82.

Scott, David L., and Kay W. Scott. 2012. *The Complete Guide to National Park Lodges.* 7th ed. Guilford, Conn.: Globe Pequot Press.

Secretariat of the Antarctic Treaty. 2020. The Antarctic Treaty. https://www.ats.aq/index_e.html.

Seidl, Andy, and Stephan Weiler. 2001. *Economic Impact of National Park Designation of the Black Canyon of the Gunnison on Montrose County, Colorado.* Agriculture and Resource Policy Report, Colorado State University.

Sellars, Richard West. 1997. *Preserving Nature in the National Parks: A History.* New Haven, Conn.: Yale University Press.

———. 2005. "Pilgrim Places: Civil War Battlefields, Historic Preservation, and America's First National Military Parks, 1863–1900." *CRM: The Journal of Heritage Stewardship* 2 (1): 23–52.

———. 2007. "A Very Large Array: Early Federal Historic Preservation—the Antiquities Act, Mesa Verde, and the National Park Service Act." *Natural Resources Journal* 47 (2): 267–328.

Shankland, Robert. 1951. *Stephen Mather of the National Park Service.* New York: Alfred A. Knopf.

Sheail, John. 2010. *Nature's Spectacle: The World's First National Parks and Protected Places.* London: Earthscan.

Sherwonit, Bill. 2013. *Denali National Park: The Complete Visitors Guide.* Seattle: Mountaineer Books.

Skillen, James R. 2009. *The Nation's Largest Landlord: The Bureau of Land Management in the American West.* Lawrence: University Press of Kansas.

Slobodchikoff, C. N., William R. Briggs, Patricia A. Dennis, and Anne-Marie C. Hodge. 2012. "Size and Shape Information Serve as Labels in the Alarm Class of Gunnison's Prairie Dogs *Cynomys gunnisoni.*" *Current Zoology* 58 (5): 741–48.

Smith, Duane A. 2002. *Mesa Verde National Park: Shadows of the Centuries.* Boulder: University Press of Colorado.

Smith, Rex Alan. 1985. *The Carving of Mount Rushmore.* New York: Abbeville Press.

Smith, Timothy B. 2006. *This Great Battlefield of Shiloh: History, Memory, and the Establishment of a Civil War National Military Park.* Knoxville: University of Tennessee Press.

Snell, Charles W., and Sharon A. Brown. 1986. *Antietam National Battlefield and National Cemetery, Sharpsburg, Maryland: An Administrative History*. Washington, D.C.: National Park Service.

Snyder, David K. 2013. *Geography of Southcentral Alaska*. Anchorage, Alas.: Picea Graphics.

Solop, Frederic I., Kristi K. Hagen, and David Ostergren. 2003. *Ethnic and Racial Diversity of National Park System Visitors and Non-visitors*. http://www .npshistory.com/publications/social-science /comprehensive-survey/ethnic-racial-diversity.pdf.

Spence, Mark David. 1999. *Dispossessing the Wilderness: Indian Removal and the Making of the National Parks*. Oxford: Oxford University Press.

Spennemann, Dirk H. R. 2004. "The Ethics of Treading on Neil Armstrong's Footprints." *Space Policy* 20 (4): 279–90.

———. 2006. "Out of This World: Issues of Managing Tourism and Humanity's Heritage on the Moon." *International Journal of Heritage Studies* 12 (4): 356–71.

———. 2007. "Extreme Cultural Tourism: From Antarctica to the Moon." *Annals of Tourism Research* 34 (4): 898–918.

Sproul, David Kent. 2001. *A Bridge between Cultures: An Administrative History of Rainbow Bridge National Monument*. Washington, D.C.: National Park Service.

Squier, Ephraim G., and Edwin H. Davis. 1848. *Ancient Monuments of the Mississippi Valley*. Washington, D.C.: Smithsonian Institution.

Squillace, Mark. 2006. "The Antiquities Act and the Exercise of Presidential Power: The Clinton Monuments." In *The Antiquities Act: A Century of American Archaeology, Historic Preservation, and Nature Conservation*, edited by David Harmon, Francis P. McManamon, and Dwight T. Pitcaithley, 106–36. Tucson: University of Arizona Press.

Stanfield, Rebecca, Robert Manning, Megha Budruk, and Myron Floy. 2006. "Racial Discrimination in Parks and Outdoor Recreation: An Empirical Study." In *Proceedings of the 2005 Northeastern Recreation Research Symposium (Newtown Square, Pennsylvania, U.S. Department of Agriculture)*, 247–53. https:// www.fs.usda.gov/ne/newtown_square/publications /technical_reports/pdfs/2006/341%20papers /stanfield341.pdf.

Stanislawski, Dan. 1946. "The Origin and Spread of the Grid-Pattern Town." *Geographical Review* 36 (1): 105–20.

———. 1947. "Early Spanish Town Planning in the New World." *Geographical Review* 37 (1): 94–105.

Steen, Harold K. 2004. *The U.S. Forest Service: A History*. Seattle: University of Washington Press.

Stein, Mark. 2008. *How the States Got Their Shapes*. New York: Harper Collins.

Stevens, Stan. 2014. "Indigenous Peoples, Biocultural Diversity, and Protected Areas." In *Indigenous Peoples, National Parks, and Protected Areas*, edited by Stan Stevens, 3–12. Tucson: University of Arizona Press.

Stodolska, Monica, K. Shinew, and M. Li. 2010. "Recreation Participation Patterns and Physical Activity among Latino Visitors to Three Urban Outdoor Recreation Environments." *Journal of Park and Recreation Administration* 28 (2): 36–56.

Stroud, George. 1985. *History of the Concession at Denali National Park (Formerly Mount McKinley National Park)*. Anchorage, Alas.: National Park Service.

Stuart, David E. 2000. *Anasazi America*. Albuquerque: University of New Mexico Press.

Sugden, David. 1982. *Arctic and the Antarctic: A Modern Geographical Synthesis*. Oxford: Basil Blackwell.

Sultana, Selima, Josh Merced, Joe Weber, Ridwaana Allen, and Gregory Carlton. 2023. "Great Smoky Mountains National Park and its Missing Black Visitors: A Preliminary Analysis on the Hidden Architecture of Landscape." *Southeastern Geographer* 63 (1): 15-35.

Summers, Adam B. 2005. *Funding the National Park System: Improving Services and Accountability with User Fees*. https://reason.org/wp-content/uploads/files /0b6c6302bfcc621638fcdbdebec8b63a.pdf.

Swain, Donald C. 1970. *Wilderness Defender: Horace M. Albright and Conservation*. Chicago: University of Chicago Press.

Sweet, Lynn C., Tyler Green, James G. C. Heintz, Neil Frakes, Nicholas Graver, Jeff S. Rangitsch, Jane E. Rodgers, Scott Heacox, and Cameron W. Barrows. 2019. "Congruence between Future Distribution Models and Empirical Data for an Iconic Species at Joshua Tree National Park." *Ecosphere* 10 (6). https:// esajournals.onlinelibrary.wiley.com/doi/epdf/10 .1002/ecs2.2763.

Szasz, Ferenc Morton. 1977. "Wheeler and Holy Cross: Colorado's 'Lost' National Monuments." *Journal of Forest History* 21 (3): 133–44.

———. 1984. *The Day the Sun Rose Twice: The Story of the Trinity Site Nuclear Explosion July 16, 1945*. Albuquerque: University of New Mexico Press.

Taylor, Bron, and Joel Geffen. 2004. "Battling Religions in Parks and Forest Reserves: Facing Religion in Conflicts over Protected Places." *George Wright Forum* 21 (2): 56–68.

Taylor, Patricia A., Burke D. Grandjean, and Bistra Anatchkova. 2011. *National Park Service Comprehensive Survey of the American Public 2008–2009: Racial and Ethnic Diversity of National Park System Visitors and Non-visitors*. Natural Resource Report NPS/NRSS/ SSD/NRR-2011432. National Park Service, Fort Collins, Colo.

Thompson, Tom L. 2007. *A Historical Analysis of Consolidation at the National Forest and District Level in the U.S. Forest Service*. https://forestservicemuseum.org /wp-content/uploads/2020/07/Historical-Analysis -of-Consolidation-2007.pdf.

Toogood, Anna Coxe. 1973. *George Washington Carver*

National Monument, Diamond, Missouri Historic Resource Study and Administrative History. Denver: National Park Service.

Tooman, L. Alex. 1995. "The Evolving Economic Impact of Tourism on the Greater Smoky Mountain Region of East Tennessee and Western North Carolina." PhD diss., University of Tennessee.

Trimble, Stephen. 1989. *The Sagebrush Ocean: A Natural History of the Great Basin.* Reno: University of Nevada Press.

Tweed, William C. 2010. *Uncertain Path: A Search for the Future of National Parks.* Berkeley: University of California Press.

Ullman, Edward L. 1954. "Amenities as a Factor in Regional Growth." *Geographical Review* 44 (1): 119–32.

UN Environment Programme World Conservation Monitoring Centre. 2020. Protected Planet. https://www.protectedplanet.net/c/about.

UNESCO (United Nations Educational, Scientific and Cultural Organization). 2020a. "Man and the Biosphere (MAB) Programme." https://en.unesco.org/mab.

———. 2020b. UNESCO World Heritage Centre. https://whc.unesco.org/en/.

United States Congress. 1964. Wilderness Act. 16 U.S.C. 1131.

United States Department of Commerce. 1966. *A Proposed Program for Scenic Roads and Parkways.* Washington, D.C.: Government Printing Office.

United States Forest Service. 2006. *Where Is the Tree You Can Drive Through?* https://www.fs.usda.gov/Internet/FSE_DOCUMENTS/fsbdev3_058751.pdf.

———. 2012. *Establishment and Modification of National Forest Boundaries and National Grasslands.* https://www.fs.fed.us/land/staff/Documents/Establishment%20and%20Modifications%20of%20National%20Forest%20Boundaries%20and%20National%20Grasslands%201891%20to%202012.pdf.

United States Geological Survey. 2003. "Physiographic Regions." https://www.americangeosciences.org/sites/default/files/education-EarthComm-UYE-c1a7q2r1.pdf.

Upchurch, Jonathan. 2015. "Zion National Park, Utah: Enhancing Visitor Experience through Improved Transportation." *Transportation Research Record* 2499 (1): 40–44.

Vale, Thomas R. 2005. *The American Wilderness: Reflections on Nature Protection in the United States.* Charlottesville: University of Virginia Press.

Valliere, William, Robert Manning, Megha Budruk, Steven Lawson, and Benjamin Wang. 2002. "Transportation Planning and Social Carrying Capacity in the National Parks." In *Proceedings of the 2001 Northeastern Recreation Research Symposium, (Newtown Square, Pennsylvania, U.S. Department of Agriculture)*, 36–39. https://www.nrs.fs.usda.gov/pubs/gtr/gtr_ne289/gtr_ne289_036.pdf.

Vaske, Jerry J., and Katie M. Lyon. 2014. *Linking the 2010 Census to National Park Visitors.* Natural Resource Technical Report NPS/WASO/NRTR-2014/880. https://irma.nps.gov/DataStore/Reference/Profile/2210640.

Vaughn, Jacqueline, and Hanna J. Cortner. 2013. *Philanthropy and the National Park Service.* New York: Palgrave Macmillan.

Verne, Jules. 1886. *Robur-le-Conquérant [Robur the Conqueror].* Paris: Pierre-Jules Hetzel.

Vincent, Carol Hardy, and Pamela Baldwin. 2004. *National Monuments: Issues and Background.* New York: Novinka Books.

Visa. 2023. Western U.S. National Parks Popular with Foreign Tourists. https://usa.visa.com/partner-with-us/visa-consulting-analytics/western-us-national-parks-popular-with-foreign-tourists.html.

Visitor Services Project. 1997. *Virgin Islands National Park Visitor Survey Spring 1997.* https://sesrc.wsu.edu/doc/93_VIIS_rept.pdf.

von Ehrenfried, Manfred. 2018. *Apollo Mission Control: The Making of a National Historic Landmark.* New York: Springer.

Wagner, Frederic H., Ronald Foresta, Richard Bruce Gill, Dale Richard McCullough, Michael R. Pelton, William F. Porter, and Hal Salwasser. 1995. *Wildlife Policies in the U.S. National Parks.* Washington, D.C.: Island Press.

Waite, Thornton. 2006. *Yellowstone by Train: A History of Rail Travel to America's First National Park.* Missoula, Mont.: Pictorial Histories Publishing Co.

Wakild, Emily. 2009. "Border Chasm: International Boundary Parks and Mexican Conservation, 1935–1945." *Environmental History* 14 (3): 453–75.

Walker, Lawrence R., and Frederick H. Landau. 2018. *A Natural History of the Mojave Desert.* Tucson: University of Arizona Press.

Walklet, Keith S. 2004. *The Ahwahnee: Yosemite's Grand Hotel.* Yosemite, Calif.: Yosemite Association.

Warf, Barney. 2008. *Time-Space Compression: Historical Geographies.* New York: Routledge.

Warner, Thomas T. 2004. *Desert Meteorology.* Cambridge: Cambridge University Press.

Washburne, Randel F. 1978. "Black Under-Participation in Wildland Recreation: Alternative Explanations." *Leisure Sciences* 1 (2): 175–89.

Watt, Laura Alice. 2017. *The Paradox of Preservation: Wilderness and Working Landscapes at Point Reyes National Seashore.* Berkeley: University of California Press.

Watts, Raymond D. 2005. *Distance to Nearest Road in the Conterminous United States.* USGS Fact Sheet 2005-3011. https://pubs.usgs.gov/fs/2005/3011/report.pdf.

Weber, Joe. 2016. "America's Lost National Park Units: A Closer Look." *George Wright Forum* 33 (1): 59–69.

———. 2019. "The Three American Wests." *Professional Geographer* 17 (2): 239–52.

Weber, Joe, and Selima Sultana. 2013. "Why Do So Few Minority People Visit National Parks? Visitation and the Accessibility of 'America's Best Idea.'" *Annals of the Association of American Geographers* 103 (3): 437–64.

Webster, Gerald R., and Jonathan I. Leib. 2011. "Living on the Grid: The U.S. Rectangular Public Land Survey System and the Engineering of the American Landscape." In *Earth Engineering*, edited by S. D. Brunn, 2123–38. Dordrecht, the Netherlands: Springer.

Weiler, Stephan. 2006. "A Park by Any Other Name: National Park Designation as a Natural Experiment in Signaling." *Journal of Urban Economics* 60 (1): 96–106.

Weiler, Stephan, and Andy Seidl. 2004. "What's in a Name? Extracting Econometric Drivers to Assess the Impact of National Park Designation." *Journal of Regional Science* 44 (2): 245–62.

Wheat, Frank. 1999. *California Desert Miracle: The Fight for Desert Parks and Wilderness*. San Diego: Sunbelt Publications.

Whisnant, Anne, David Whisnant, and Tim Silver. 2011. *Shenandoah National Park Official Guidebook*. Washington, D.C.: National Park Service.

Whisnant, David E., and Anne Mitchell Whisnant. 2007. *Small Park, Large Issues: De Soto National Memorial and the Commemoration of a Difficult History*. Washington, D.C.: National Park Service.

Whiteley, Lee, and Jane Whiteley. 2003. *The Playground Trail: The National Park to Park Highway: To and through the National Parks of the West in 1920*. Boulder, Colo.: Johnson Printing.

Widder, Keith R. 1975. *Mackinac National Park, 1875–1895*. Reports in Mackinac History and Archaeology No. 4. Mackinac State Historic Parks.

Wilkinson, Charles F. 1992. *Crossing the Next Meridian: Land, Water, and the Future of the West*. Washington, D.C.: Island Press.

Williams, Chuck, Linda McCollum, Dan Wray, Cam Camburn, Crystalaura Jackson, Norm Kresge, and Sharon Schaaf. 2015. *Seekers, Saints and Scoundrels: The Colorful Characters of Red Rock Canyon*. Salt Lake City: Paragon Press.

Williams, Lizzie. 2005. *Nigeria*. Buckinghamshire: Bradt.

Williams, Richard, and Jane G. Ferrigno. 2008. *Glaciers of North America*. U.S. Geological Survey Professional Paper 1386-J. https://pubs.usgs.gov/pp/p1386j/.

Wilson, Janet. 2004. "A New Campfire Song: Help!" *Los Angeles Times*, April 6, 2004. https://www.latimes.com/archives/la-xpm-2004-apr-06-os-newfamily6-story.html.

Wilson, Randall K. 2016. *America's Public Lands: From Yellowstone to Smokey Bear and Beyond*. Lanham, Md.: Rowman & Littlefield.

Wines, Abby. 2019. "Death Valley Hosts Record 1,678,660 Visitors in 2018." https://www.nps.gov/deva/learn/news/2018-visitation.htm.

Winks, Robin. 1997. *Laurance S. Rockefeller: Catalyst for Conservation*. Washington, D.C.: Island Press.

Winter, Patricia L., Woo Jeong, and Geoffrey Godbey. 2004. "Outdoor Recreation among Asian Americans: A Case Study of San Francisco Bay Area Residents." *Journal of Park and Recreation Administration* 22 (3): 114–36.

Wirth, Conrad L. 1980. *Parks, Politics, and the People*. Norman: University of Oklahoma Press.

Wolch, Jennifer, and Jin Zhang. 2004. "Beach Recreation, Cultural Diversity and Attitudes toward Nature." *Journal of Leisure Research* 36 (3): 414–43.

World Conservation Union. 1992. *Protected Areas of the World: A Review of National Systems, Volume 1: Indomalaya, Oceania, Australia, and Antarctic*. Gland, Switzerland: IUCN.

Worster, Donald. 1992. *Under Western Skies: Nature and History in the American West*. New York: Oxford University Press.

Yard, Robert Sterling. 1919. *The Book of the National Parks*. New York: Charles Scribner's Sons.

Yochim, Michael J. 2009. *Yellowstone and the Snowmobile: Locking Horns over National Park Use*. Lawrence: University Press of Kansas.

Young, Terence. 2006. "False, Cheap and Degraded: When History, Economy and Environment Collided at Cades Cove, Great Smoky Mountains National Park." *Journal of Historical Geography* 32 (1): 169–89.

———. 2009. "A Contradiction in Democratic Government: W. J. Trent, Jr., and the Struggle to Desegregate National Park Campgrounds." *Environmental History* 14 (4): 651–82.

———. 2017. *Heading Out: A History of American Camping*. Ithaca, N.Y.: Cornell University Press.

Young, Terence, Alan MacEachern, and Lary Dilsaver. 2021. "From Conservation to Cooperation: A History of Canada-US National Park Relations." *Environment and History* 27 (4): 607–34.

Youngs, Yolanda, Dave White, and Jill Wodrich. 2008. "Transportation Systems as Cultural Landscapes in National Parks: The Case of Yosemite." *Society and Natural Resources* 21 (9): 797–811.

Ziesler, Pamela S., and David Pettebone. 2018. "Counting on Visitors: A Review of Methods and Applications for the National Park Service's Visitor Use Statistics Program." *Journal of Park & Recreation Administration* 36 (1): 39–55.